Fundamentals of C# Programming

for Information Systems

George C. Philip
The University of Wisconsin—Oshkosh

Chapter 13 contributed by Dr. Jakob Iversen

The University of Wisconsin–Oshkosh

Copyright © 2017 Prospect Press, Inc. All rights reserved.

No part of this publication may be reproduced, stored in a retrieval system or transmitted in any form or by any means, electronic, mechanical, photocopying, recording, scanning or otherwise, except as permitted under Sections 107 or 108 of the 1976 United States Copyright Act, without either the prior written permission of the Publisher, or authorization through payment of the appropriate per-copy fee to the Copyright Clearance Center, Inc. 222 Rosewood Drive, Danvers, MA 01923, website www.copyright.com. Requests to the Publisher for permission should be addressed to the Permissions Department, Prospect Press, 47 Prospect Parkway, Burlington, VT 05401 or email to Beth.Golub@ProspectPressVT.com.

Founded in 2014, Prospect Press serves the academic discipline of Information Systems by publishing innovative textbooks across the curriculum including introductory, emerging, and upper level courses. Prospect Press offers reasonable prices by selling directly to students. Prospect Press provides tight relationships between authors, publisher, and adopters that many larger publishers are unable to offer. Based in Burlington, Vermont, Prospect Press distributes titles worldwide. We welcome new authors to send proposals or inquiries to Beth.golub@prospectpressvt.com.

The programs and other materials presented in this book have been edited carefully, but they are not guaranteed for any particular purpose. The publisher does not accept any liabilities with respect to the materials.

Cover Illustration and Design: Maddy Lesure
Production Management: Kathy Bond Borie, Rachel Paul
Editor: Beth Lang Golub

eTextbook, ISBN: **978-1-943153-15-2**
Available from Redshelf.com and VitalSource.com

Black & white paperback, ISBN: **978-1-943153-16-9**
Available from Redshelf.com and CreateSpace.com

Full-color paperback: ISBN: **978-1-943153-17-6**
Available from CreateSpace.com

To Mary, my wife of more than 40 years,
for her invaluable love, patience and support.

To my children, Samantha and Joshua,
for the fulfillment they have brought to my life, and for their love.

To my nephews and nieces,
Finney, Vava, Philipkuttan, Julie, Georgie, Wesley, Jerrymon,
Kochumon, Kochumol, Faith, Titty, Tony, Sarah, Sharon and Molsey,
who give me faith in the next generation.

Brief Table of Contents

Chapter 1: Introduction to Visual Studio and Programming 1

Chapter 2: Data Types 56

Chapter 3: Decision Structures and Validation 86

Chapter 4: Iteration Structure: Loops 121

Chapter 5: Methods 144

Chapter 6: Graphical User Interface Controls 176

Chapter 7: Arrays 208

Chapter 8: Sequential Files and Arrays 230

Chapter 9: Collections 279

Chapter 10: Graphical User Interface: Additional Controls 302

Chapter 11: Multi-form Applications and Menus 316

Chapter 12: Databases 339

Chapter 13: Object Oriented Programming (Online Only*)

Appendix A: Binary Files 415

Appendix B: Creating Databases 418

Appendix C: Answers to Review Questions 420

Index 432

* Chapter 13 is available as a free PDF download at http://www.prospectpressvt.com/titles/c-sharp-programming/. Click on the Student Resources link in the gray box on the right side of the page.

Expanded Table of Contents

Chapter 1: Introduction to Visual Studio and Programming 1

   1.1 Introduction to Programming 1

   1.2 Introduction to Visual Studio 5

   Tutorial 1: Ice Cream Cost 8

   1.3 Visual Studio Environment 10

   1.4 Adding Controls to a Form 16

   1.5 The Code Editor Window 20

   1.6 Introductory Programming Concepts 25

   1.7 Data Types 26

   1.8 Working with Variables 29

   Tutorial 2: Developing the Code 33

   1.9 Doing Calculations 37

   1.10 Formatting and Displaying Output in a Label 38

   1.11 Additional Concepts: Constants, Scope of Variables and ListBox ... 43

   1.12 Exception Handling: Catching Errors 51

   Exercises 54

Chapter 2: Data Types 56

   2.1 Working with Different Numeric Data Types 56

   Tutorial 1: Ice Cream Total with Data Types 58

   2.2 Casting 59

   2.3 Working with Strings and Characters 68

   Tutorial 2: String Operations 69

   2.4 Working with Dates and Times 74

   Tutorial 3: DateTime Operations 74

   Exercises 85

Chapter 3: Decision Structures and Validation 86

   3.1 Flowchart for Decision Structure 86

   3.2 if else Statement 87

   Tutorial 1: Ice Cream Total with Discount 88

   3.3 if else if Statement 96

   3.4 Nested if Statements 98

Tutorial 2: Compute Grade 99

3.5 The switch Statement 102

Tutorial 3: switch Statement 104

3.6 Data Validation 106

Tutorial 4: Late Validation 108

Tutorial 5: Immediate Validation Using Validating Event 111

Checking for Null Values 114

Exercises 118

Programming Assignment 1 120

Chapter 4: Iteration Structure: Loops 121

4.1 The while Loop 121

4.2 Console Applications 123

Tutorial 1: while Loop to Verify Password 123

4.3 Incrementing Variables in a Loop 125

Tutorial 2: while Loop: Financial Planning Application 128

4.4 The do-while Loop 134

4.5 The for Loop 135

Tutorial 3: for Loop: Financial Planning Application 138

4.6 The break and continue Statements 140

Exercises 143

Chapter 5: Methods 144

5.1 Introduction to Methods 144

Tutorial 1: Methods: Compute Heart Rate 147

5.2 Passing Values to Methods 148

5.3 Passing Arguments by Value 154

5.4 Passing Arguments by Reference 157

5.5 Top-Down Design 161

5.6 Methods That Return a Value 168

5.7 Enumerations: Limiting Parameter Values 170

Tutorial 2: Enumeration: Enumeration Type Parameter 172

Exercises 174

Chapter 6: Graphical User Interface Controls 176

6.1 Working with ScrollBars 176

Tutorial 1: Theater Tickets Sale Application 177

6.2 Working with CheckBoxes 180

6.3 Working with RadioButtons 182

6.4 Validating Input 184

6.5 Getting User Input from Controls 185

6.6 Working with ComboBoxes 188

6.7 Working with ListBoxes 198

6.8 Windows Presentation Foundation (WPF) 204

Chapter 7: Arrays 208

7.1 Introduction to One-Dimensional Arrays 208

Tutorial 1: Working with 1-D Arrays: Test Scores Application 211

7.2 Assigning a Value to an Element of an Array 214

7.3 Accessing Elements of an Array Using the Index 215

7.4 Accessing Elements of an Array Using the foreach Loop 219

7.5 Copying an Array 221

7.6 Looking Up Values in an Array 225

Tutorial 2: Lookup Units Sold Using 1-D Arrays 226

Exercises 228

Chapter 8: Sequential Files and Arrays 230

8.1 Introduction to Text Files 230

Tutorial 1: Reading Text Files 233

8.2 Splitting a Row: Split Method 235

8.3 Reading Data from Files into Arrays 238

8.4 Additional Methods of Arrays 246

Tutorial 2: Looking Up Phone Numbers Using Arrays 246

Tutorial 3: Using a ComboBox to Select the Search Name 250

8.5 Writing to Text Files 253

Tutorial 4: Writing Scores to Text File 253

8.6 SaveFileDialog Control 260

8.7 Passing Arrays to Methods 264

Tutorial 5: Passing Arrays to Compute Average 264

8.8 Two-Dimensional (2-D) Arrays 268

Tutorial 6: 2-D Array to Find the Sales for a Product 272

Exercises 276

Programming Assignment 2 277

Chapter 9: Collections 279

9.1 Introduction to Collections 279

9.2 List<T> Collection 281

Tutorial 1: Looking Up Sales Data Using a List 283

9.3 Dictionary<TKey, TValue> Collection 289

Tutorial 2: Dictionary with Product# as Key and Price as Value 291

Tutorial 3: Dictionary with Product# as Key and Sales Array as Value 297

Exercises 301

Chapter 10: Graphical User Interface: Additional Controls 302

10.1 ListView: Features 302

10.2 Adding Items to ListView 303

Tutorial 1: Selecting a Show from a ListView 306

10.3 Accessing Data from a ListView 307

10.4 Forms with Tab Pages 312

Trutorial 2: Multipage Form with Tab control 313

Exercises 314

Chapter 11: Multi-form Applications and Menus 316

11.1 Introduction 316

11.2 Accessing an Existing Form from Another 317

Tutorial 1: Multi-form Financial Planning Application 319

11.3 Sharing a Method between Multiple Forms 321

Tutorial 2: Sharing the LoadDictionary Method between Forms 323

11.4 Closing a Parent Form 328

11.5 Menus 331

11.6 ToolStrips 336

Exercises 338

Chapter 12: Databases 339

12.1 Introduction 339

12.2 Creating Database Objects 340

Tutorial 1: Display Employee Records Using the Wizard 340

12.3 Displaying Data in a DataGridView 343

12.4 Accessing Data Items from a DataGridView 350

12.5 Displaying Data in Details View 353

12.6 Finding the Record for a Selected Key 355

Tutorial 2: Find Employee Records Using the Wizard 355

12.7 Selecting a Group of Records Using the Binding Source 360

12.8 Selecting Records Using the Table Adapter 366

Tutorial 3: Parameter Query to Select Records 367

12.9 Untyped Datasets: Displaying Records 370

Tutorial 4: Display Records Using Untyped Datasets 371

12.10 Untyped Datasets: Selecting Records 380

Tutorial 5: Select Records Using Untyped Datasets 382

12.11 Untyped Datasets: Add/Edit/Delete Records 389

Tutorial 6: Add/Edit/Delete Using Untyped Datasets 390

12.12 Command Object and DataReader 399

Tutorial 7: SqlCommand and SqlDataReader to Work with Tables 401

Exercises 410

Programming Assignment 3 412

Chapter 13: Object Oriented Programming (Online Only*)

13.1 Introduction to Objects and Classes

13.2 Classes vs. Objects

13.3 Information Hiding (Encapsulation)

Tutorial 1: Create an Employee Class

13.4 Properties

Tutorial 2: User Interface and Properties

13.5 Calling Methods (Sending Messages to Objects)

Tutorial 3: Working with Multiple Forms as Objects

13.6 Introduction to Inheritance

13.7 Implementing Inheritance

Tutorial 4: Creating Subclasses

13.8 Using Subclasses and Superclasses

13.9 Overriding Methods

Tutorial 5: Implementing the GrossPay Method

13.10 Polymorphism

Appendix A: Binary Files 415

A.1 Writing to Binary Files 415

Tutorial: Working with Binary Files 415

A.2 Reading Binary Files 416

Appendix B: Creating a Database 418

Tutorial: Creating HR Database 418

Appendix C: Answers to Review Questions 420

Index 432

* Chapter 13 is available as a free PDF download at http://www.prospectpressvt.com/titles/c-sharp-programming/. Click on the Student Resources link in the gray box on the right side of the page.

Preface

Mission of the Text

Welcome to *Fundamentals of C# Programming for Information Systems*. This book teaches the fundamentals of programming in C# to provide a solid foundation to build business and other real-world applications. Programming concepts are discussed in the context of familiar practical applications that use graphical interfaces.

Target Audience

This book is designed for introductory programming courses in IS/MIS, CIS and IT. This book also would fit into a computer science curriculum with an introductory course that uses a GUI-based application-oriented approach to teach programming concepts. The breadth and depth of coverage makes this book suitable for a two-course sequence in programming, particularly when students come to the first course with no programming background and a slower pace is desired. An approach in a two-course sequence would be to do in-depth coverage of topics like collections, databases, object oriented programming, and others presented in later chapters only in the second course.

Key Features

A key feature of the book is that programming concepts are introduced in small chunks through examples and illustrations accompanied by hands-on tutorials. The tutorials, which are interspersed with the concepts, help students apply and explore what they learn immediately. Additionally, review questions and exercises within the chapters enhance student interest and learning.

Although the book is written for beginners, it is thorough and concise. Graphical illustrations and screenshots are used throughout the book to enhance learning for both beginners and experienced students.

Windows forms are used from the beginning to provide GUI-based as opposed to console-based interface.

This book builds graphical user interfaces and code in the .Net environment using Visual Studio. You may use the current free version, Visual Studio Community 2015, or the earlier versions, including Visual Studio Express, Visual Studio 2013, Visual Studio 2012 and Visual Studio 2010.

Supplements

For Students: Tutorial_Starts.zip file that contains

- o Partially completed projects for tutorials
- o Data files/databases used in projects

You may download the Tutorial_Starts.zip file from
http://www.prospectpressvt.com/titles/c-sharp-programming/student-resources/

For Instructors: Instructor resources include

- o Completed tutorials
- o PowerPoint slides for all chapters
- o Test bank
- o Partially completed projects for tutorials
- o Data files/databases used in projects

To access instructor resources, please complete the request form at
http://www.prospectpressvt.com/faculty-resources/instructors-material/

Installing Visual Studio

You may download Visual Studio Community 2015 from the following website:
https://www.visualstudio.com/downloads/download-visual-studio-vs
To install Visual Studio, open the downloaded .exe file and run it.

Overview of the Content and Organization

Every possible sequence of topics seems to put constraints on the quality of illustrative examples and applications that can be used in a chapter. The organization of chapters in this book attempts to minimize such problems and to enhance the ability to build on prior chapters. However, except for the foundational Chapters 1–5, there is significant flexibility in choosing specific topics and the depth of coverage. As suggested by the dependencies summarized below, there is some flexibility in the sequencing too.

Chapters 1–5 cover the Visual Studio environment and introductory programming concepts, including methods. These chapters, which provide the foundational knowledge, should be covered in sequence before other chapters, though certain topics like working with dates and times (section 2.4) may be postponed or skipped.

Chapter 6 presents the application of the GUI controls ScrollBars, RadioButtons, CheckBoxes, ComboBoxes and ListBoxes.

Chapter 7 provides a detailed presentation of one-dimensional arrays, and Chapter 8 presents accessing sequential files and using arrays in combination with files. Chapter 6 is not a prerequisite for Chapters 7 or 8. GUI is presented early on in Chapter 6 to motivate students with more interesting graphical interfaces. It should be noted that the comprehensive assignment (Assignment 2) specified at the end of Chapter 8 requires the use of several GUI controls.

Chapter 9 introduces collections, and discusses the List and Dictionary collections in more detail. Chapter 8 ("Sequential Files and Arrays") is a prerequisite for this chapter. Because of the close relationship between collections and arrays, collections are presented in this book immediately following Chapter 8 on arrays.

Chapter 10 discusses the application of ListView and TabControl. The dependency of this chapter on Chapter 9 ("Collections") is very low. The prerequisite for this chapter includes Chapters 6 and 8.

Chapter 11 presents multiform applications, Menus and ToolStrips. This chapter has some dependency on previous chapters, except Chapter 10.

Chapter 12 provides in-depth coverage of accessing databases from C# programs. Chapter 6 is a prerequisite for this chapter. In addition, the ListView control presented in Chapter 10 is used in an example in the last part of this chapter, and it is required in the third comprehensive assignment (Assignment 3) at the end of this chapter. Assignment 3 also requires the use of MainMenu control discussed in Chapter 11. Other than that, the dependence of Chapter 12 on Chapters 7–11 is relatively low.

Chapter 13 provides an introduction to object oriented programming (OOP) principles and techniques. The initial part of this chapter may be used for an introduction to OOP early in the semester. The latter part of the chapter on inheritance relies on collections from Chapter 9. Chapter 13 is available as a free PDF download at http://www.prospectpressvt.com/titles/c-sharp-programming/. (Click on the Student Resources link in the gray box on the right side of the page.)

Acknowledgments

I am thankful for the valuable assistance provided by many people in the preparation of this book. I was fortunate to work with Beth Lang Golub, editor and president of Prospect Press, who was flexible and supportive of my goal to offer a good quality programming textbook at a reasonable price. Special thanks go to Susan Hegedus, Kathy Bond Borie and Rachel Paul for their painstaking attention to detail in editing this book, and to Maddy Lesure for the cover design.

I wish to acknowledge the contributions of the following reviewers for their valuable guidance in improving the presentation and contents of this book:

> Janet Bailey, University of Arkansas at Little Rock
> Wei Kian Chen, Champlain College
> Clinton Daniel, University of South Florida
> Joni L. Jones, University of South Florida
> David Pumphrey, Colorado Mesa University
> Manonita M. Ratwatte, University of Oklahoma (Retired)
> Theadora Ross, University of Arkansas at Little Rock
> David M. Weber, Northern Arizona University

Thanks are also due to Dr. Jakob Iversen, The University of Wisconsin–Oshkosh, for authoring Chapter 13, "Object Oriented Programming."

About the Author

Dr. George Philip is Professor Emeritus of Information Systems at the College of Business, The University of Wisconsin–Oshkosh. He has more than twenty-five years of teaching and consulting experience in the information systems field, including computer programming in multiple languages. He also served as chair of the Information Systems Team, and director of the M.S. in Information Systems program. He has published numerous articles in this field.

Chapter 1

Introduction to Visual Studio and Programming

Welcome to programming in C# language. In this chapter, you will learn to develop simple programs in the Visual Studio development environment, and to work with different types of data.

Topics

1.1	Introduction to Programming	1.8	Working with Variables
1.2	Introduction to Visual Studio	1.9	Doing Calculations
1.3	Visual Studio Environment	1.10	Formatting and Displaying Output in a Label
1.4	Adding Controls to a Form	1.11	Additional Concepts: Constants, Scope of Variables and ListBox
1.5	The Code Editor Window		
1.6	Introductory Programming Concepts	1.12	Exception Handling: Catching Errors
1.7	Data Types		

1.1 Introduction to Programming

Programming is the process of developing computer programs. If a computer program seems like a mystery to you, it is just a set of instructions telling the computer how to do a task, like looking up the price of an item or finding the Chinese restaurants in a city.

Unfortunately, computers cannot understand normal English. So, programs have to be written using special commands and statements according to strict rules. A key aspect of programming is breaking down what you want the computer to do, into detailed instructions. Like the directions that a GPS gives you to get to a place, the instructions in a program need to be precise.

Typically, a program uses one or more data items to produce some results. For example, a program that processes an order might use the item number and order quantity to compute the subtotal, sales tax and total cost, as represented in Figure 1-1.

Figure 1-1: Inputs, process and outputs of a program

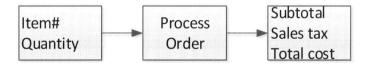

The program, represented by the block "Process Order," may include multiple subtasks like look up the unit price, check inventory and compute results.

The data that are used by a program are called the **input** to the program, and the results produced are called the **output** of the program. In addition to processing input data to produce the output, a program might write data to and read data from **storage** devices like a flash drive or a hard drive, as represented in Figure 1-2.

Figure 1-2: A general representation of a software system

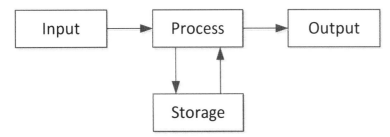

How Do I Go About Developing a Program?

To understand the process of developing a program, let's use a simplified order-processing system as an example. To develop good programs and do so efficiently, follow these steps:

1. Define the purpose, and identify the input, process and output of the program.
2. Design and develop the graphical user interface (GUI).
3. Identify the components and logic of the program.
4. Design and develop files/databases, if any.
5. Write and test the code.

Let's look at these steps in more detail.

1. Define the purpose, and identify input/process/output

Before you can write the program, you need to lay some groundwork. In this step, you identify what the user wants the program to do, including the input, process and output of the program.

Depending on the size and complexity of the program, this could involve extensive analysis of the requirements, including interviews with users; examination of current forms, reports and transactions; and identifying processes like checking inventory and looking up price in an order.

Here is an example of a simplified statement of the **purpose** of the order processing program:

    Purpose: Compute and display the subtotal, sales tax and total cost for an order

The **output** of the program often follows from the purpose. For this example, the output would be

    Output: subtotal, sales tax, total cost

The **process** specifies not only **what** the program should do (e.g., compute total cost) but also **how** it should be done (e.g., how to compute total cost), as follows:

    Process: (What?)   Look up unit price, look up sales tax rate,
                             compute subtotal, sales tax and total cost
           (How?)   subtotal: unit price * order quantity
                             sales tax: subtotal * sales tax rate
                             total cost: subtotal + sales tax

Specifying the process also would include identifying the sources of data, like the product file to get the unit price and sales tax file to get the sales tax rate.

The **input** specifies the data items that are needed to carry out the process to produce the output. The input for this order-processing system would be

    Input: item number, order quantity

Note that unit price and sales tax rate are not included in the input because the program looks them up. A real-world system would be a lot more complex. Typically, the process would include additional subtasks like handling orders when inventory is insufficient, and output may include various reports. In such systems, graphical methods like **Data Flow Diagrams** and **UML diagrams** are used to represent the processes and the data accessed by them.

2. Design and develop the graphical user interface (GUI)

After identifying the input, process and output of a program, you design and create the user interface—that is, how the user would interact with the program, and how the program would communicate with the user. This is the fun part where you bring in your creativity.

Typically, you use forms to interact with the program. As you will learn in the next section, forms have various types of objects, called **controls**, such as Button, TextBox and Label. In this step, you identify the type of controls to be used, specify their names and captions as appropriate, and design the layout. Figure 1-3 shows an example of the GUI for a simplified order-processing system where the user doesn't provide the unit price and sales tax rate.

Figure 1-3: GUI for an order-processing system

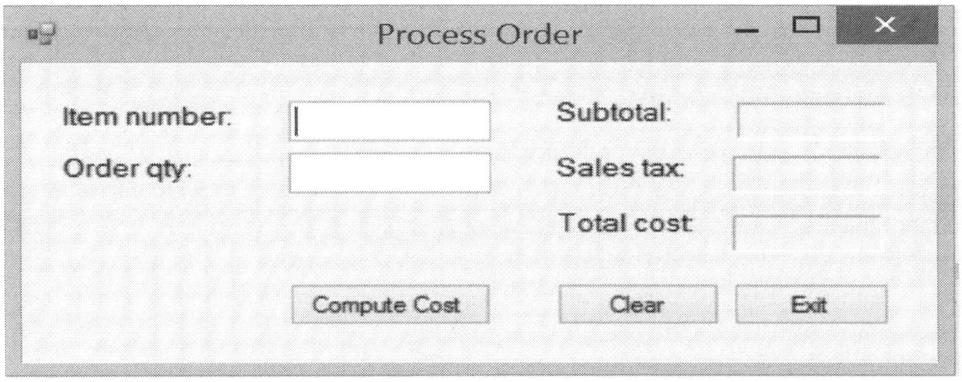

3. Identify the components and logic of the program

This step identifies the major subtasks of the program. For example, in order to process an order, the program needs to do the following subtasks:

>   Get Item# and quantity
>   Look up unit price and tax rate
>   Compute subtotal, sales tax and total cost
>   Display subtotal, sales tax and total cost

Again, a real-world program may have to do additional subtasks, like checking the inventory to make sure there is sufficient quantity on stock.

For relatively simpler programs, after identifying the subtasks, you may go directly to writing the program for each subtask. However, for tasks involving more complex logic, it might help to develop an outline of the logic of performing the subtasks. The representation of the logic of a program in plain English is called **pseudo code**. You also may represent the logic graphically using a **flowchart**, as discussed in Chapter 3.

4. Design and develop files/databases

If data is read from or written to files and/or databases, these are designed and developed prior to writing the program. Depending on the application, this step may have to be done in parallel with previous steps.

5. Write and test the code

The final step is to write and test the code. You can program in a variety of languages. C#, Java, Visual Basic, Python and PHP are among the popular languages. You will use C#, which is a popular language for developing desktop and web applications.

Programing may involve iteratively developing an application by going through the above steps multiple times.

Syntax, Logic and Runtime Errors

The programming statements you write have to follow strict rules of the language, called the **syntax**. The program wouldn't **compile** if it had any **syntax error**, like a missing semicolon at the end of a statement or a misspelled key word. Compiling is the process of translating the program you write, called the **source code**, to another language before running a program, as described in the next section. So, your first task is to make sure that there are no syntax errors. The good news is that Visual Studio provides a lot of help in identifying syntax errors.

After the syntax errors are eliminated, the program may run. But, it's still too early to celebrate because the results could be incorrect due to errors in the program logic, just like you can write a grammatically correct sentence that doesn't convey the intended message. Errors that cause a program to produce incorrect or unintended results are called **logic errors**. A tax-filing software using the wrong tax rate and a billing software overcharging a customer are examples of logic errors.

There are errors other than logic errors that can occur at runtime. These are called **runtime errors**. Runtime errors cause the program to crash (unless the program catches and handles such errors) because the program asks the computer to do something it is unable to do, like accessing a file with an invalid path or dividing a number by zero.

The process of identifying errors (bugs) is called **debugging**. Testing programs to identify and eliminate errors is an extremely important part of developing software.

Review Questions

1.1 Consider Google as a software system. What would be the input, process and output for Google?

1.2 Consider a software system that enrolls students into classes. Identify some key inputs that the system needs every time a student enrolls in a class, and the subtasks (process) that need to be performed. What are some outputs the system should produce for students and instructors?

1.3 List the major steps in developing a program.

1.4 Incorrect punctuation in a program is an example of what type of error?

1.5 A payroll program uses the wrong formula to compute overtime pay. What type of error is it?

1.6 True or false: A program that doesn't have any syntax errors should produce the correct results.

1.2 Introduction to Visual Studio

In this section, you will learn how to use Visual Studio (VS) to create the user interface and write C# programs. Visual Studio is an integrated development environment (IDE) for developing applications in a variety of languages, including C#, Visual Basic, C++, Python and HTML/JavaScript. VS supports development of desktop and web applications for Windows, Android and iOS. A major strength of Visual Studio is that it provides a user-friendly environment for developing applications.

Installing Visual Studio

You may download Visual Studio Community 2015 from the following website:

https://www.visualstudio.com/downloads/download-visual-studio-vs

To install Visual Studio, *open* the downloaded **vs_community.exe** file. The installation software will guide you through the process. The first time you start Visual Studio, you will be asked to choose the *Development Settings* and a *Color Theme*. Choose *Visual C# for* settings. You may change these settings later, as described in Tutorial 1.

Components of Visual Studio

Though you have the choice to write programs in a variety of languages, the CPU can understand only **Machine Language,** which is extremely difficult for humans to understand. Machine Language requires detailed instructions that consist of patterns of bits (0 and 1), like 10001010, and are dependent on the machine (the specific type of computer). Because of the need to write detailed machine-dependent instructions, Machine Language is called a **low-level language**.

Except in special cases, programs are written in **high-level languages,** like C#, which require fewer statements, are less dependent on the hardware and are easier to understand because they use words rather than patterns of bits.

Programs written in high-level languages are translated to Machine Language before they are run. Different languages use different methods to translate and run programs. Many languages use a special software called a **compiler** to translate the source code to Machine Language. The compiler typically produces a separate executable Machine Language program that can be run any number of times without having to compile every time it is run.

Some programming languages use an **interpreter** that translates each statement to Machine Language and runs it without producing an executable program. So, every time the program is run, it needs to be translated to Machine Language.

Visual Studio uses a compiler, but it translates the source code to an **intermediate language**, which is translated to Machine Language and run using another software, as described later in this section. Thus, Visual Studio translates your statements to Machine Language in two steps:

First, a **compiler** converts the source code into a language called Microsoft Intermediate Language (IL). The compiled code, along with references to prebuilt programs (called classes), is stored in an executable file called **Microsoft Intermediate Language (IL) Assembly**. Such files have the extension .exe or .dll.

Next, another software called **Common Language Runtime (CLR)** translates the assemblies from Intermediate Language to Machine Language and executes the programs. The process of translating and running the source code may be represented as follows:

The products that support developing and running programs within the Visual Studio family include the following:

1. An **Integrated Development Environment (IDE)**
   An IDE provides an environment to develop programs, which includes code editors for **Visual C#, Visual Basic, Visual J#, Visual C++, HTML and XML**, and designers for **Windows forms** and **web forms.**
   In Visual Studio, a software application typically is organized into **Projects** that may contain one or more **forms.**
   Forms provide the **user interface** that allows users to input data for the program, to interact with the program and to display results.

2. **A compiler** that translates the source code into **Microsoft Intermediate Language (MSIL)**

3. **.Net Framework**, which includes
   a. the **Common Language Runtime (CLR)** that translates the assemblies from Intermediate Language to Machine Language and executes the programs, and
   b. .Net Framework **Class Library** that includes a large number of prebuilt programs called **classes.**

Running a program, as described earlier, consists of (1) the compiler translating the source code (the project) into Microsoft Intermediate Language Assembly and (2) the Common Language Runtime translating the assemblies from Intermediate Language to Machine Language and executing the program to produce the output.

Next, we will look at how to work with Visual Studio to develop C# programs.

Review Questions

1.7   What is the only programming language that the CPU can understand?

1.8   What is a compiler?

1.9   What is an interpreter?

1.10   What is Microsoft Intermediate Language Assembly?

1.11   What is the function of Common Language Runtime in Visual Studio?

Creating an Application in Visual Studio

To become familiar with the Visual Studio environment, let's create an application for an ice cream parlor, consisting of a simple Windows form to compute and display the cost for an order of ice cream. We will follow the five steps identified earlier to develop it:

1. Define the purpose

Purpose:	Compute and display the total cost for an order of ice cream
Input:	Unit price, number of scoops
Process:	Compute total cost
	Total cost: unit price * number of scoops
Output:	Total cost

2. Design and develop the Graphical User Interface (GUI)

Figure 1-4 presents the design of the form that shows the user interface to let the user enter the number of scoops and unit price to display the cost. You will create the form in Tutorial 1.

Figure 1-4: Windows form to compute ice cream cost

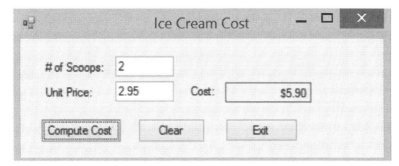

3. Identify the components and logic of the program

The major subtasks include

>Get number of scoops and unit price
>Compute cost
>Display cost

4. Design and develop files/databases

This application doesn't involve the use of files or databases.

5. Write and test the code

To understand the process of developing the entire application, including the code, let's create the form presented in step 2 and write the code for the tasks identified in step 3.

Creating a Windows Form

In Visual Studio, forms are created within a **Project**. Typically, all forms within a Project relate to a common task. Creating a project also creates a **Solution,** which is a container for one or more projects. Each project we discuss in this book is in a separate solution that has the same name as the project.

Tutorial 1 creates the project and the form to compute the cost of an order of ice cream.

Tutorial 1: Creating a Form: Ice Cream Cost

Step 1-1: Open Visual Studio Community 2015. You will see the start page, shown in Figure 1-5.

Figure 1-5: Visual Studio Start Page

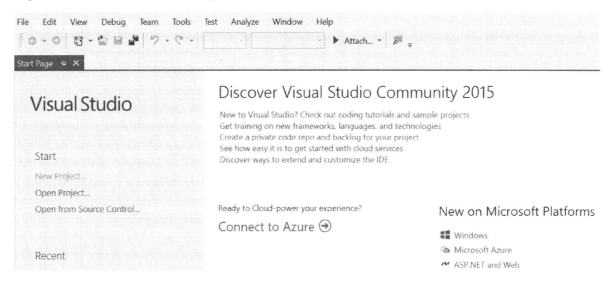

First, you will change some settings.

Changing default settings in Visual Studio

Step 1-2: Change the default folder to save your projects:

Select *Tools, Options*. Expand *Projects and Solutions* and select *General*.

For *Projects Location*, select a folder (e.g., your flash drive) where you will save your projects.

Step 1-3: Change the setting to automatically save projects when they are created:

Make sure that the CheckBox for *Save new projects when created* is checked. Click *OK*.

You may change the *color theme* by selecting *Environment, General* from the *Options* window. Similarly, you may reset your settings by selecting *Tools, Import and Export Settings, Reset all Settings*.

Creating a new project

Step 1-4: Create a new project:

Click *New Project*. You will see the New Project window displayed in Figure 1-6.

Figure 1-6: *New Project* window

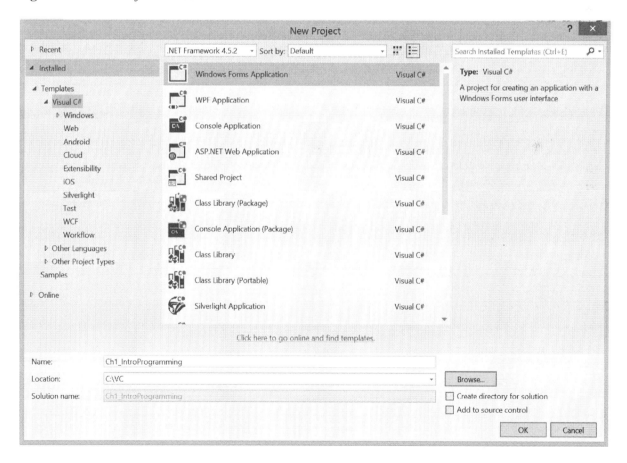

Select *Visual C#* from *Templates, Other Languages.* For the type of application, select *Windows Forms Application*, as shown in Figure 1-6.

For *Name*, enter Ch1_IntroProgramming.

Select the *Location*, if different from the default folder.

Leave the CheckBox for *Create directory for solution* unchecked so that both the Project and the Solution will be created in the same folder. (If there are multiple Projects within a Solution, it would be better to have separate directories for the Solution and Projects.) Note that the default name for the Solution is the same as that of the Project.) Click *OK*.

The Visual Studio development environment appears, as shown in Figure 1-7.

1.3 Visual Studio Environment

This section introduces the various windows and the environment within Visual Studio.

Figure 1-7 shows the default form named **Form1** in the **Designer** window, the **Toolbox** on the left, and the **Solution Explorer** window and the **Properties** window on the right.

The appearance of the form will vary with the version of Visual Studio and Windows you use. For example, the title of the form is centered in Windows 8 and 10, whereas it is left justified in Windows 7.

The **Designer** window allows you to create the user interface by adding controls to a form.

The **Toolbox** contains the components to build the user interface. The items within the Toolbox (e.g., ComboBox, TextBox and Label) are called **controls**.

The **Solution Explorer** window shows the name of the Solution, which is a container of projects; the name of the project (or projects, if there are multiple projects); and the names of the files within each project. If the name of the Solution is not displayed, select

> *Tools, Options, Projects and Solutions, General* and check *Always Show Solution*

The **Properties** window displays and lets you set the properties of the currently selected object in the *Designer* window. For example, the *Name* field in the Properties window shows that the current name of the form is **Form1**.

Figure 1-7: Visual Studio environment

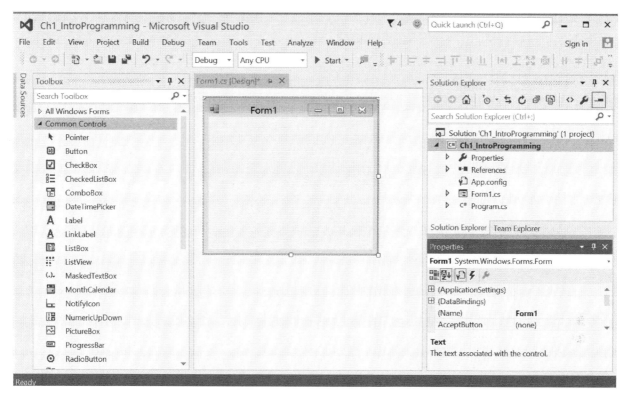

A brief description of each property is displayed at the bottom of the properties window when you click on the property, as shown in Figure 1-8. Make sure you expand the description area to make the description visible. If the description is not displayed, right click the property and check *Description*.

Figure 1-8: Description of properties

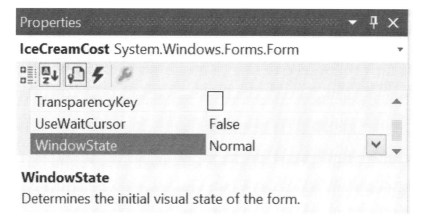

Displaying a closed window

If any window is not displayed, select *View* from the menu, and select the corresponding window.

Changing the docking position of a window

You may change the docking position of a window, such as the Properties window, by grabbing it with the pointer and then dropping it in one of the four positions (top, bottom, left, right), shown in Figure 1-9.

Figure 1-9: Docking positions

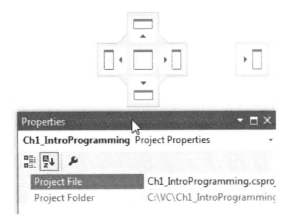

Undocking (floating) a window

By default, all windows are attached to a side of the Visual Studio window. To undock a window, right click its title bar, and select float. To dock a floating window, grab it with the pointer and drop it in one of the positions shown in Figure 1-9.

Autohide

The pushpin icon on a window, as shown in Figure 1-10, allows you to hide the window and show it as a tab on the side.

Figure 1-10: Pushpin for autohide

Click the pushpin to turn autohide on and reduce the window to tab, as shown in Figure 1-11. Note that the Solution Explorer is now a vertical bar to the right of the Properties window.

Figure 1-11: Solution Explorer with autohide on

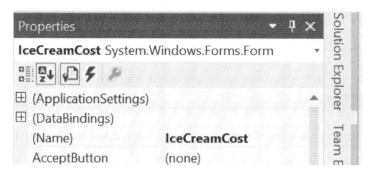

To turn autohide off, click on the tab and click the pushpin.

Menu and toolbar items

The menu and toolbar, shown in Figure 1-12, provide access to an extensive set of functions.

Figure 1-12: Menu and toolbar

The function of each toolbar item is displayed in a **ToolTip** box when you hover the mouse pointer over the item. For a general understanding of the functions, you may view the descriptions of the Toolbar items by displaying the Tooltips. You will become more familiar with them through the projects in this book.

Adding and deleting forms

You may add a form to the project as follows:

Select: *Project, Add Windows Form*. You will see the *Add New Item* window.

Make sure Visual C# Items and Windows Forms are selected.

Type in a name for the form. Click *Add*.

To delete a form,

Right click the name of the file (.cs) in the Solutions window, and select *Delete*.

Deleting a file deletes the form and all files associated with the form permanently.

Excluding a form from the project

You may exclude a form from the Solution Explorer but still keep the files associated with the form, in the project folder, as follows:

Right click the file name (.cs) in the *Solution Explorer*, and select *Exclude from Project*.

To bring back a form that was excluded from the project,

> Select *Project, Add Existing Item*.
>
> Select the file (.cs) from the project folder.

The form will be added to the Solution Explorer.

Changing the name and title of a form

The *Name* field in the Properties window shows that the current name of the form is **Form1**.

The Solution Explorer also displays a file named **Form1.cs**, which is the file that contains the code. Expanding this file, by clicking the triangle to its left, displays two other files,

> **Form1.Designer.cs** that contains the generated code representing the design of the form and
> **Form1.res** that contains additional resources associated with the form,

as shown below:

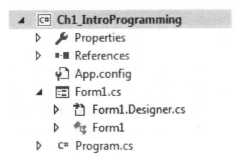

Guidelines to name a form

Names of forms (and classes, in general) must start with a letter, and can contain only letters, numbers or the underscore character (_). However, the use of the underscore character is not recommended.

The convention in C# is to use the **Pascal Case** to name forms: The first letter of the form name and the first letter of subsequent words, if any, are capitalized, as in IceCreamCost and FinancialPlanner.

Use a meaningful name and avoid using abbreviations or acronyms that are not widely recognized. A noun or noun phrase is recommended for a form name, though you will find a few form names in this book that violate this guideline.

Now that you are familiar with the environment, let's develop the application using the default form, Form1, in the project that you created in Step 1-4.

Step 1-5: Change the form name and title:

> Right click *Form1.cs,* and rename it **IceCreamCost.cs**. Notice that the name of the form in the Properties window and the names of all three files under Form1.cs are changed.
>
> The title of the form still remains Form1. To change the title, select the Text property of the form in the Properties window and change it to **Ice Cream Cost**.

Specifying the startup form

A project typically contains multiple related forms. The Program.cs file contains the startup program called Main that includes the following statement, where *FormName* represents the name of the startup form.

Application.Run(new *FormName*());

Step 1-6: Double click on Program.cs in the Solution Explorer window, and make sure that the name of the startup form is IceCreamCost, as shown below:

Application.Run(new IceCreamCost());

Running a form

Step 1-7: Run the form by clicking the *Start* button (shown below) from the Toolbar.

Visual Studio displays the empty form. Because you haven't added any controls to the form, there is not much you can do with it. Close the form by clicking the Close button ("x") at the top right of the form.

Additional properties of forms

Let's look at a couple of properties that apply to the form only when it is run.

Click on the *start position* property of IceCreamCosts.cs, and view its description displayed at the bottom of the Properties window. If the description is not visible, expand the description area to make it visible.

Click the dropdown arrow to the right of StartPosition to view the options. Select *Center Screen* from the list. Run the form and see how the property impacts the position of the form. Change the property back to *WindowsDefaultPosition*.

Repeat for *WindowState* property, setting it to *Maximized*.

Review Questions

1.12   What property of a form would you use to change the title displayed on the title bar of the form?

1.13   When you run a project containing multiple forms, which form does C# run?

1.4 Adding Controls to a Form

The term **control** means objects like TextBoxes, Labels and Buttons. Typically, **TextBoxes** are used to enter data when you run the form, **Labels** to display a label or computed results and **Buttons** to run a program segment.

Labels and TextBoxes

As stated above, a TextBox allows you to enter data into it when you run the form, whereas a Label control can only display data. Generally, TextBoxes are used to enter input data, and Labels to display output or labels. The Label and TextBox controls are represented in the Toolbox, as follows:

    A   Label   [abl]   TextBox

Text property of Label

To display text in a Label in the Design window, type the text in the box to the right of the **Text property** of the Label.

AutoSize property of Label

When the AutoSize property of a Label is set to its default value, True, you cannot adjust the size of the Label; the size of the Label is determined by what is displayed in it. If you want to adjust the size, you need to set AutoSize to False.

Naming controls

To make programs easy to understand, it is important to give meaningful names to all controls that are used in the code. In addition, use standard prefixes for naming controls on a form. Table 1-1 shows standard prefixes for some commonly used controls.

Table 1-1: Prefixes for controls

Control	Prefix	Example
TextBox	txt	txtCourseNumber
Label	lbl	lblTotalSales
Button	btn	btnComputeTotal
ComboBox	cbo	cboProduct
ListBox	lst	lstIngredient
ListView	lvw	lvwPrice

We will introduce prefixes for other controls later when we use them.

To create the form, IceCreamCost, we need to add several controls from the Toolbox window. First, you need to make sure the Toolbox is displayed.

Step 1-8: Display controls in the Toolbox as follows:

> If the Toolbox tab is not displayed, select *View, Toolbox*.
> If the Toolbox appears as a tab, click on the Toolbox tab, and then click the Auto Hide button at the top-right of the window.
> Expand *Common Controls*. See Figure 1-13.

Figure 1-13: Toolbox controls

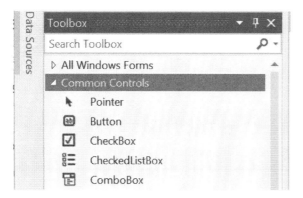

Next, you will add the controls shown in Figure 1-14 to the form. Note that if you use Windows 7, the title of the form won't be centered.

Figure 1-14: The Designer window with the form

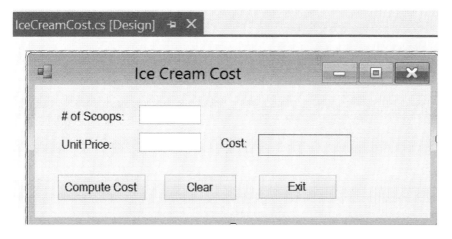

You may want to **resize** a form before or after adding controls to it. To resize a form, move the cursor to the border of the form, and click and drag the cursor when it turns to a double-headed arrow.

You may add a control by dragging and dropping it on the form, or by double clicking it.

Step 1-9: Add a TextBox to enter the number of scoops, as follows:

> Drag and drop a TextBox from the Toolbox menu to the form and position it at the location shown in Figure 1-14, and resize it as necessary.

Click the ellipsis (...) to the right of Font property to set the FontSize to 10 and the BorderStyle to *Fixed3D*.

Change Name property of the TextBox to **txtScoops**.

Step 1-10: Add a Label to display the label, "# of Scoops":

Drag and drop a Label and position it to the left of the TextBox.

Change its AutoSize and Text properties, as follows:

AutoSize:	False
Text:	# of Scoops

Step 1-11: Add the remaining seven controls and set their properties, as specified in Table 1-2:

Unless specified otherwise, set the FontSize to 10 for all controls, and BorderStyle to None for Labels, and Fixed3D for TextBoxes. Note that the only Label that is given a name (lblCost) is the one that is used in the code to dislay the cost. Other Labels that are not used in the code keep their default names.

Table 1-2: Controls on the form IceCreamCost and their properties

Purpose	Type of control	Property	Setting
Enter number of scoops	TextBox	Name	txtScoops
Display label for scoops	Label	Text	# of Scoops:
Enter unit price	TextBox	Name	txtUnitPrice
Display label for unit price	Label	Text	Unit Price:
Display cost	Label	Name	lblCost
		BorderStyle	FixedSingle
		Text	(blank)
Display label for cost	Label	Text	Cost:
Compute cost	Button	Name	btnComputeCost
		Text	Compute Cost
Clear controls	Button	Name	btnClear
		Text	Clear
Close the form	Button	Name	btnExit
		Text	Exit

Closing and opening the Designer window and the project

To save the changes to the form, click the **Save** button on the Toolbar. The **Save All** button would save changes to all forms within a project. The Save and Save All buttons appear on the Toolbar as follows:

To close the Designer window, click the **Close** button (x) on the tab above the form (see Figure 1-14), or select **File, Close**.

To open the Designer window, double click the file name IceCreamCost.cs in the Solution Explorer, or right click the file name and select **View Designer**.

To close the project, Select: **File, Close Solution.**

To open the project and the solution from within Visual Studio,

> Select File, Open, Project/Solution,
> Open the project folder, Ch1_IntroProgramming,
> Select Ch1_IntroProgramming.csproj (or, Ch1_IntroProgramming.sln), as shown below, and
> Click Open.
> If file extensions are not shown along with file names, please see the next subsection, **Display File Extensions**.

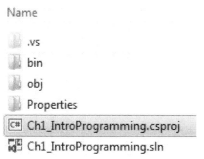

To open the project from outside Visual Studio, open the project folder Ch1_IntroProgramming from File Explorer (in Windows 10), or Windows Explorer (in earlier versions of Windows). The project folder displays the following folders and files:

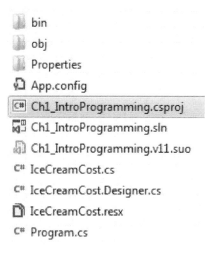

Double click the file Ch1_IntroProgramming.csproj or Ch1_IntroProgramming.sln to open the project and the solution.

Display file extensions

In File Explorer (Windows 10), select View, and check the CheckBox, *File name extensions*.

In Windows Explorer (Windows 8.1), select *Organize, Folders and search options, View*.
Uncheck the CheckBox, *Hide extensions for known file types*.

Phases completed

In Steps 1-1–1-11 of Tutorial 1, you developed the Graphical User Interface, which is phase 2 of the program development process. Earlier, you also identified the subtasks of the program in phase 3:

>   Get number of scoops and unit price
>   Compute cost
>   Display cost

This application doesn't involve the development of files or databases (phase 4). The next phase of program development is phase 5: Write and test the code.

Review Questions

1.14   What control would you use to let the user enter a data item when you run a form?

1.15   What property of a Label do you set to display a data item in a Label?

1.5 The Code Editor Window

Before you write and test the code, let's take a look at the Code Editor window (or Code window) that includes some initial code already generated by C#.

Step 1-12: Open Code window, as follows:

>   Select IceCreamCost.cs in Solution Explorer, and
>   click *View Code* button (<>) on the top right
>   (or right click IceCreamCost.cs and select *View Code*).
>   See Figure 1-15.

Figure 1-15: Opening Code window

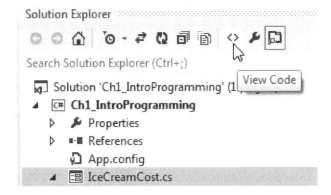

Figure 1-16 shows the Code window.

Figure 1-16: The Code window

```
IceCreamCost.cs*   IceCreamCost.cs [Design]*
Ch1_IntroProgramming.IceCreamCost                    IceCreamCost()
 1  using System;
 2  using System.Collections.Generic;
 3  using System.ComponentModel;
 4  using System.Data;
 5  using System.Drawing;
 6  using System.Linq;
 7  using System.Text;
 8  using System.Threading.Tasks;
 9  using System.Windows.Forms;
10
11  namespace Ch1_IntroProgramming
12  {
13      public partial class IceCreamCost : Form
14      {
15          public IceCreamCost()
16          {
17              InitializeComponent();
18          }
19      }
20  }
```

You may **display line numbers** in Code Editor window as follows:

> Select *Tools, Options*.
> Expand *Text Editor* by clicking the triangle on its left.
> Expand *C#*.
> Select *General* and check the CheckBox *Line numbers* as shown below:

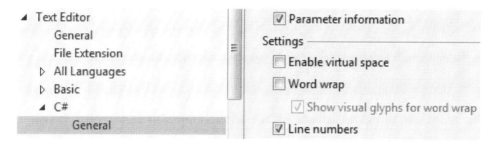

Let's take a look at what the statements mean.

Lines 1–9: The statements in lines 1–9 that start with the key word *using* (e.g., using System.Windows.Forms) are called directives. Each *using* directive specifies a namespace, like **System.Windows.Forms**. A **namespace** is just a container for a group of prebuilt programs (called **classes**) from the .Net Framework Class Library.

For example, the **System.Windows.Forms** is a namespace that contains the classes that define the TextBox, Label, Button and other controls that are commonly used in a form. The directive, using System.Windows.Forms, makes those classes available within the form. The C# compiler looks in the specified namespaces for classes used in the program.

What Is a Class?

Without going into a lot of detail, a **class** may be viewed as a group of program units called **methods,** and data items that are represented by **properties**. A method is a program unit that performs a particular task. For example, a class representing customer orders may have one method (program unit) to compute the sales tax and another method to compute the total cost. The properties of the class may include product number and order quantity.

Thus, namespaces contain classes, and classes contain methods. A method consists of a group of statements that perform a task.

Line 11: namespace Ch1_IntroProgramming

This statement specifies the project's namespace, which is essentially the group of classes within the project. The beginning brace on line 12 and ending brace on line 20 indicate that everything in between is part of the project's namespace.

Line 13: public partial class IceCreamCost : Form

This statement shows the name of the class IceCreamCost that specifies the form. The key word **Public** means the code units (methods) and properties of this form are accessible to other classes, if the code units are public. **Partial** means that the class may be split to different files.

Everything between the starting brace (line 14) and the closing brace (line 19) is part of this class.

Line 15: public IceCreamCost()

This statement is the header of a method named IceCreamCost within the class IceCreamCost.

Developing the Code

As you observed in the above code, C# uses a pair of left (opening) and right (closing) **braces**, {}, to enclose a block of statements.

Before we write the code, let's look at the input, process and output structure of the program, shown in Figure 1-17. The output that the user wants is the cost of ice cream. The cost is computed using unit price and the number of scoops, which are the inputs. The processing consists of computing the cost (= unit price x number of scoops).

Figure 1-17: Input, process and output of the program

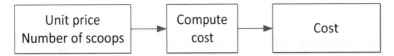

Thus, developing the code involves the following subtasks that were identified in Phase 3, earlier:

    Get number of scoops and unit price
    Compute cost
    Display cost

You need to develop code to do each of the above subtasks.

Event-Driven Programs and Event Handlers

In the .Net environment, code often is connected ("attached" or "wired") to specific events, like clicking a button, loading a form or pressing a key. Therefore, programs developed in .Net are called **event-driven programs**. A method that runs in response to an event such as clicking a button is called an **event handler**.

For example, to compute cost when the user clicks a button, we create a program unit (method) that is connected to the click event of the button. Because the method runs in response to the click event, it is called the click event handler of the button.

Creating event handlers

Let's look at how an event handler is created.

Step 1-13: Create click event handler method for btnComputeCost:

Switch back to the Design view by clicking the designer tab, IceCreamCost.cs[Design].

Double click on the Compute Cost button to reopen the Code window and to create the **click event handler**, shown below:

```
private void btnComputeCost_Click(object sender, EventArgs e)
{


}
```

The first line is the header of this method, which shows the method's name, **btnComputeCost_Click**. The key word **private** indicates that this event handler method can be accessed only from this form, and **void** means that this method doesn't return any value to the calling program. These terms and the concept of parameters within the parentheses are explained in later chapters.

Deleting event handlers

Suppose you created an event handler that you don't need. For example, you created the Click event handler of a button and decided to delete the button. Before you delete the button, you need to delete the event handler to avoid an error.

If you manually delete the event handler method from the code window, you will get an error when you try to open the form in the Design window because of the reference to the event handler from the generated code. For example, if you create the click event handler for btnExit, and delete the code for the event handler, you will get the following error message when you try to display the form.

1.5 The Code Editor Window

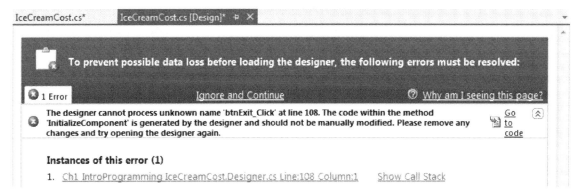

To remove the error, click on the "Go to Code" button on the error window. The **Designer.cs** file, which contains the generated code, opens with the pointer at the line that refers to the click event handler, as shown below:

```
108        this.btnExit.Click += new System.EventHandler(this.btnExit_Click);
```

You may manually delete this statement from the Designer.cs file to get rid of the error and display the form. **Be careful not to change or delete any other line**. Close the Designer.cs file to display the form.

How do you properly delete such an event handler? Use the following steps:

Select the button on the form.

Click the **Events** button (⚡) in the Properties window to display the events for the button, as shown below:

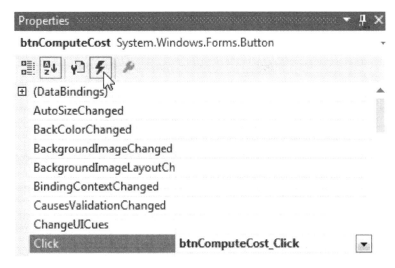

Select the Click event.

Click the dropdown arrow and select the event handler you want to delete (you need to do it even if the event handler already is displayed).

Right click the event handler in the Properties window and select Delete.

This process will delete the event handler if you haven't entered any code within the event handler. If you have entered any code, you still need to select Delete in the Properties window and then manually delete the event handler from the code window.

Step 1-14: Double click btnExit to create the Click event handler. Then delete the event handler by first deleting it from the Properties window, and then deleting the code for the event handler, including the header, the braces (and code, if any, within the braces) from the Code window.

Now you are ready to move to phase 5: Write and test the code.

Review Questions

1.16   What is a namespace?

1.17   What does the *using* directive specify?

1.18   What is a method?

1.19   What is an event handler?

1.6 Introductory Programming Concepts

To help you understand some basic programming concepts, let's first look at the completed code, shown in Figure 1-18, to compute the cost for an order of ice cream.

Figure 1-18: Code to compute cost

```
20          private void btnComputeCost_Click(object sender, EventArgs e)
21          {
22              // Compute the total cost for an order of icecream:
23
24              // Declare variables to hold input data and computed results
25              int scoops;
26              double unitPrice, cost;
27
28              //Get input data from Text Boxes:
29              scoops = int.Parse(txtScoops.Text);
30              unitPrice = double.Parse(txtUnitPrice.Text);
31
32              // Compute result
33              cost = unitPrice * scoops;
34
35              // Display Result
36              lblCost.Text = cost.ToString("C");
37          }
```

As you can see from the code, the method includes three segments corresponding to the three subtasks that were identified earlier:

   Get number of scoops and unit price (lines 29 & 30)
   Compute cost (line 33)
   Display cost (line 36)

Note that the line numbers are not part of the code. They are inserted using a setting in Visual Studio for ease in reference to a statement (select *Tools, Options, Text Editor, C#* and check *Line Numbers*).

Let's look closely at the statements in this program.

Comments

Line 22: //Compute the total cost for an order of ice cream:
Lines starting with // are comments for the programmers and are ignored by the compiler.
You may comment multiple lines by placing the lines between /* and */, as in Figure 1-19.

Figure 1-19: Commenting multiple lines

```
22          /*Compute the total cost for an order of icecream
23             using the formula: cost = unitPrice * NumberOfScoops:
24             Declare variables to hold input data and computed results*/
```

You may comment out a set of lines by highlighting the lines (drag the mouse pointer over them) and then clicking the Toolbar button *Comment out the selected lines*, shown in the following image:

The button to the right would uncomment the selected lines.

1.7 Data Types

Lines 25 and 26 declare variables and specify what types they are:

    Line 25: int scoops;
    Line 26: double unitPrice, cost;

In order to understand the above statements and the rest of the program, we will take a closer look at data types and variables.

Data used in a program can be grouped broadly into the following types:

    Integer: An integer is a whole number (e.g., 255677, -54).
    Nonintegers or reals: A real number has one or more decimal places (e.g., 255677.75; -0.255).
    Text: Text data include **character** type that consists of an individual character (e.g., 'A'; 'B') and **string** type that consists of a group of characters, including alphabetical characters, numeric digits and special characters (e.g., "John Smith"; "P410-2023-3180-05").
    Notice that a character is enclosed in single quotes, and a string is enclosed in double quotes.
    Boolean: A Boolean has the value *true* or *false*.
    Date/Time: Date and time are enclosed in a pair of pound signs, as in #05/25/2016 02:30:00 PM#.

Let's look at the different C# types for integers and reals.

Integer types

Table 1-3 shows the different integer types in C#. Note that C# is case sensitive. That means the key words for data types, like short and int, are case sensitive. So, "Short" and "Int" are invalid key words.

Table 1-3: Built-in C# data types for Integers

C# key word	Size (memory used)	Range of values
sbyte	8 bits	-128 to 127
short	16 bits	-32,768 to 32,767
int	32 bits	-2,147,483,648 to 2,147,483,647
long	64 bits	-9,223,372,036,854,775,808 to 9,223,372,036,854,775,807
char	16 bits	0 to 65535

There are additional "unsigned" data types, **byte, ushort, uint** and **ulong,** which can store only positive, but larger, integers with maximum size close to double that of signed integers. For example, **uint** type variables can hold values up to 4,294,967,295, whereas the maximum for **int** type is only 2,147,483,647.

The size shows the memory needed to store the data, and the range shows the range of values that can be stored in the variable. Larger data types take more memory, as reflected in the size.

Note that **character** data (denoted by the C# key word **char**) is represented by integers using one of the popular standards called Unicode that allows representation of 65,535 different characters from different languages.

C# versus .Net Types

The names of basic data types in C# are different from those in .Net Framework. However, there is direct correspondence between them. For example, **int** type in C# is the same as **System.Int32** in .Net, and **long** in C# is the same as **System.Int64**.

Noninteger (real) types

Table 1-4 shows the data types for non-integers.

Table 1-4: C# noninteger types

C# key word	Size	Range of values	Precision	How the number 123456789012.345678901234567890 may be represented
float	32 bits	$\pm 1.5 \times 10^{-45}$ to $\pm 3.4 \times 10^{38}$	7 digits	0.1234568×10^{12}
double	64 bits	$\pm 5.0 \times 10^{-324}$ to $\pm 1.7 \times 10^{308}$	15 digits	$0.123456789012346 \times 10^{12}$
decimal	128 bits	$\pm 1.0 \times 10^{-28}$ to $\pm 7.9 \times 10^{28}$	28–29 digits	123456789012.345678901234567800

As indicated by the ranges, float and double can store larger numbers than decimal type. They also take less memory. But float and double cannot represent all noninteger numbers exactly (due to the method used by them to represent numbers), making them less suitable than decimals for financial calculations.

Decimal type, in general, can represent real numbers exactly, and it has better precision than float and double. Hence, as stated earlier, decimal is better suited for financial calculations where exact representation is important. However, decimal type takes more memory and is slower to work with than float and doubles. Because decimal has a lower range, it may not be appropriate for some scientific calculations.

The C# types, float and double, use a method of representation called **floating point** method to approximate a wide range of values to a fixed number of significant digits. Therefore, float and double also are referred to as floating point types.

string type

The C# type **string** represents a combination of zero or more characters. "John Smith" and "P410-2023-3180-05" are examples of string type data. Note that string type data is enclosed in quotes.

bool type

The C# type bool represents Boolean data that include the values **true** and **false**.

DateTime type

C# doesn't have a data type for dates and times. The .Net base type DateTime is used to represent dates and/or times.

Review Questions

1.20 What is the appropriate data type to represent student GPA?

1.21 Which real data type is best suited for financial calculations where exact representation of numbers is important?

1.22 What are the advantages of float and double types compared to decimal?

1.8 Working with Variables

Programs typically use variables to hold data. A simple variable holds one piece of data, and more complex types like an array can hold a set of data. In general, we will use the term "variable" to denote a simple variable.

A variable is a location in computer memory where it holds data while the program is running. The program accesses the location using the name given to the variable.

For variables, C# requires that you specify what type of data will be stored in each variable.

Creating (Declaring) Variables

Variables are created by specifying the type and a name of the variable. The statement shown below from Line 25 in Figure 1-18 creates a variable named **scoops** that is of **int** type:

    int scoops;

When the program is run, the above statement creates a variable named *scoops* in computer memory to hold an int type integer data. Because we haven't assigned any value to scoops, by default, the value zero is stored in the variable, as represented below:

scoops

You may declare two or more variables of the same type in a statement, as in line 26 in Figure 1-18:

    double unitPrice, cost;

This statement creates two variables, unitPrice and cost, to hold double type **floating point** data.

Note that a **semicolon** is required to denote the end of each C# statement.

Guidelines for naming variables

Variable names must start with a letter, and can contain only letters, numbers or the underscore "_" character.

The convention in C# is to use the **Camel Case:** The first letter of a variable is lowercase, and the first letter of subsequent words, if any, are capitalized, as in totalCost and averageRetailPrice.

Use meaningful names for variables to make programs easy to read. You should avoid using abbreviations or acronyms that are not widely recognized. For example, use totalQuantitySold rather than tqs.

Variables to hold other types of data

The example of code in Figure 1-18 that we are looking at does not include variables to hold other types of data, like character and string data. Every type of variable is declared using the same syntax. Here are examples of statements to create character and string variables:

    char middleInitial;
    string lastName;

Assigning Values to Variables

After you create a variable, you can assign a value to it using a statement of the form

    variable name = value (or an expression that computes a value)

For example, the variable scoops declared in the program as type **int** may be assigned the value 3, using the statement

    scoops = 3

That is, the value 3 will be stored in a memory location called scoops, as represented below:

    | 3 |
   scoops

Similarly, the statement

    scoops = (10/5) + 1

would assign the value 3 to scoops.

Lifetime of a variable

You cannot use a variable that doesn't exist! That means you cannot use a variable before it is declared, because the variable exits only after it is created using the declaration statement.

Similarly, a variable ceases to exist after completion of execution of the method where it is declared. Thus, the **lifetime** of a variable that is declared within a method is the time period between its creation in memory during execution of the method and the time when the execution of the method is completed. In general, the lifetime represents the time period between the creation of a variable and the destruction of the variable from memory.

A related term is **scope** of a variable, which is the section of a program where the variable can be accessed. The scope of a variable is the block of statements between the variable's declaration and the next right curly brace. The scope of variables is discussed further in section 1.10.

Initializing a variable

You may declare a variable and also assign it a value (initialize the variable) in one statement, using the syntax,

>   type variable name = value

Here are some examples:

>   int scoops = 3;
>   double unitPrice = 3.95;
>   char middleInitial = 'C';
>   string lastName = "Philip";
>   bool validData = true;

Caution! The statement **"3 = scoops"** will not do the same thing as **"scoops=3"**; it is meaningless in a program. The variable name must be on the left side of the equal sign, and the value on the right side.

Default Values of Variables

If a variable is declared without initializing it, the variable will have no value until a value is assigned to it. However, if you use the **new** key word to create a variable, as in

>   **int scoops = new int(),**

the variable will have a default value depending on its type.

The use of the default constructor, new, is explained in Chapters 2, 7 and 13. The default values of a few different types of variables are shown in Table 1-5.

Table 1-5: Default values of data types

Type	Default value
Integers and reals, except decimal types	0
decimal	0m ("m" represents decimal type)
bool	false
DateTime	# 01/01/0001 12:00:00AM #
String	null reference

Thus, the statement **int scoops = new int()** has the same effect as **int scoops = 0**.

Representing Numbers

Scientific notation

You may represent very large numbers using scientific notation:

    3.95e8

e8 represents 10 to the power 8. Thus, 3.95e8 is equal to 3.95 times 10^8—that is, 395,000,000.

Similarly, 3.95e-8 represents 3.95 times 10 to the power -8—that is, 0.0000000395.

Numbers can have only numeric digits, decimal point, minus and plus signs, and the letter "e" or "E" when using scientific notation. Commas and dollar signs ($) are invalid within a number. So, the following statements are invalid:

```
double unitPrice = $1.50;      //Invalid; dollar sign is not allowed in a number
double cost = 3,520.00;        //Invalid; comma is not allowed in a number
double amount = "15.45";       //Invalid; "15.45" is a string
```

Caution! In general, the compiler does not like it when you assign the wrong type of value to a variable. The following statement would result in an error because scoops is declared as type int, which is a type of integer, but 1.5 is a real number.

```
int scoops = 1.5        //Invalid; 1.5 is not an integer
```

We will look at specific rules on mixing different types of integer and real types in Chapter 2.

Assigning the Value of One Variable to Another Variable

You may assign the value of one variable to another variable, as in the following example:

```
int scoops = 3;
int temp;
temp = scoops;
```

The statement "**temp = scoops**" copies the value of scoops into the variable temp. After execution of this statement, the storage locations for both variables will contain the same value 3:

What happens if you change the value of scoops by adding the following statement after the above code?

```
scoops = 5;
```

Does it change the value of temp? The answer is no, because each variable has its own storage location. After execution of the statement "**scoops = 5,**" the two variables will have the following values:

Caution! The statement

> temp = scoops;

which assigns the value of scoops to temp, is not the same as

> scoops = temp

which assigns the value of temp to scoops.

Tutorial 2: Developing the Code

In this tutorial, you enter the code shown in Figure 1-18 into the click event handler of the button that computes the cost.

Step 2-1: Declare variables scoops, unitPrice and cost:

> Make sure the Code window is open.
> Enter the following two statements after the opening brace within the btnComputeCost_Click method that was created in Step 1-13, as shown in Figure 1-20. If you don't have this method, double click the btnComputeCost button to create it.
> > int scoops;
> > double unitPrice, cost;

Figure 1-20: Click event handler for btnComputeCost button

```
15      public IceCreamCost()
16      {
17          InitializeComponent();
18      }
19
20      private void btnCompteCost_Click(object sender, EventArgs e)
21      {
22          int scoops;
23          double unitPrice, cost;
24      }
```

As stated earlier in secton 1.6, C# is case sensitive. That means the key words for data types, like int and double are case sensitive. So, if you type "Int scoops" instead of "int scoops," you will get an error when you run the project.

IntelliSense

As you may have noticed, when you type code, the IntelliSense list box pops up with a list of items you can choose from to help you complete the code. For example, when you type the letter "i" at the beginning of a statement, the following list box is displayed.

```
private void btnComputeCost_Click(object sender, EventArgs e)
{
    i
}
    InstallerTypeAttribute
    InstanceCreationEditor
    InsufficientExecutionStackException
    InsufficientMemoryException
    int
    Int16
```

You may select "int" by pressing the Tab key.

Next, do Step 2-2 to see what happens if you assign the wrong type of data to a variable.

Step 2-2: Experiment with assigning different types of data to the variables:

Add each of the following statements, identify the error in each (hover the cursor over the number), write down in your own words what the problem is and how to correct it, and delete the statement.

scoops = 1.5;
cost = $3.95;
unitPrice = 1,575.50;
cost = "3.95";

What's special about double types?

If you are wondering why we used only double type (not float or decimal) for variables that store real numbers like unit price and cost, it is because all real numbers, by default, are treated as double type. For example, 3.5, 1575.0 and 0.6 are double types. As we will see in Chapter 2, it is invalid to assign such double type numbers to a float or decimal type variable without appropriate conversion of the double type.

Getting Values from a TextBox

The data you type into a TextBox, whether it is a number or a string, is represented as a string. So, if you enter 3.95 into the TextBox, txtUnitPrice, the Text property of the TextBox, txtUnitPrice.Text, holds the string "3.95," not the number 3.95.

Because the variable unitPrice is of double type, the value that is assigned to it cannot be a string; the value (txtUnitPrice.Text) must be converted to double before assigning it to the variable unitPrice.

Parse Method: Converting String Type Data to a Numeric Type

Every data type has a set of built-in programs called **methods** to do different tasks related to the data type. Numeric data types have a method called **Parse** that converts a string data to the numeric data type by parsing each character.

Let's look at the statements in lines 29 and 30 in Figure 1-18. The statement in line 29,

    29   scoops = int.Parse(txtScoops.Text);

uses the **Parse()** method of int type to convert the string, txtScoops.Text, to an int type number. The converted number is assigned to the variable scoops.

Similarly, in line 30,

    30   unitPrice = double.Parse(txtUnitPrice.Text);

double.Parse(txtScoops.Text) converts the string value txtUnitPrice.Text to double data type, which is assigned to unitPrice.

The Parse method would result in error at runtime if the string to be converted is empty. For example, if the TextBox, txtUnitPrice, is empty, the program would break at line 30 with an error message. Further, the Parse method can covert only a string to numeric data types. It cannot convert one numeric data type, like int, to another numeric data type, like double.

Convert Class: An Alternative to Parse Method

Convert is a .Net class that has different methods that convert to/from base .Net data types. As discussed earlier under "C# vs. .Net Types," the data types in C# are different from the base .Net data types. However, there is direct correspondence between them. For example, the int type in C# is the same as **Int32** in .Net, and double in C# is the same as **Double** in .Net. Because Convert is a .Net class, it can be used in other .Net languages like Visual Basic as well.

The following statement converts the string, txtScoops.Text, to int type number, using the ToInt32() method.

        scoops = Convert.ToInt32(txtScoops.Text);

You may use the above statement in place of the current statement in line 29 that uses the Parse method.

Similarly, the ToDouble() method of Convert class converts the string, txtScoops.Text, to Double type.

        unitPrice = Convert.ToDouble(txtScoops.Text);

The Parse method can convert only a **string** to numeric data types, but the methods of Convert class also can be used to covert a numeric data type to another numeric data type.

Step 2-3: Add the following code to convert the strings, txtScoops.Text and txtUnitPrice.Text, to numeric values, and to store them in variables:

    scoops = int.Parse(txtScoops.Text);
    unitPrice = double.Parse(txtUnitPrice.Text);

Review Questions

1.23 What is the Camel Case convention to name variables in C#?

1.24 What are the requirements for variable names?

Correct errors, if any, in the following statements:

1.25 int orderQty = 4.5;

1.26 char lastName = "Jones";

1.27 string firstName = 'Jack';

1.28 double totalAmount = "145.50";

1.29 double totalCost = 2,145.00;

1.30 double unitPrice = $15.50;

1.31 int quantity;
     15 = quantity;

1.32 int quantity = txtQuantity.Text; (txtQuantity is the name of a TextBox)

1.33 int units = double.Prase(txtUnits.Text); (txtUnits is the name of a TextBox)

What would be the values of the variables currentSales and previousSales after execution of the following set of statements?

1.34 int currentSales = 110;
     int previousSales;
     currentSales = previousSales;

1.35 int currentSales = 110;
     int previousSales;
     previousSales = currentSales;

1.9 Doing Calculations

Common mathematical operators used in calculations are shown in Table 1-6.

Table 1-6: Math operators to do calculations

Operator	Operation	Example
*	multiply	Multiply 3.95 by 3 (3.95*3)
/	divide	Divide 11.85 by 3 (11.85/3)
+	add	Add 11.85 and 0.65 (11.85+0.65)
-	subtract	Subtract 1.25 from 12.5 (12.5-1.25)
-	unary minus, for negative numbers; e.g., -350.50	Negative 2.5 (-2.5)

The statement shown below from Line 33 in Figure 1-18 multiplies the unit price and number of scoops to compute the cost.

33   cost = unitPrice * scoops;

Step 2-4: Type in the code to compute cost:

    cost = unitPrice * scoops;

Creating Expressions: Precedence of Operators

In an expression, multiplications and divisions are done before additions and subtractions, and then calculations take place from left to right. Consider the following expression:

    5+4*3-2/2

The above expression would evaluate to 5+(4*3)-(2/2), yielding 16.

Here is another example:

    12/4*2

This expression would be evaluated from left to right as (12/4)*2, yielding 6.

It is always a good practice to use parentheses to make the order of calculations clear and to avoid unintended errors.

Review Questions

Evaluate the following expressions:

1.36   10/2+3*4

1.37   12-6/2*3+1

1.10 Formatting and Displaying Output in a Label

What you see in a Label is the value of its Text property. Earlier in Step 1-10, you displayed text in a Label in the Designer window using its Text property. Now we look at how to display a text in a Label using code at runtime. To do this, you assign the text to the Text property of the Label.

The statement in line 36,

>   lblCost.Text = cost.ToString("C");

displays the value of cost in the Label, lblCost, by assigning it to the Text property of lblCost.

The value of cost, which is a double type, is converted to a string using the ToString() method of the variable, described next.

Formatting Numbers Using the ToString() Method

Each numeric data type has a method (program unit) called ToString() that converts and formats the number to a string. For example,

>   cost.ToString()

would convert the value of cost to a string so that it can be displayed in a Label.

The ToString() method also allows a format code (string) to be specified, as in

>   cost.ToString("C")

where "C" specifies currency format.

Examples of using additional ToString() format codes are presented in Table 1-7.

Table 1-7: ToString() format codes

Number	Format string	ToString() would format the number as
1145.857	null	1145.857
1145.857	"C"	$1,145.86
1145.857	"N"	1,145.86
1145.857	"F"	1145.86
0.156	"P"	15.60%
Optionally, you may specify the number of decimal places. Here are some examples:		
1145.857	"C0"	$1,146
-1145.857	"C1"	($1,145.9)
-1145.857	"N1"	-1,145.9
1145	"N3"	1,145.000
Format code for integers only		
157	"D5"	00157
157	"D"	157
157	"D1"	157

If the format code doesn't specify the number of decimal places, as in **cost.ToString("C")**, the default number of decimal places is two.

Step 2-5: Add the statement to display cost:

> lblCost.Text = cost.ToString("C");

The btnComputeCost_Click event handler should look as shown in Figure 1-21.

Figure 1-21: Click event handler of btnComputeCost button

```
20      private void btnComputeCost_Click(object sender, EventArgs e)
21      {
22          /*Compute the total cost for an order of icecream
23             using the formula: cost = unitPrice * NumberOfScoops:
24             Declare variables to hold input data and computed results*/
25          int scoops;
26          double unitPrice, cost;
27
28          //Get input data from Text Boxes:
29          scoops = int.Parse(txtScoops.Text);
30          unitPrice = double.Parse(txtUnitPrice.Text);
31
32          // Compute result
33          cost = unitPrice * scoops;
34
35          // Display Result
36          lblCost.Text = cost.ToString("C");
```

Step 2-6: Run the form:

> Double click **program.cs** to open the code editor window.
> Specify IceCreamCost as the name of the form to be run, as shown:
>
>> Application.Run(new IceCreamCost());

Click the *Start* button on the Toolbar to run the form.
Type in 3 for scoops and 3.95 for unitPrice. Click the Compute Cost button to see the result.
Close the window.

Return to the Code Editor window by clicking on the IceCreamCost.cs tab.
Change the format string "C" to "C0," run the program and observe the difference in format.

Repeat with format strings "C1," "N" and "F," and observe the format in each case.

Run the form with no data for unit price. The program breaks at line 30, with an error message as shown below, because txtUnitPrice.Text is an empty string.

```
unitPrice = double.Parse(txtUnitPrice.Text);

// Compute result                 ⚠ FormatException was unhandled
cost = unitPrice * scoops;        An unhandled exception of type 'System.FormatException' occurred in mscorlib.dll

// Display Result                 Additional information: Input string was not in a correct format.
```

Clearing Controls

TextBoxes have a Clear() method that clears their content. The Labels, however, are cleared by setting the Text property to null (which is nothing) or an empty string represented by two quotes with nothing in between ("").You also may use this method for clearing TextBoxes.

Step 2-7: Double click the Clear button to create the click event handler of the button, and type in the code to clear the controls:

```
private void btnClear_Click(object sender, EventArgs e)
{
   txtScoops.Clear();
   txtUnitPrice.Clear();
   lblCost.Text = null;
}
```

Note that Clear is the name of a method (a program unit), and methods require the pair of parentheses at the end of their names.

Closing a Form

To close a form that is running, use the Close() method of the form.

Step 2-8: Double click the Close button, and type in the code within the click event handler:

```
private void btnExit_Click(object sender, EventArgs e)
{
   this.Close();
}
```

In the above code, "this" refers to the current form.

The Complete Code for the IceCreamCost Form

Figure 1-22 shows the complete code.

Figure 1-22: The complete code for IceCreamCost

```
using System;
using System.Collections.Generic;
using System.ComponentModel;
using System.Data;
using System.Drawing;
using System.Linq;
using System.Text;
using System.Threading.Tasks;
using System.Windows.Forms;
```

```csharp
namespace VCintro
{
    public partial class IceCreamCost : Form
    {
        public IceCreamCost()
        {
            InitializeComponent();
        }

        private void btnComputeCost_Click(object sender, EventArgs e)
        {
            //Compute the total cost for an order of ice cream:

            //Declare variables to hold input data and computed results
            int scoops;
            double unitPrice, cost;

            //Get input data from Text Boxes:
            scoops = int.Parse(txtScoops.Text);
            unitPrice = double.Parse(txtUnitPrice.Text);

            // Compute result
            cost = unitPrice * scoops;

            // Display Result
            lblCost.Text = cost.ToString("C");
        }

        private void btnExit_Click(object sender, EventArgs e)
        {
            this.Close();
        }

        private void btnClear_Click(object sender, EventArgs e)
        {
            txtScoops.Clear();
            txtUnitPrice.Clear();
            lblCost.Text = null;
        }
    }
}
```

Step 2-9: Run the form:

Test the Clear and Exit buttons.

Experimentation with Expressions

To see for yourself how C# evaluates an expression, do Step 2-10.

Step 2-10: To give $1 off on each scoop, change the expression for computing the cost as follows:

cost = unitPrice-1 * scoops;

Click the Start button to run the program. Enter 2 for scoops and 3.5 for unit cost.
Before clicking the button, compute the value of cost manually.
Note: If you manually computed 3.5-1 * 2 to be 5, you are wrong.

Next, click the Compute Cost button to display the actual value.
Why is it 1.5? Because multiplication is done before addition, the expression is evaluated as 3.5-(1*2), which yields 1.5.

To specify that the expression, (unitPrice-1), should be multiplied by scoops, use parentheses around the expression, as in (unitPrice-1)*scoops. Run the program again with the same values for scoops and unitPrice (2 and 3.5, respectively). Check the value of cost. Now, it should be evaluated as (3.5-1)*2, yielding 5.

Change the expression back to cost = unitPrice * scoops.

Invalid numbers

To see for yourself errors due to invalid data, do Step 2-11.

Step 2-11: Run the program, and enter 2 for scoops and $3.95 for unitPrice.

Click the Compute Cost button.
The program breaks and highlights the statement that contains the error
    unitPrice = double.Parse(txtUnitPrice.Text);
and displays the message

 FormatException was unhandled
Input string was not in a correct format.

The error occurs because $3.95 is not a valid number; hence, it cannot be converted to a decimal.

The Parse method, however, accepts numbers with commas. For example 2,450 is valid. To test it, run the program, enter 2 for scoops and 2,450 for unitPrice (it is some expensive ice cream!) and compute the cost. But, remember, it is invalid to assign a number that contains a comma to a variable, as in

cost = 2,450;   // invalid!

It's time to practice! Do Exercise 1-1 (uses integer data) and Exercise 1-2 (uses real type data), provided at the end of the chapter.

1.11 Additional Concepts: Constants, Scope of Variables and ListBox

Now we expand on the concepts you learned in previous sections and introduce constants, scope of variables, and formatting multiple lines of output using ListBox. We use an expanded version of the application developed in Tutorials 1 and 2 to help understand these topics.

Tutorial 3: Ice Cream Cost with Sales Tax

This tutorial expands the application developed in Tutorials 1 and 2 by adding sales tax to compute the total cost. The subtotal, sales tax and total cost are formatted and displayed in a ListBox on a new form called **IceCreamTotal,** as shown in Figure 1-23.

Figure 1-23: IceCreamTotal form with sales tax

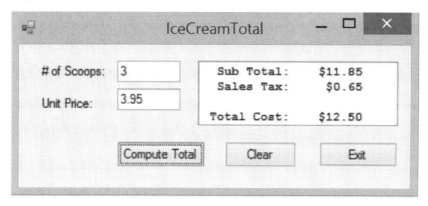

Figure 1-24 shows the input, process and output of the program. The output now consists of the subtotal and sales tax in addition to the cost.

Figure 1-24: Input, processing and output for IceCreamTotal

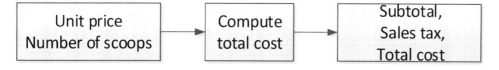

The process, compute total cost, involves computing the following expressions:

    Subtotal = Unit price * Number of scoops
    Sales tax = Subtotal * Sales tax rate
    Total cost = Subtotal + Sales tax

Note that the sales tax rate, which is needed to compute sales tax, is not a part of the input. Then where does the program get it from? The value of sales tax rate is "hard-coded" into the program, as shown in Figure 1-25—that is, it is a part of a program statement, which keeps the example simple but makes it difficult to change the rate.

Step 3-1: To create this application, add a new form to the project, Ch1_IntroProgramming:

   Select: *Project, Add Windows Form*
Change the name of the file form1.cs in Solution Explorer to **IceCreamTotal.cs**.
Add the controls shown in Figure 1-23, as follows:

   Rather than adding the Labels, TextBoxes and Buttons to the form, you may copy the following controls from IceCreamCost and paste them to the current form:

   Labels and TextBoxes for number of scoops and unit price
   Buttons for compute total, clear and exit

To copy multiple controls from a form, you may select the controls by dragging the pointer over an area that contains the controls, or you may select individual controls while the shift key is pressed. Right click any of the selected controls, and select *Copy*.

To paste the control on IceCreamTotal, right click the form, and select *Paste*.

In addition, add a ListBox and rename it **lstOutput**.

Step 3-2: Double click btnComputeTotal to create the Click event handler of the button, as shown below:

   private void btnComputeTotal_Click(object sender, EventArgs e)
   {
   }

The complete code required to compute and display the total cost is shown in Figure 1-25. You will enter each segment of the code after we discuss the code.

Figure 1-25: Code to compute and display total cost

```
20          private void btnComputeTotal_Click(object sender, EventArgs e)
21          {
22              // Compute the total cost for an order of icecream, including sales tax.
23
24              // Declare variables to hold input data and computed results
25              int scoops;
26              double unitPrice, subTotal, salesTax, totalCost;
27
28              //Declare a constant to hold the tax rate
29              const double TAX_RATE = 5.5;
30
31              //Get input data from Text Boxes:
32              unitPrice = double.Parse(txtUnitPrice.Text);
33              scoops = int.Parse(txtScoops.Text);
34
35              // Compute results
36              subTotal = unitPrice * scoops;
37              salesTax = subTotal * TAX_RATE / 100;
38              totalCost = subTotal + salesTax;
39
40              // Display Results with decimals aligned
41              lstOutput.Items.Add(string.Format("{0,12}{1,10:C}", "Sub Total:", subTotal));
42              lstOutput.Items.Add(string.Format("{0,12}{1,10:C}", "Sales Tax:", salesTax));
43              lstOutput.Items.Add("");   // Add a blank line
44              lstOutput.Items.Add(string.Format("{0,12}{1,10:C}", "Total Cost:", totalCost));
45          }
```

Lines 25 and 26 declare variables as in Tutorial 2.

Constants

Line 29 declares a **constant**. The key word **const** in line 29,

> const double TAX_RATE = 5.5;

specifies that TAX_RATE is a constant (not a variable).

Constants are used to represent literal values that do not have to be changed at runtime. The use of a constant for tax rate assumes that the tax rate is a fixed value. However, if the tax rate is to be read from a file or a TextBox at runtime, a variable would be more appropriate.

The constant has the following characteristics:

1. The value of a constant cannot be changed at runtime.
   The following code that assigns a value to the constant, numberOfScoops, at runtime, is invalid:

   > const int numberOfScoops = 0;
   > numberOfScoops = 2;   // Invalid

2. A constant must be initialized as part of the declaration by assigning a literal value or a constant. For example, the statement,

   > const int numberOfScoops = int.Parse(txtScoops.Text); // Invalid

   is invalid because the value assigned to the constant is not a constant or a literal value. Similarly, the following statement also is invalid because the constant is not initialized as part of the declaration:

   > const int numberOfScoops;   // Invalid

3. A constant is replaced by its value at compile time.

Step 3-3: Double click btnComputeTotal to create its click event handler, and add the comments and statements to declare the variables and a constant, as shown in Figure 1-26.

Figure 1-26: Code to declare variables and constant

```
22          // Compute the total cost for an order of icecream, including sales tax.
23
24          // Declare variables to hold input data and computed results
25          int scoops;
26          double unitPrice, subTotal, salesTax, totalCost;
27
28          //Declare a constant to hold the tax rate
29          const double TAX_RATE = 5.5:
```

Figure 1-27 shows the code to get input data from TextBoxes (the same as in Tutorial 2), and to compute subtotal, sales tax and total cost.

Figure 1-27: Code to get input data and compute results

```
31              //Get input data from Text Boxes:
32              unitPrice = double.Parse(txtUnitPrice.Text);
33              scoops = int.Parse(txtScoops.Text);
34
35              // Compute results
36              subTotal = unitPrice * scoops;
37              salesTax = subTotal * TAX_RATE / 100;
38              totalCost = subTotal + salesTax;
```

The statement in line 32 converts the data typed into txtUnitPrice to double type because unitPrice is a double. Similarly, in line 33, data from txtScoops is converted to int type because scoops is an int type variable.

Note that when an expression has multiplication and division, as in line 37, the two operations are treated as equal in deciding the order of computation. The calculations take place from left to right, as in (subTotal * TAX_RATE) / 100, which is okay here because it doesn't make any difference whether multiplication is done first or division is done first.

Step 3-4: Enter code from Figure 1-27.

Displaying Data in a ListBox

In this application, the output is displayed in a ListBox in two columns, the first column for the labels and the second column for the numbers, as shown in Figure 1-28. The decimals are aligned automatically by right justifying the numbers in the second column and formatting them with two decimal places.

Figure 1-28: Formatted output

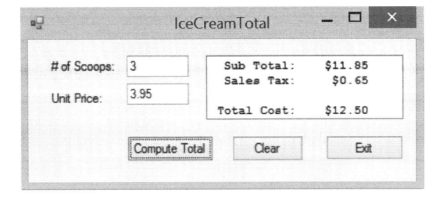

A ListBox can be used to display items and also let users select one or more items from the list. Here we use the ListBox just to display output. We will discuss ListBox in more detail in Chapter 5.

string.Format() method

First, we look at formatting data using the Format() method of string type, which has the following syntax:

    string.Format(format string, data items separated by commas)

The Format method formats multiple data items and combines them into a single string. For example, the following statement formats the first line of the output shown in Figure 1-28:

    string.Format("{0,12}{1,10:C}", "Subtotal:", subTotal)

This statement lists two data items, a literal and a variable, to be formatted: "Subtotal:" and subTotal. These two items are identified by indices 0 and 1, respectively.

The format string, "{0,12}{1,10:C}," specifies that

    parameter 0 ("Subtotal") should be displayed in a column that is 12 characters wide, and
    parameter 1 (subtotal) should be displayed in a column that is 10 characters wide.

The "C" in {1,10:C} specifies currency format (default is two decimal places).

By default, each data item is right justified within its column.

Displaying output in a ListBox

Each line (item) in a ListBox is a member of the **Items collection** of the ListBox. The **Add() method** adds a new line to the Items collection of the ListBox. For example,

    lstOutput.Items.Add("")

would add a blank line to the ListBox.

The following statement would display a title in the ListBox:

    lstOutput.Items.Add("Ice cream cost with sales tax");

We will discuss collections and its methods in detail in Chapter 9.

To display the first line of output, you use the string.Format method, as follows:

    lstOutput.Items.Add(string.Format("{0,12}{1,10:C}", "Subtotal:", subTotal));

The following statements from lines 43 and 44 use the same format string as the one used to display the subtotal. Because all three values (subtotal, sales tax and total cost) are right justified (by default) and displayed using the same column width, the decimal points of the three numbers are automatically aligned.

    43   lstOutput.Items.Add(string.Format("{0,12}{1,10:C}", "Sales Tax:", salesTax));
    44   lstOutput.Items.Add(string.Format("{0,12}{1,10:C}", "Total Cost:", totalCost));

Additional formatting codes presented earlier in Table 1-7 for the ToString() method also apply to string.Format(). These include N, F, and P and others.

Font for ListBox

For the decimals to align, it also is important that the font for the ListBox is set to a fixed-width font like Courier New.

Step 3-5: Change the font type for the ListBox to *Courier New*. Add the following code from Figure 1-29 to display results in the ListBox.

Figure 1-29: Code to display formatted output in ListBox

```
40          // Display Results with decimals aligned
41          lstOutput.Items.Add(string.Format("{0,12}{1,10:C}", "Sub Total:", subTotal));
42          lstOutput.Items.Add(string.Format("{0,12}{1,10:C}", "Sales Tax:", salesTax));
43          lstOutput.Items.Add("");  // Add a blank line
44          lstOutput.Items.Add(string.Format("{0,12}{1,10:C}", "Total Cost:", totalCost));
```

The complete code should look as shown in Figure 1-25.

Step 3-6: Run the form:

Double click **program.cs** to open the code window.
Specify IceCreamTotal as the name of the form to be run:

Application.Run(new IceCreamTotal());

Click the *Start* button to run the form and test the code you entered. Run the form with no data for number of scoops. Run it again with $3.90 for unit price.
Note that in both cases, the program crashes with an error message, and you will have to rerun the program. Later in this chapter you will learn how to have the program deal with errors more gracefully.

Clearing a ListBox

The Clear method of the Items collection clears all lines (items) from the ListBox. Figure 1-30 shows the code to clear the TextBoxes and the ListBox.

Figure 1-30: Code to clear controls

```
private void cmdClear_Click(object sender, EventArgs e)
{
  txtScoops.Clear();
  txtUnitPrice.Clear();
  lstOutput.Items.Clear();
}
```

Step 3-7: Add the code from Figure 1-30 to the click event handler of Clear button.

Scope of Variables

Variables like scoops and unitPrice that are declared within a method are called **local variables.** A local variable is available for use only within the method where it is declared, and only in the statements that follow the declaration.

As discussed earlier, the part of a program where a variable is accessible is called the **scope** of the variable. The scope of a variable is the block of statements between the variable's declaration and the next right curly brace. For example, the scope of the variable unitPrice is lines 27–44, which is part of the btnComputeTotal_Click event handler method following the declaration of unitPrice.

A variable is not known outside its scope. That means, it would be an error to use scoops in any method other than btnComputeTotal_Click where it is declared

Step 3-8: Use the variable scoops inside btnClear_Click method as follows:

```
private void cmdClear_Click(object sender, EventArgs e)
{
    txtScoops.Clear();
    txtUnitPrice.Clear();
    lstOutput.Items.Clear();
    scoops = 0;  // Error! - Invalid use of scoops.
}
```
Note that the use of scoops is invalid, as indicated by the error.

Fields: Variables That Are Declared Outside Methods

Variables that are declared within a class but outside all methods are called fields. For example, the code shown in Figure 1-31 declares scoops as a field (line 19) within the class IceCreamTotal that represents the form, rather than as a local variable inside btnComputeTotal_Click method (line 29). The rest of the code remains the same as what was previously discussed. Fields are declared at the form level rather than inside a method.

Figure 1-31: Declaring fields

```csharp
13      public partial class IceCreamTotal : Form
14      {
15          public IceCreamTotal()
16          {
17              InitializeComponent();
18          }
19          int scoops;      // scoops declared outside the methods
20          private void btnComputeTotal_Click(object sender, EventArgs e)
21          {
22              // Compute the total cost for an order of icecream, including sales tax
23
24              // Declare variables to hold input data and computed results
25              // int scoops;
26              double unitPrice, subTotal, salesTax, totalCost;
27
28              //Declare a constant to hold the tax rate
29              const double TAX_RATE = 5.5;
30
31              //Get input data from Text Boxes:
32              unitPrice = double.Parse(txtUnitPrice.Text);
33              scoops = int.Parse(txtScoops.Text);
34
35              // Compute results
36              subTotal = unitPrice * scoops;
37              salesTax = subTotal * TAX_RATE / 100;
38              totalCost = subTotal + salesTax;
39
40              // Display Results with decimals aligned
41              lstOutput.Items.Add(string.Format("{0,12}{1,10:C}", "Sub Total:", subTotal));
42              lstOutput.Items.Add(string.Format("{0,12}{1,10:C}", "Sales Tax:", salesTax));
43              lstOutput.Items.Add("");   // Add a blank line
44              lstOutput.Items.Add(string.Format("{0,12}{1,10:C}", "Total Cost:", totalCost));
45      }
```

Because scoops is declared at the class level, it is available within each method and outside the methods. That means the variable can be assigned a value within btnComputeTotal_Click event handler method and set to zero within the btnClear_Click method. Variables declared at the form level exist until the form is closed.

Step 3-9: Declare scoops at the class level, and comment out its declaration within the method, as shown in Figure 1-31.

Note that the statement scoops = 0, within btnClear_Click method, does not give any error because scoops is declared at the class level. (It is not necessary to set scoops to zero; it is done to demonstrate the use of fields.)

Now, comment out the class level declaration in line 19, and uncomment the declaration of unitPrice in line 27. Comment out the statement unitPrice = 0 that you inserted into the btnClear_Click event handler.

When to use fields in a form

A variable may be declared as a field (i.e., declared at the class level) if it is to be used in more than one event handler method. We will see such use of fields in later chapters.

Why not declare every variable at the class level (as fields)?

First, a variable declared as a field stays in the memory until the form is closed. If the variable is needed only in one method, declaring it as a local variable inside the method will remove it from memory when execution of that method is completed—that is, fields consume memory for longer periods of time.

Second, when a variable is declared at the class level when it is needed only in one method, any other method may inadvertently change its value.

So, if a variable is to be used only in one method, then it should be declared inside that method—that is, declare it as a local variable, except in some special cases.

Review Questions

1.38   What would the Label lblCost display after execution of the following set of statements?

   doubleextendedCost = 4525.457;
   lblCost.Text = extendedCost.ToString("N");

1.39   A program needs to read the sales tax rate at runtime from a database and use it to compute the sales tax for customer orders. Would you use a constant or a variable to hold the sales tax rate while the program is running?

1.40   What are the special requirements of a constant?

1.41   What is a local variable?

1.42   How is a field different from a variable?

1.12 Exception Handling: Catching Errors

It would be annoying if you have to rerun a program just because you made a simple mistake like including a "$" sign in the dollar amount for unit price, as you found out in Step 3-6. How do you prevent the program from crashing if it encounters an error when it is run?

There is special code, called **exception handler**, which you can insert to catch such errors (exceptions). When a program encounters an error, it creates an object that contains information about the error and hands it to the system that runs the program. This process is called throwing an exception. Let's look at how to use the **try-catch** statement to catch exceptions.

Figure 1-32 shows how you would use a simple form of the try-catch method in the btnComputeTotal_Click event handler.

1.12 Exception Handling: Catching Errors

An error in any statement within the **try** clause will cause the method to throw an exception, skip the rest of the statements and jump to the catch clause. If there are no errors, the statements in the catch clause, called the catch block, are skipped.

Figure 1-32: Try-catch statement

```
private void btnComputeTotal_Click(object sender, EventArgs e)
{
    try
    {
        // Compute the total cost for an order of ice cream, including sales tax.
        // Declare variables to hold input data and computed results
        int scoops;
        ...
        lstOutput.Items.Add(string.Format("{0,12}{1,10:C}", "Total Cost:", totalCost));
    }
    catch (Exception err)
    {
        MessageBox.Show(err.Message);
    }
}
```

You assign the name **err** (or any meaningful name) to the exception object using the catch clause as follows:

catch (Exception err)

A commonly used property of the exception object is the **Message** property, which contains the error message that describes the error that caused the exception—that is, **err.Message** contains the specific error message.

MessageBox

MessageBox is an object available in .Net to display messages. The **Show** method of the MessageBox allows any string to be displayed as the message. For example, a message could be displayed using the statement

MessageBox.Show("Invalid data; please re-enter").

We use MessageBox to display the message contained in err.Message. In the catch clause, we use the statement

MessageBox.Show(err.Message)

to display err.Message whose value is a string that describes the error.

Step 3-10: Incorporate a try-catch statement in the btnComputeTotal_Click event handler, as shown in Figure 1-32.

Step 3-11: Run the form. Click the Compute Total button after entering unit price but no data for number of scoops. Note the error message. Repeat with both data entered, but include a "$" sign in unit price.

Figure 1-33 shows the complete code that includes computing the total, clearing the controls and closing the form.

Figure 1-33: Complete code for IceCreamTotal form

```csharp
20      private void btnComputeTotal_Click(object sender, EventArgs e)
21      {
22          try
23          {
24              // Compute the total cost for an order of icecream, including sales tax.
25
26              // Declare variables to hold input data and computed results
27              int scoops;
28              double unitPrice, subTotal, salesTax, totalCost;
29
30              //Declare a constant to hold the tax rate
31              const double TAX_RATE = 5.5;
32
33              //Get input data from Text Boxes:
34              unitPrice = double.Parse(txtUnitPrice.Text);
35              scoops = int.Parse(txtScoops.Text);
36
37              // Compute results
38              subTotal = unitPrice * scoops;
39              salesTax = subTotal * TAX_RATE / 100;
40              totalCost = subTotal + salesTax;
41
42              // Display Results with decimals aligned
43              lstOutput.Items.Add(string.Format("{0,12}{1,10:C}", "Sub Total:", subTotal));
44              lstOutput.Items.Add(string.Format("{0,12}{1,10:C}", "Sales Tax:", salesTax));
45              lstOutput.Items.Add("");   // Add a blank line
46              lstOutput.Items.Add(string.Format("{0,12}{1,10:C}", "Total Cost:", totalCost));
47          }
48          catch (Exception err)
49          {
50              MessageBox.Show(err.Message);
51          }
52      }
53
54      private void cmdClear_Click(object sender, EventArgs e)
55      {
56          txtScoops.Clear();
57          txtUnitPrice.Clear();
58          lstOutput.Items.Clear();
59          //unitPrice = 0;  // Error! - Invalid use of unitPrice.
60      }
61
62      private void cmdExit_Click(object sender, EventArgs e)
63      {
64          this.Close();
65      }
```

It's time to practice! Do Exercise 1-3 (uses constant and try-catch) and Exercise 1-4 (uses ListBox and try-catch).

Exercises

Exercise 1-1: Maximum heart rate

The maximum heart rate during exercise is given by the formula: (220 - age). Create a form named HeartRate, within a project named Exercises_Ch1, to let the user enter an integer age into a TextBox, and compute and display the maximum rate in a Label by clicking a button.

Exercise 1-2: MPG

The miles per gallon is given by dividing the miles driven by the gallons of gas used. Assume both are nonintegers. Create a form named MilesPerGallon, within a project named Exercises_Ch1. The form should let the user enter the miles driven and gallons used into two TextBoxes, and compute and display the miles per gallon in a Label by clicking a button.

Exercise 1-3: Total cost

Create a form, TotalCost, within a project named Exercises_Ch1, to let the user enter the labor and material costs of a job into two TextBoxes, and display the subtotal, sales tax, and total as follows:

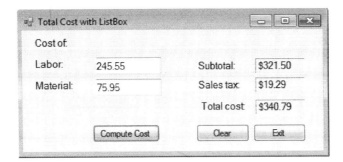

Use a try-catch statement to catch unexpected errors, like a missing data item. Use a constant for sales tax rate (6%). Subtotal = labor cost + material cost. Total cost is the sum of subtotal and sales tax.

Exercise 1-4: Total cost with ListBox

Create a form, TotalCostWithListBox, within a project named PrExercises_1, to let the user enter the labor cost and material cost of a job, and display the subtotal, sales tax and total cost in a ListBox as follows:

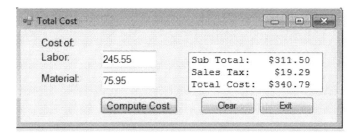

Use a try-catch statement to catch unexpected errors, like a missing data item. Assume 6% sales tax. Subtotal = labor cost + material cost. Total cost is the sum of subtotal and sales tax.

Chapter 2

Data Types

In Chapter 1, we looked at different types of integer, noninteger (real) and text data that are used in programs. There are certain rules you need to follow when you mix different types of numbers, like a decimal type and double type, in an expression or statement.

In this chapter, first we look at the differences between different types of numeric data and the rules on mixing them. Next, we discuss string type, followed by DateTime type.

Topics

2.1	Working with Different Numeric Data Types	2.3	Working with Strings and Characters
2.2	Casting	2.4	Working with Dates and Times

2.1 Working with Different Numeric Data Types

To refresh your memory, Table 2-1 shows commonly used integer types of numbers, their sizes and range of values. You are already familiar with the int type that you used in Chapter 1.

Table 2-1: Integer types

C# key word	Size (memory used)	Range of values
sbyte	8 bits	-128 to 127
short	16 bits	-32,768 to 32,767
int	32 bits	-2,147,483,648 to 2,147,483,647
long	64 bits	-9,223,372,036,854,775,808 to 9,223,372,036,854,775,807

As stated in Chapter 1, there are additional "unsigned" data types, **byte, ushort, uint** and **ulong,** which can store only positive, but larger, integers with maximum size close to double that of signed integers. For example, **uint** type variables can hold values up to 4,294,967,295, whereas the maximum for **int** type is only 2,147,483,647. There is also a special integer type, denoted by C# key word char, which can represent 65,535 characters from different languages.

The size shows the memory needed to store the data, and the range shows the range of values that can be stored in each type. For integers, larger size also means larger range.

One of the basic rules discussed in the next section is that **you cannot store a data type with a larger range in a variable of data type with a smaller range without explicitly converting the larger type to the smaller type.**

For example, the second statement in the following code is invalid:

```
int count = 150;          //valid; 150 is well within the range for int type
short recordCount = count;     // Invalid; count is int type, which is larger than short type
```

Note that the actual value of count is well within the range of short type, but the assignment is still invalid because count is of a larger type than recordCount. You will learn how to convert data item of one type to another type in the next section.

Next, we will look at the noninteger types, shown in Table 2-2.

Table 2-2: Real types

C# key word	Size	Range of values	Precision	How the following number may be represented: 123456789012.345678901234567890
float	32 bits	$\pm 3.4 \times 10^{38}$	7 digits	0.1234568×10^{12}
double	64 bits	$\pm 1.7 \times 10^{308}$	15 digits	$0.123456789012346 \times 10^{12}$
decimal	128 bits	$\pm 7.9 \times 10^{28}$	28–29 digits	123456789012345678901234567800

In addition to size and range, noninteger types have a precision property, with float type having only 7 digits of precision.

It is important to note that although the decimal type uses more memory (i.e., it has a larger size) than float and double, it has a smaller range (10^{28}). Decimals also are slower to work with. Hence, decimals are less suited than float and double for scientific applications involving extremely large numbers.

Why does decimal type take more memory although its range is smaller? Decimal type uses a method of internal representation that allows exact representation of real numbers with higher precision than float and double.

Although float and double have larger ranges and still take less memory, they cannot represent all noninteger numbers exactly (due to the method used by them to represent numbers), making them less suitable than decimals for financial calculations. For example, you would expect the following calculation to yield a value of 84.83 for cost:

```
float cost = (float) 4.99 * 17;
```

However, the actual value would be 84.82999. The meaning of "(float)" on the right side will be explained later in this section.

So, decimals are better suited for financial calculations than float and double when exact representation of numbers is important.

The C# types, float and double, use a method of representation called the **floating point** method to approximate a wide range of values to fixed number of significant digits. Therefore, float and double also are referred to as floating point types.

In the following sections, we examine the use of different types through a modified version of the IceCreamTotal form that you developed in Chapter 1.

Tutorial 1: Ice Cream Total with Data Types

This tutorial creates a new project named Ch2_DataTypes, and a modified version of IceCreamTotal form from Chapter 1. The modified form is named IceCreamTotalDataTypes. Figure 2-1 shows the form, which looks identical to IceCreamTotal.

The only change is that in place of double type variables, we use **decimal** type that is more suitable for financial calculations.

Figure 2-1: The IceCreamTotalDataTypes form

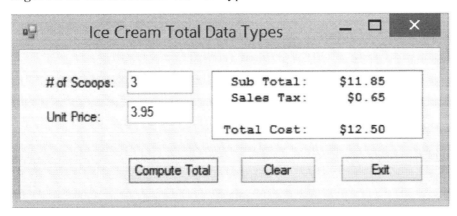

The primary goal of this tutorial is to learn how to work with different numeric data types.

Step 1-1: Copy the zipped file, Tutorial_Starts.zip. Extract the files (right click and select *Extract All*). Open the project, Ch2_DataType. Open the form **IceCreamTotalDataTypes.**

Because there are multiple forms in this project, you need to specify which form you want to run. Double click the Program.cs file to open it, and specify IceCreamTotalDataTypes as the form to be run:

Application.Run(new IceCreamTotalDataTypes());

Step 1-2: Open the code window to view the code that computes the total cost, as shown in Figure 2-2.

The code shown in Figure 2-2 differs from the code in the **IceCreamTotal** form only in lines 26 and 29. In line 26, we use **decimal** type variables to represent the unit price, subtotal, sales tax and total cost. Similarly, in line 29, we use decimal type constant for tax rate.

First, let's look at the statement in line 29 that assigns a value to the constant TAX_RATE,

const decimal TAX_RATE = (decimal) 5.5;

What does "**(decimal) 5.5**" mean? The word **decimal** within the parentheses converts the number 5.5 to decimal type, as explained in the next section.

Figure 2-2: Modified code to compute total cost

```
20      private void btnComputeTotal_Click(object sender, EventArgs e)
21      {
22          //Compute the total cost for an order of icecream:
23
24          //Declare variables to hold input data and computed results
25          int scoops;
26          decimal unitPrice, subTotal, salesTax, totalCost;
27
28          //Declare a constant to hold the tax rate
29          const decimal TAX_RATE = (decimal) 5.5;
30
31          //Get input data from Text Boxes:
32          unitPrice = decimal.Parse(txtUnitPrice.Text);
33          scoops = int.Parse(txtScoops.Text);
34
35          // Compute results
36          subTotal = unitPrice * scoops;
37          salesTax = subTotal * TAX_RATE / 100;
38          totalCost = subTotal + salesTax;
39
40          // Display Results with decimals aligned
41          lstOutput.Items.Add(string.Format("{0,12}{1,10:C}", "Sub Total:", subTotal));
42          lstOutput.Items.Add(string.Format("{0,12}{1,10:C}", "Sales Tax:", salesTax));
43          lstOutput.Items.Add("");   // Add a blank line
44          lstOutput.Items.Add(string.Format("{0,12}{1,10:C}", "Total Cost:", totalCost));
45      }
```

2.2 Casting

The process of explicitly converting one data type to another type is called **casting.** Let's look at when casting is required and what the rules are.

Literals

Numbers like 5.5 that are used in programs are called **literal values** or **literals. All noninteger literal values, by default, are treated as double type.** Thus, 5.5 is double type. But, TAX_RATE is decimal type. Recall that double type has a larger range (10^{308}) than decimals (10^{28}).

Rule for Assigning Literal Values to Variables and Constants

You cannot store a literal of a data type that has a larger **range** in a variable of a data type that has a smaller range without explicitly converting (casting) the literal to the smaller type. Here are the data types sorted by increasing range of values:

sbyte, short, int, long, decimal, float, double

Thus, a data type cannot be assigned to any of the data types on its left. For example, double type cannot be assigned to any of the other types without explicit conversion. This is true even if the actual value of the data item is small; it is the **range of the data type** that matters.

So, the following code will result in an error because 5.5 is double type, which has a larger range than decimal, although double uses less memory than decimal:

    const decimal TAX_RATE = 5.5;   // invalid

Similarly, float cannot be assigned to decimal or integer types, but it can be assigned to double, which has a larger range.

To correct the problem, the literal 5.5 needs to be explicitly converted to decimal type by specifying the data type enclosed in parentheses as follows:

    const decimal TAX_RATE = **(decimal)** 5.5;   // valid

Another way to cast a literal to decimal type would be to add the letter "m" or "M" to the literal:

    const decimal TAX_RATE = 5.5m;   // valid

Except for decimals, data types of smaller range can be stored in data types of larger range. So, the following assignment is valid because the literal 5 is an integer. All integer types have ranges smaller than that of decimal:

    const decimal TAX_RATE = 5;   // valid

decimal type: The exception

The exception is that **decimal type cannot be stored in float and double (though they have large ranges) without explicitly converting it**. Remember, decimal needs more memory than float and double, though it has a smaller range. So, the following is invalid, because decimal type cannot be assigned to double type:

    const double TAX_RATE = (decimal) 5.5;   // invalid

Step 1-3: Change the statement in line 29 to: const decimal TAX_RATE = 5.5;
    Hover the pointer over the number to view the error message, "Literal of double type cannot be implicitly converted to type 'decimal'..."

Step 1-4: Change the statement in line 29 to

    const decimal TAX_RATE = (decimal) 5.5;

Note that the error disappears.
Next, change the statement to

    const decimal TAX_RATE = 5;

Still there is no error.
Now, you may change it back to

    const decimal TAX_RATE = 5.5m;

The above rules imply that **casting is needed when storing a double type in any other type,** because all other types have smaller ranges than double.

Casting is needed also when storing a decimal type in any other type, because of the special nature of decimals—decimal needs more memory than float and double, though it has a smaller range.

Rule for Casting

You can cast a literal or variable of one type to another type only if the value of the literal/variable is within the range of the type it is cast to.

For example, casting 40000 to short in the following statement would fail because it is outside the range (32767) for short:

    short sales = (short) 40000;   // Cast would fail

Casting a real to integer

You may cast a real to an integer type if the value of the real is within the range of the integer type, but you will lose the decimal part of the value. For example, the following statement is valid, but it will result in losing the decimal part (0.5) of the number:

    int quantity = (int)10.5;

Casting to float

Caution! When you cast a real number with more than 7 significant digits to float, you lose some precision. Consider the following:

    float price = 123456789.55**F**;

The number 123456789.55, which is a double by default, is cast to a float and assigned to the float type variable price. The value stored in unitPrice would have only 7 digits of precision, because float type has only 7 digits of precision.

Integer literals

With integer numbers, the value of the number determines its type. For example, the literal 120 is considered by the compiler as sbyte, because it is less than the sbyte's limit of 127. Similarly, 32000 is short type, and 2,000,000,000 is int type.

Debugging Programs: Putting Break Points in a Program

If you encounter unexpected results or errors in a program, a good way to determine what is going on is to examine the values of variables at different statements of the program. To do this, you can stop execution of a program at a statement by putting a break point at that statement.

To put a break at a statement, click on the margin to the left of the statement. You will see a break represented by a red dot in the margin.

Locals window

When the program breaks, you can view the values of all **local variables** (variables declared within a method) by displaying the Locals window (select Debug, windows, Locals). You also can move the pointer over each variable to see its value.

Suppose you inadvertently declared tax rate as int type, which results in truncation of the number 5.5 to 5, and the wrong output. Let's look at how to identify the problem.

Step 1-5: Check the effect of casting a double to int by putting a break in the program:
Change the statement in line 29 to

const int TAX_RATE = (int) 5.5;

and run the form with 1 scoop and a unit price of 10. The sales tax should be 0.55 (10 x 0.055), but the output shows 0.50.
To identify the problem, put a break point at the closing brace "}" on line 45 by clicking on the margin to the left of the brace. C# inserts a red circle in the margin, as shown below:

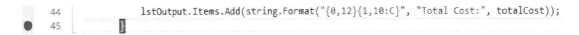

```
44        lstOutput.Items.Add(string.Format("{0,12}{1,10:C}", "Total Cost:", totalCost));
45    }
```

Run the program. Program execution stops at the brace, indicated by a yellow arrow over the red circle.
Display the Locals window by selecting Debug, Windows, Locals. The Locals window appears, as shown in Figure 2-3.

Figure 2-3: Locals window

Name	Value	Type
this	{Ch2_DataTypes.IceCreamTota	Ch2_DataTypes.
sender	{Text = "Compute Total"}	object {System.\
e	{X = 20 Y = 15 Button = Left}	System.EventArg
scoops	1	int
unitPrice	10	decimal
subTotal	10	decimal
salesTax	0.5	decimal
totalCost	10.5	decimal
TAX_RATE	5	int

Examining the values of variables helps identify that TAX_RATE has the wrong value of 5. You also may move the pointer over TAX_RATE or any variable to display its value, one at a time.
Click the Continue button on the Toolbar to continue running.
Close the form, and correct the problem by changing the statement in line 29 back to the original form: const decimal TAX_RATE = (decimal) 5.5;

Next we will use the break to examine the effect of casting a double to float.

Chapter 2: Data Types   63

Step 1-6: Insert the following statement in line 39, which is currently blank:

    float testNumber = 123456789.45F;

Run the form. Enter values for scoops and unit price, and click the Compute Total button. When the program breaks, hover the pointer over testNumber to view its value, 123456792.0, which is different from 123456789.45 due to the lower precision (7 digits) of floats compared to doubles. Close the form, and delete the newly inserted statement.

Rule for Assigning One Variable to Another Variable of a Different Type

The rule for assigning a variable of one type to a variable of another type is essentially the same as the rule for assigning a literal to a variable: You cannot store a variable (or constant) of a data type that has a larger **range** in a variable of a data type that has a smaller range. Here are the data types sorted by increasing range of values:

    sbyte, short, int, long, decimal, float, double

The second statement shown below is invalid because price is of double type, which has a large range (10^{308}) compared to decimal (10^{28}).

    double price = 25;   // valid; all integer types have ranges smaller than double
    decimal currentPrice = price; // invalid; double has a larger range than decimal

To correct the problem, you need to cast price to decimal as follows:

    decimal testNumber = (decimal) price; // valid

Except for decimals, data types of smaller range can be stored in data types of larger range. Decimal type cannot be stored in float or double (though they have large ranges) without explicitly converting it.

The second statement shown below is invalid because price is of decimal type, which cannot be assigned to any other type.

    decimal price = 45500;   // valid!
    double currenPrice = price; // invalid

Again, to correct the problem, you need to cast price to double, as follows:

    double currentPrice = (double) price; // valid

Thus, both double and decimal cannot be assigned to any other type without explicit conversion.

Step 1-7: To observe the role of casting when a variable of larger type is assigned to another variable of smaller type, enter the following two statements into the click event handler of ComputeTotal button, and note the error in the second statement:

    decimal price = 2.95m;   // "m" represents decimal
    double testNumber = price; // invalid

To correct the error, cast price to double.

    double testNumber = (double) price; // valid

Delete both statements. Repeat the process by adding the following statements and observing the error:

>   double price = 2.95;
>   decimal testNumber = price;   // invalid!

To correct the error, cast price to decimal.

>   decimal testNumber = (decimal) price;   // valid

Casting: Narrowing and Widening Conversion

Explicit casting and narrowing conversions

As discussed above, **explicit casting** is needed when you store a data type with a larger range in a data type with a smaller range, and also when you store a decimal in other types. Such a conversion is called a **narrowing conversion.**

Implicit casting and widening conversion

Explicit casting is not needed when storing any data type, other than decimal, in another data type that has a larger range.

Thus, literal or variable of any type other than decimal can be assigned to a double type variable without explicit casting. In such cases, C# performs an **implicit cast** to do a **widening conversion**.

Consider the following code that assigns the literal 1234567890 of int type (determined by the value) to a float type variable, costPerUnit, and then assigns costPerUnit to another variable, price.

>   float costPerUnit = 1234567890;   // valid
>   double price = costPerUnit;   // valid

In the first statement, C# performs an implicit cast from integer to float that has a larger range to do widening conversion. The second statement assigns the value of costPerUnit, which is float, to unitPrice, which is a double. Again, C# performs an implicit cast from float to the larger double.

Caution! Float has only 7 digits of precision. Hence, the value of costPerUnit will have only 7 digits of precision.

Step 1-8: Make sure that there is break at the closing brace. Add the following two statements to the click event handler of Compute Total button, run the program and note the value of costPerUnit and price.

>   float costPerUnit = 1234567890;   // valid, but loses precision.
>   double price = costPerUnit;   // valid

Delete the statements.

Note that the values of costPerUnit and price are different from 1234567890 due to lack of precision for float.

Mixing data types in an expression

When an operation involves two different data types, then the data type with the smaller range, except for decimal, is implicitly converted to the larger type, and the result would be of the same type as the larger type.

So, mixing different integer types in an expression, as in the following code, is valid:

```
short weight = 60;
int score = 1000;
int weightedScore = weight * score;  // valid to mix different integer types
```

The short type is implicitly converted to the larger int type.

Similarly, mixing integer types with a real type like float, double or decimal is valid, as in the following code:

```
float weight = 60;
int score = 1000;
float weightedScore = weight * score;  // valid to mix integer types with any real type.
```

Again, the int type is implicitly converted to the larger float type.

Mixing float and double types also is valid. The float type will be implicitly converted to the larger double type.

```
float weight = 60;
double score = 950.50;
double weightedScore = weight * score;  // valid to mix float with double (and integer types)
```

Decimal type: Again, the troublemaker!

While mixing float and double types is valid, **mixing decimal with float or double is invalid.** Decimal is not implicitly converted to float or double. Similarly, double and float are not implicitly converted to decimal.

The expression in the following statement is invalid because subTotal is declared as decimal, but the literal 5.5 is double:

```
salesTax = subTotal * 5.5 / 100;        // invalid!
```

So, the literal 5.5 must be explicitly converted to decimal, as follows, so that the result will be decimal type:

```
salesTax = subTotal * (decimal) 5.5 / 100;
```

or

```
salesTax = subTotal * 5.5m / 100;
```

Caution! Division of an integer by another integer would be an integer, resulting in loss of the fractional part, if any. For example, the value of average in the following statement would be 2, though average is float type, because division of 25 by 10 is represented as an integer:

```
float average = 25/10;   // yields the result 2
```

Casting at least one of the two integers to float type would yield the correct result:

>   float average = 25f/10;

The result of 25f/10 would be float type, which is the larger of the two types, yielding the correct result, 2.5.

So, when you divide an integer by another integer, at least one of the two must be cast to a real type to get the result with the fractional part, if any.

Step 1-9: Add the following statement to the click event handler of Compute Total button, run the program and observe the value of average:

>   float average = 25 / 10;   //yields 2

Change the expression to 25f/10f, and see the difference in value (2.5).
Change the expression to 25f/10, which yields 2.5.
Delete the statement.

Multiplication with large integers

Multiplication with large integers can result in unexpected consequences:

>   int price = 3000;
>   int quantity = 1000000;
>   long cost = price * quantity;

The product 3000 * 1000000 yields 3 billion. However, because the product of the two int types (price * quantity) is represented as int, whose limit is approximately 2.15 billion, the result is outside the limit and yields incorrect results.

Step 1-10: Add the following statements, run the program and note the value of cost.

>   int price = 3000;
>   int quantity = 1000000;
>   long cost = price * quantity;
>   Delete the statements.

The value of cost is incorrect because the result is int type, whose limit is about 2.15 billion, which is too small for the result of 3000,000,000 (3 billion).

An oddity in C#

When an operation involves short or sbyte types, these types are automatically converted to int type before performing the operation. So, the result will be int type.

>   short exam1 = 80;
>   sbyte exam2 = 90;
>   short total = exam1 + exam2; // Error! The result is int type (odd!), whereas total is short type

You may cast the result to short, as in

>   short total = (short) (exam1 + exam2);

Because of this odd behavior when using sbyte and short, in general, **you may use int type for all integers that are within the range for int**, except in applications where memory usage is of concern.

It's time to practice! Do Exercise 2-1.

Review Questions

2.1 Which real data type is best suited for financial calculations where exact representation of numbers is important?

2.2 What are the advantages of float and double types compared to decimal?

2.3 Which data type has the largest range?

2.4 Which data type has the largest size (memory)?

Correct errors, if any, in the following statements:

2.5 double sales = 2560.50;

2.6 float cost = 2.50;

2.7 float units = 45000;

2.8 sbyte quantity = 200;

2.9 decimal price = 300;
float currentPrice = price;

2.10 short units = 2500;
int quantity = units;

2.11 int qty1 = 1000000000;
int qty2 = 2000000000;
long total = qty1 + qty2;

2.12 float total = 255.50 + 500;

2.13 The following code is intended to compute the average:
int totalScore = 25;
int count = 4;
float average = total/count;

2.14 decimal price = 100;
float quantity = 5.5;
decimal netCost = price*quantity - 10;

2.3 Working with Strings and Characters

In this section, you will learn how to work with string type and character type data and variables. The C# type **string** represents a combination of zero or more characters. "John Adams" and "P410-2023-3180-05" are examples of string type data. Note that strings are enclosed in quotes.

Value and Reference Types

The integer and real types we discussed in Chapter 1 are called **value types** because each variable of these types holds the actual value t in the memory location for that variable.

By contrast, strings (and other objects like arrays) are called **reference types** because each variable of these types holds only a reference to the data—that is, the memory location or address of the data, not the actual data. Reference type is discussed later in Chapter 7.

To understand working with strings, let's look at an application that creates a user name by combining the following three pieces of string type data:

1. First 5 characters of the last name
2. First character of the first name
3. Last two digits of the id

Figure 2-4 shows the user interface form named UserNameString.

Figure 2-4: UserNameString form

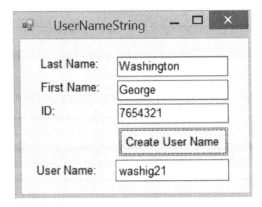

Tutorial 2: String Operations

This tutorial creates the UserNameString form shown in Figure 2-4. The primary objective is to learn string operations.

Step 2-1: Open the project, Ch2_DataTypes, from Tutorial_Starts folder. Open the form, UserNameString. To make it easy to understand the code, the names of controls that are used in the code are shown in Table 2-3.

Table 2-3: Controls on the form, UserNameString

Purpose	Type of control	Property	Setting
Enter last name	TextBox	Name	txtLname
Enter first name	TextBox	Name	txtFname
Enter id	TextBox	Name	txtId
Create user name	Button	Name	btnUserName
Display user name	Label	Name	lblUserName

Figure 2-5 shows the code to create the user name.

Figure 2-5: Code to create the user name

```
20      private void btnUserName_Click(object sender, EventArgs e)
21      {
22          string lName = txtLname.Text;   // Get input data
23          string fName = txtFname.Text.Trim();
24          string id = txtId.Text.Trim();
25          string userName;
26          // Combine three strings to get the user name
27          userName = lName.Substring(0, 5).ToLower() + fName.Substring(0, 1).ToLower() +
28                                      id.Substring((id.Length - 2), 2);
29          // Alternate way using the Concat method;
30          userName = string.Concat(lName.Substring(0, 5).ToLower(), fName.Substring(0, 1).ToLower(),
31                                      id.Substring((id.Length - 2), 2));
32          lblUserName.Text = userName;
33      }
```

Let's take a look at what the statements do.

Lines 22–24 get the input data from TextBoxes and save the data in variables.

Line 27 extracts the required substrings from the last name, first name and id; converts them to lowercase; and combines them to form the user name.

The string type has a large number of methods and properties to help work with strings. We first look at the methods/properties used in the above program.

Trim() method

The Trim() method removes all leading and trailing blank (white) spaces from the string. For example,

    "   George   ".Trim()

would return the string, "George".

In the statement in line 22,

    string lName = txtLname.Text.Trim();

the Trim() method of the string txtLname.Text removes trailing and leading blank spaces, if any, inserted by the user into the TextBox.

Additional options to trim include

    TrimStart()—removes leading blanks, if any
    TrimEnd()—removes trailing blanks, if any

Step 2-2: Double click the Create User Name button, and add lines 22 through 25 to the click event handler of the button. Put a break on the closing brace of the event handler. Run the form after entering the data with leading and trailing blank spaces. Make sure the last name is at least 5 characters long, because the code doesn't allow for shorter names.

Substring(i, n) method

The Substring(i, n) method returns n characters from the string, starting with index position i. Index position 0 refers to the first character, 1 refers to the second character, 2 refers to the third character and so on.

For example, in line 27, **lName.Substring(0, 5)** returns a string that consists of 5 characters, starting with the character in position zero—that is, it gives the first 5 characters of the last name.

Similarly, **fName.Substring(0, 1)** returns a string that consists of the first character of the first name.

id.Substring((id.Length - 2), 2) gives the two digits of the id starting with the second position from the end—that is, the last 2 digits of id. For example, for a 7-digit id, (id.Lenth - 2) would be 5, thus making the expression equivalent to **id.Substring(5,2)**, which gets the 2 characters starting with index position 5—that is, starting with the sixth character.

Length property

The Length property gives the number of characters in the string.

In line 27, **id.Length** uses the Length property of the string, id, to get the number of characters in the id to help extract the last two digits of id:

27   userName = lName.Substring(0, 5).ToLower() + fName.Substring(0, 1).ToLower() +
              id.Substring((id.Length - 2), 2);

ToLower() method

The ToLower() method returns a copy of the string in lowercase.

lName.Substring(0, 5).ToLower() converts the first five characters of the last name to lowercase. Similarly, **ToUpper()** converts to uppercase.

Concatenating strings

Two or more strings may be concatenated together using the "+" operator, as in

    String fullName = "G " + "C " + "Philip"

The value of fullName would be "G C Philip." This is the method used in computing username in Line 27:

    userName = lName.Substring(0, 5).ToLower() +
        fName.Substring(0, 1).ToLower() + id.Substring((id.Length - 2), 2);

Step 2-3: Add the statement that computes user name in line 27 & 28. Run the form to test it.

    Change the parameters of Substring method for fName to 3 and 1, as in

        fName.Substring(3, 1).

    Observe the change in user name.

    Type in the last name with uppercase and with leading and trailing blanks, and observe the output. Type in id of a different lengths.

Another option to concatenate strings is to use the **Concat()** method of string, as in

    String fullName = String.Concat("G ", "C ", "Philip");

which yields "G C Philip."

Line 30 uses the Concat method to combine strings:

30   userName = string.Concat(lName.Substring(0, 5). ToLower(),
        fName.Substring(0, 1).ToLower(),id.Substring((id.Length - 2), 2));

Step 2-4: Add lines 30 through 32 from Figure 2-5 to complete the program, and test the code.

    Make each of the following changes, run the program, and describe what impact it has on the output and why.

    Change the parameters (0, 5) in lName.Substring(0, 5) to (0,4).

    Change the same set of parameters as above to (2, 3).

    Change the parameters ((id.Length - 2), 2)) to ((id.Length - 1), 1)).

Additional Methods of the String() Method

Replace (string, string) method: Replaces the first string, if any, with the second string.

Example: lName = lName.Replace(" ", null);

The above statement would replace blank spaces, if any, by null, in essence removing any spaces within the name.

Step 2-5: Insert the statement
    lName = lName.Replace(" ", null);
immediately after line 22 that has the statement
    string lName = txtLname.Text.Trim();
Test the program by including spaces in the last name that you type in.

Some of the commonly used methods and properties of string are given in Table 2-4.

Table 2-4: Methods and properties of string type variables

String methods & properties	Description
Methods	
IndexOf(*ch*)	Returns the index of the first occurrence of the character, *c*, if it is found; if not, -1 is returned
IndexOf(*str*)	Returns the index of the first occurrence of the string, *str*, if it is found; if not, -1 is returned.
Insert(*i, str*)	Inserts the string, *str*, at index position i, and returns the resulting string
Remove(*i, n*)	Removes *n* characters starting at index position i, and returns the resulting string
Replace(*str1, str2*)	Replaces every occurrence of *str1* by *str2* and returns the resulting string
SubString(*i,n*)	Returns *n* characters starting with index position *i*
ToLower()	Returns a copy of the string in lowercase
ToUpper()	Returns a copy of the string in uppercase
Trim()	Removes all leading and trailing blank (white) spaces, and returns the resulting string
Property	
Length	Gives the number of characters in the string

Additional methods of string variables include
    Contains(*ch*), Contains(*str*)
    IndexOf(*string, i*), IndexOf(*string, 1, n*), IndexOf(*ch, i*), IndexOf(*ch, i, n*), LastIndexOf(*ch*)
    Insert(i, str)
    StartsWith(*str*), EndsWith(*str*)
    SubString(i)
    TrimStart(), TrimEnd()

String versus. Character

A string that consists of only one character is not the same as a character type value. As stated earlier, a string variable is a reference type, but a character is a value type.

The string methods are applicable to strings of all sizes, including a size of one character, **but they are not applicable to character type**.

Combining strings with characters

If you concatenate a char type to a string, C# will implicitly convert the char type to string, as in the following statement:

```
char initial = 'g';
lblUserName.Text = "Phili" + initial;
```

Step 2-6: Insert the above two statements to the end of the click event handler. Comment out the rest of the code, and run the form. The string "phili" and character 'g' are combined to display the string "Philig." Delete the two statements.

The methods of string type variables, like ToUpper and ToLower, are not valid for character type variables. Hence, the following statement would give an error:

```
char initial = 'g';
lblUserName.Text = "Phili" + initial.ToUpper(); //error
```

You may convert the character type variable, initial, to string type and then use the ToUpper method of string variables.

Step 2-7: Add the above two statements to the end of the code in the click event handler, and observe the error message.

Correct the error by converting the value of initial to string type using the ToString() method of char type, as shown below:

```
lblUserName.Text = "Phili" + initial.ToString().ToUpper();
```

You also may use the ToUpper method of the char type, as in

```
lblUserName.Text = "Phili" + char.ToUpper(initial);
```

An alternative is to declare initial as a string instead of a character, and then use the ToUpper method of the string.

Step 2-8: Add the two statements shown below to declare initial as string, and then use the ToUpper() method. Run the code. Observe that there is no error.

```
string initial = "g";
lblUserName.Text = "Phili" + initial.ToUpper().
```

Delete the above two statements.

It's time to practice! Do Exercise 2-2.

Review Questions

2.15 A string variable named firstName contains the first name, and another string variable named lastName contains the last name of a person. Write an expression to create a string that consists of the first 2 characters of the first name and the entire last name, with a space between them, and convert the string to all capital letters.

2.16 A student id consisting only of numeric digits is stored in a string variable named studentId. Write the code to display the last 2 digits of the id in a ListBox named lstDisplay.

2.4 Working with Dates and Times

Working with dates and times is important in a variety of applications, like computation of due dates, late fees, duration of events and rental fees. This section takes a detailed look at how to work with dates and times. However, the materials presented in this section are not prerequisites for the chapters that follow.

As stated in Chapter 1, C# does not have a data type for dates and times. The .Net base type DateTime is used for dates and/or times. DateTime data are enclosed in pound signs, as in #12/25/2017 10:30:00 AM#.

Tutorial 3: DateTime Operations

To explore various operations with dates and times, you will create an application that uses your graduation date and time. Different pieces of information related to the graduation date/time are displayed on a form named GraduationDate, as shown in Figure 2-6.

Figure 2-6: GraduationDate form

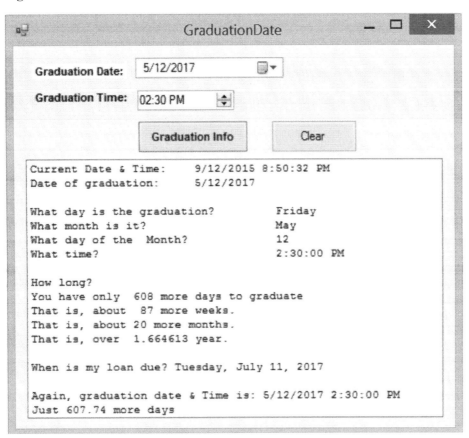

Step 3-1: Open the form named GrduationDate from Ch2_DataTypes projects in Tutorial_Starts folder.

Step 3-2: The controls on the form are shown in Table 2-5.

Chapter 2: Data Types

Table 2-5: Controls on the form GraduationDate

Purpose	Type of control	Property	Setting
Select date	DateTimePicker	Name	dtpGradDate
		Format	short
		Value	Delete the time, leaving just the date
Select time	DateTimePicker	Name	dtpGtadTime
		Format	Custom
		CustomFormat	hh:mm tt (use HH:MM tt for 24-hour military time)
		ShowUpDown	True
Display prompt for GradDate	Label	Text	Graduation Date:
Display prompt for GradTime	Label	Text	Graduation Time:
Compute results	Button	Text	Graduation Info
Clear controls	Button	Text	Clear
Display results	ListBox	Name	lstDisplay
		Font	Courier New, 8.25pt.

Using DateTimePicker to select date and times

Rather than hardcoding dates as strings and converting to DateTime, a DateTimePicker may be used to let users select a date or time. By default, the DateTimePicker, shown below with Format set to **short**, lets the user select a date.

The selected date, which also includes the current time, is returned by the **Value** property. To remove the current time, delete the time from the Value property. C# will add the default time 12:00:00 to the selected date. In the GraduationDate form, the current time has been removed from the Value property of dtpGradDate. So, the date includes the default time 12:00:00.

To let users select a Time from DateTimePicker, you need to set the Format, CustomFormat and ShowUpDown properties, as shown in Table 2-6.

Table 2-6: Settings for DateTimePicker to display time

Format	Custom
CustomFormat	hh:mm tt
ShowUpDown	True

The time selected by the user is returned by the Value property that also includes today's date. The above properties of dtpGradTime are already set to the values specified in Table 2-6.

Step 3-3: Double click the Graduation Info button to open the click event handler of the button.

Working with Dates

Here is an example of creating a date and storing it in a DateTime variable by hard-coding the date:
    DateTime gradDate = DateTime.Parse("05/12/17");

You also may specify a time:
    DateTime gradDateTIme = DateTime.Parse("05/12/17 02:30 PM");

Or, you may specify a time with no date:
    DateTime gradTime = DateTime.Parse("02:30 PM");

When no date is specified, today's date is added to the time, by default.
Similarly, if no time is specified, the default time 12:00 AM is added to the date.

The date must be specified in one of the standard date formats, 05-12-2017, May 12, 2017 or 2017-05-12.

The general syntax to create a DateTime is
    DateTime variableName = DateTime.Parse(string);

Step 3-4: Enter the following statements in the click event handler of Graduate Info button to convert strings to DateTime.

    DateTime gradDate = DateTime.Parse("05/12/17");
    DateTime gradDateTime = DateTime.Parse("05/12/17 02:30 PM");
    DateTime gradTime = DateTime.Parse("02:30 PM");

Put a break at the last line of the click event handler that has the closing brace.
Run the program, click on the Graduation Info button and observe the value each of the three variables by moving the cursor over it. Pay particular attention to the default value of date and time.
Delete the statements. Remove the break from the closing brace.

Creating a date/time using the "new" key word

Alternatively, you may use the **new** key word, using the syntax

    DateTime variableName = new DateTime(year, month, day, hour, minute, second, millisecond)

where hour, minute, second and millisecond are optional.

The **new** key word creates a structure of DateTime type in memory with the default value of DateTime (#01/01/0001 12:00:00 AM#). The default value is assigned to the variable specified on the left side of the = operator. The new key word, which is called the default constructor, is discussed in more detail in Chapters 7 and 13.

Here is an example:

    DateTime gradDate = new DateTime(2017, 5, 12);
    // The value of gradDate would be 5/12/2017.

Alternately, you can include the time 2:30 PM represented by 14, 30, 0:
    DateTime gradDateTime = new DateTime(2017, 5, 12, 14, 30, 0);
    // The value of gradDateTime would be 5/12/2017 2:30 PM.

Step 3-5: Specify date/time using the new key word, by adding the following statements:

```
DateTime gradDate = new DateTime(2017, 5, 12);
DateTime gradDateTime = new DateTime(2017, 5, 12, 14, 30, 0);
```

Run the program, and observe the values of gradDate and gradDateTime.
Delete the statements.

Step 3-6: Add the following code to declare variables and get graduation date and time from DateTimePickers.
Run the form, and check the values of the variables gradDate and gradTime.

```
22
23          DateTime gradDate, gradDateTime, gradTime;
24          float numberOfDays;
25
26          //Get date of graduation:
27          gradDate = dtpGradDate.Value;
28          // Get time of graduation:
29          gradTime = dtpGradTime.Value;
```

Properties of DateTime

DateTime structure has a variety of properties. Some of the commonly used properties are shown in Table 2-7. Column 3 shows the value of the property for gradDate created as follows:
DateTime graduationDate = DateTime.Parse("5/12/17 02:30 PM");

Table 2-7: Properties of DateTime structure

Property	Description	Value of property for graduationDate (5/12/2017 02:30 PM)
Day	Returns the day as an integer	graduationDate.Day: 12
Month	Returns the month as an integer	graduationDate.Month: 5
Year	Returns the year as an integer	graduationDate.Year: 2017
Now	Returns the current date and time	DateTime.Now: (today's date and time)
Today	Returns the current date	DateTime.Today: (today's date with the default time 12:00:00 AM)
Hour	Returns the hour as an integer	graduationDate.Hour: 2
TimeOfDay	Returns the duration that has elapsed since midnight, as a TimeSpan (Hours, Minutes, Seconds, …)	TimeOfDay.Hours: 2 TimeOfDay.Minutes: 30 TimeOfDay.Seconds: 0
Additional properties include Minutes, Seconds, DayOfWeek and DayOfYear		

Today versus Now

The property **Today** gives today's date with the default time 12:00:00 AM, whereas **Now** gives today's date with the current time.

Formatting DateTime

Table 2-8 shows commonly used formatting options for DateTime.

Table 2-8: Formatting options for DateTime

Format	5/12/2017 will be displayed as
ToShortDateString	5/12/2017
ToLongDateString	Friday, May 12, 2017
(No Format)	5/12/2017 12:00:00 AM
ToString	5/12/2017 12:00:00 AM

Step 3-7: Type in the code shown below to display current date & time and the graduation date:

```
31    // Format and display current date & time and graduation date:
32    lstDisplay.Items.Clear();
33    lstDisplay.Items.Add("Current Date & Time:    " + DateTime.Now);
34    lstDisplay.Items.Add("Date of graduation:     " + gradDate.ToShortDateString());
```

Run the program and observe the output. Change the format for gradDate to ToLongDateString, run the program and observe the output.
Now remove the format, and see the output with the default format.

Formatting date/time using ToString() method

Just like formatting numbers using ToString() method, you can format dates and times using ToString method of DateTime. For example, you can display the name of the day or name of the month for a date. Table 2-9 shows format codes available for ToString() method.

Table 2-9: Format codes for ToString method

Code	5/12/2017 2:30 PM will be displayed as
"dd"	12
"ddd"	Fri
"dddd"	Friday
"MM"	5
"MMM"	May
"MMMM"	May
"yy"	17
"yyyy"	2017
"hh"	02
"mm"	30
"ss"	00

You may combine the above codes to create custom formats as follows:
  "dd MMM yyyy hh:mm:ss tt"   12 May 2017 2:30:00 PM
  "MM/dd/yy hh:mm tt"   5/12/17 2:30 PM

Step 3-8: Type in the code to display day, month, day of the month from gradDate, and time of graduation from gradDate, as follows:

```
What day is the graduation?      Friday
What month is it?                May
What day of the  Month?          12
What time?                       2:30:00 PM
```

Here is the code to be added. Note how the format codes "dddd" and "MMM" are used to display the day and month from a date:

```
36    // Display the month & day of graduation:
37    lstDisplay.Items.Add("");
38    lstDisplay.Items.Add("What day is the graduation?      " + gradDate.ToString("dddd"));
39    lstDisplay.Items.Add("What month is it?                " + gradDate.ToString("MMM"));
40    lstDisplay.Items.Add("What day of the month?           " + gradDate.Day);
41
42    //Format and display the time of graduation:
43    lstDisplay.Items.Add("What time?                       " + gradTime.ToLongTimeString());
```

Modify line 38 so that the three-letter abbreviated name of day (e.g., Fri) is displayed.

Step 3-9: You may experiment with the different format codes from Table 2-8 and observe the outputs. Next, change the statement in line 43 as follows so that no format is used:
  lstDisplay.Items.Add("What time?   " + gradTime)
Run the form. Note that the time is displayed along with today's date, which is the default date.

Doing Computations with Date/Times

The DateTime structure has a number of methods to find the difference between two date/times and to add intervals to a date/time. Table 2-10 shows such methods.

Table 2-10: Methods of DateTime

Method	Description	Value returned by the method
AddDays(days)	Returns the DateTime obtained by adding the specified number of days to a DateTime	graduationDate.AddDays(3): Returns 5/15/17 02:30 PM
AddMonths(months)	Returns the DateTime obtained by adding the specified number of months to a DateTime	graduationDate.AddMonths(3): Returns 8/12/17 02:30 PM
AddHours(hours)	Returns the DateTime obtained by adding the specified number of hours to a DateTime	graduationDate.AddHours(3): Returns 5/12/17 05:30 PM
Note: Additional methods with the same syntax are available to add years, minutes and seconds.		

Subtract(DateTime)	Returns a TimeSpan obtained by subtracting a specified DateTime	graduationDate.Subtract(DateTime.Now): Returns a TimeSpan with the number of days, hours, and so on between now and graduation date.
Add(TimeSpan)	Returns the DateTime obtained by adding the specified TimeSpan to a DateTime	DateTime.Now.Add (fillTime)*: Returns the DateTime value obtained by adding 3 days, 2 hours and 30 minutes to current date and time. *fillTime: (3, 2, 30, 0, 0)

Computing the difference between two dates

.Net Framework uses a **TimeSpan** structure to represent the time interval between two DateTime values.

TimeSpan structure

To get the difference between two DateTime values, you may subtract one from the other. For example, the time to graduate may be computed and stored in a TimeSpan variable as follows:

TimeSpan timeToGraduate = gradDate-DateTime.Now;

Table 2-11 shows commonly used properties of TimeSpan Stucture.

Table 2-11: Properties of TimeSpan stucture

Properties of TimeSpan	Description
Days, Hours, Minutes, Seconds, Milliseconds, Ticks	The number of whole days and remaining whole hours, minutes, seconds, milliseconds and ticks in a time interval. For example, the TimeSpan representing (#12/19/2017 4:40:45 PM# - #12/18/2017 1:10:30 PM#) would have following values for its properties: Days - 1 Hours - 3 Minutes - 30 Seconds - 15
TotalDays, TotalHours, TotalMinutes, TotalSeconds, TotalMilliSeconds	Each property returns the entire interval expressed in a particular unit like days, hours or minutes. For example, the TimeSpan representing (#12/19/2015 4:30:45 PM# - #12/18/2015 1:20:40 PM#) would have following values for its properties: TotalDays - 1.146 (0.146 represents the fractional days) TotalHours - 27.504 (includes 27 days and .504 fractional hours) TotalMinutes - 1650.25

A TimeSpan also may be created using the new key word:

TimeSpan variableName = new TimeSpan(days, hours, minutes, seconds, milliseconds);

Here is a specific example of a TimeSpan that represents the time it takes to fill an order: 2 days, 3 hours and 30 minutes:

TimeSpan fillTime = new TimeSpan(2, 3, 30, 0, 0);

Step 3-10: Insert the following statements that compute the number of days to graduate:

```
46      // Compute the interval between today's date and graduation date
47      TimeSpan tsGraduation = (gradDate - DateTime.Today);
48      /* today includes current date, but not current time.
49         You may also use the Subtract() method of DateTime
50         TimeSpan tsGraduation = gradDate.Subtract(DateTime.Today);     */
51
52      numberOfDays = (float)tsGraduation.Days;
53      // Since graduation date and DateTime.Today both have the same default time (12:00:00),
54      // the remaining hours, minutes and seconds of the difference will be zero.
55      lstDisplay.Items.Add("");
56      lstDisplay.Items.Add("How long?");
57      lstDisplay.Items.Add("You have only " + numberOfDays + " more days to graduate");
```

Note that in the expression (gradDate - DateTime.Today), both dates have the same default time (12:00:00). Hence, the remaining hours, minutes and seconds of the difference will be zero. So, the TotalDays property of the TimeSpan would give the same value as the Days property.

Later, Step 3-14 introduces an example where there is a time difference between two DateTime values.

Step 3-11: This is an **optional** addition to the code. Skipping this code won't affect the rest of the program. Add the code to compute and display the number of weeks, months and years to graduation, as follows:

```
59      // Computing weeks, months and years:
60      float numberOfWeeks = numberOfDays / 7;
61      float numberOfMonths = ((gradDate.Year - DateTime.Today.Year) * 12) +
62                              (gradDate.Month - DateTime.Today.Month);
63      float numberOfYears = (float) (numberOfDays / 365.25);
64      // The above calculation of month and year may be imprecise under certain conditions.
65
66      // Display number of weeks, months & years
67      lstDisplay.Items.Add("That is, about " + numberOfWeeks.ToString("N0") + " more weeks.");
68      lstDisplay.Items.Add("That is, about " + numberOfMonths + " more months.");
69      lstDisplay.Items.Add("That is, over " + numberOfYears + " year.");
```

Adding years, months, days, hours, and so on to a DateTime

Table 2-10 presented the methods of DateTime to add years, months, days, hours, and so on to a DateTime. Let's use one of the methods to compute the date when the student loan is due, assuming it is 60 days after graduation.

Setp 3-12: Add the following statements to compute and display the date when the student loan is due:

```
71      // Compute the day when the loan is due - Add 60 days to graduation date:
72      lstDisplay.Items.Add("");
73      lstDisplay.Items.Add("When is my loan due? " + gradDate.AddDays(60).ToLongDateString());
74      // To subtract days, use a negative number
```

Combining a date and time

A date from a DateTime variable and the time from another DateTime variable can be combined into a single variable. For example, the date from gradDate and the time from gradTime can be combined and stored into the single variable, gradDateTime. Because you cannot add two DateTimes, the time from gradTime is converted to a TimeSpan using the TimeOfDay property of DateTime and then added to gradDate, as shown in Step 3-13.

Step 3-13: Add the code to combine the date from gradDate and time from gradTime into a single variable gradDateTime. Here is the code to do it:

```
76      // Combine the date and time and store in a single variable:
77      // Convert graduation time to a TimeSpan and add to date:
78      TimeSpan tsGradTime = gradTime.TimeOfDay;
79      // or,             = new TimeSpan(gradTime.Hour, gradTime.Minute, gradTime.Second);
80      gradDateTime = gradDate.Add(tsGradTime);
81      lstDisplay.Items.Add("");
82      lstDisplay.Items.Add("Again, graduation date & time is: " + gradDateTime);
```

Step 3-14: Add the code to compute the number of days to graduation, taking into account the difference between graduation time and current time.

```
84      // Compute the number of days to graduation, including the fractional part
85      tsGraduation = (gradDateTime - DateTime.Now);   //Now includes current date and time
86      lstDisplay.Items.Add("Just " + tsGraduation.TotalDays.ToString("N") + " more days");
```

Run the program, and observe that the number of days includes a fractional part due to the difference between graduation time and current time.

Alternatively, you may use the Days (instead of TotalDays) and Hours properties to display the whole days and hours using the following statement:
    lstDisplay.Items.Add("Just " + tsGraduation.Days.ToString("N1") + " more days" + " and " + tsGraduation.Hours + " hrs");
Stop the program.

Now you have the complete code that should look as follows (a copy of the code can be found at Tutorial_Starts/Ch2_Data Types/Ch2_GraduationDate_Code.txt):

private void btnDisplayGradInfo_Click(object sender, EventArgs e)
{
    DateTime gradDate, gradDateTime, gradTime;
    float numberOfDays, numberOfHours;

    //Get date of graduation:
    gradDate = dtpGradDate.Value;
    // Get time of graduation:
    gradTime = dtpGradTime.Value;

    // Format and display current date & time and graduation date:
    lstDisplay.Items.Clear();
    lstDisplay.Items.Add("Current Date & Time: " + DateTime.Now);
    lstDisplay.Items.Add("Date of graduation: " + gradDate.ToShortDateString());

```csharp
// Display the month & day of graduation:
lstDisplay.Items.Add("");
lstDisplay.Items.Add("What day is the graduation?    " + gradDate.ToString("dddd"));
lstDisplay.Items.Add("What month is it?              " + gradDate.ToString("MMM"));
lstDisplay.Items.Add("What day of the Month?         " + gradDate.Day);
//Format and display the time of graduation:
lstDisplay.Items.Add("What time?                     " + gradTime.ToLongTimeString());

// Compute the interval between today's date and graduation date

TimeSpan tsGraduation = (gradDate - DateTime.Today);
    // today includes current date, but not current time.
// You may also use the Subtract() method of DateTime
// TimeSpan tsGraduation = gradDate.Subtract(DateTime.Today);

numberOfDays = (float) tsGraduation.Days;
// Since graduation date and DateTime.Today both have the same default time (12:00:00),
// the remaining hours, minutes and seconds of the difference will be zero.
lstDisplay.Items.Add(" ");
lstDisplay.Items.Add("How long?");
lstDisplay.Items.Add("You have only  " + numberOfDays + " more days to graduate");

// Computing weeks, months and years:
float numberOfWeeks = numberOfDays / 7;
float numberOfMonths = ((gradDate.Year - DateTime.Today.Year) * 12) + (gradDate.Month -
           DateTime.Today.Month);
float numberOfYears = (int)(numberOfDays / 365.25);
// The above calculation of month and year may be imprecise under certain conditions.

// Display number of weeks, months & years
lstDisplay.Items.Add("That is, about  " + numberOfWeeks.ToString("N0") + " more weeks.");
lstDisplay.Items.Add("That is, about " + numberOfMonths + " more months.");
lstDisplay.Items.Add("That is, over  " + numberOfYears + " year.");

// Compute the day when the loan is due - Add 60 days to graduation date:
lstDisplay.Items.Add("");
lstDisplay.Items.Add("When is my loan due? " + gradDate.AddDays(60).ToLongDateString());
// To subtract days, use a negative number

// Combine the date and time and store in a single variable:
// Convert graduation time to a TimeSpan and add to date:
TimeSpan tsGradTime = new TimeSpan(gradTime.Hour, gradTime.Minute, gradTime.Second);
gradDateTime = gradDate.Add(tsGradTime);

lstDisplay.Items.Add("");
lstDisplay.Items.Add("Again, graduation date & Time is: " + gradDateTime);

// Compute the number of days to graduation, including the fractional part
tsGraduation = (gradDateTime - DateTime.Now);  //Now includes current date and time
lstDisplay.Items.Add("Just " + tsGraduation.TotalDays.ToString("N") + " more days");

}
```

Step 3-15: Double click the Clear button, and enter the following code into the Click event handler of the Clear button. The code shows three different statements to reset the date. Run the form and observe the effect of each statement by commenting out the others.

```
 89            private void btnClear_Click(object sender, EventArgs e)
 90            {
 91                //Clear ListBox
 92                lstDisplay.Items.Clear();
 93
 94                //Reset the DateTimePicker for date:
 95                dtpGradDate.Value = DateTime.Parse("01/01/1900");
 96                // Or,
 97                dtpGradDate.Value = new DateTime(1900, 01, 01);
 98                //Or,
 99                dtpGradDate.Value = DateTime.Today;
100
101                // Reset the DateTimePicker for Time:
102                dtpGradTime.Value = DateTime.Parse("12:00 AM");
103            }
```

It's time to practice! Do Exercise 2-3 and 2-4.

Review Questions

2.17   The DateTime structure has two related properties—**Today** and **Now**. What is the difference between them?

Refer to the following figure and code for the next two questions:

//Get rental date from DateTimePicker named dtpRentalDate
DateTime rentalDate = dtpRentalDate.Value;
// Get time of rental DateTimePicker named dtpRentalTime
DateTime rentalTime = dtpRentalTime.Value;

2.18   When a rented item is returned, it is desired to compute and display the whole number of days (without the fractional part due to time difference, if any) the item was rented. Write the code to display the number of days in the ListBox, lstDisplay.

2.19   Write the code to combine the rental date and time, and store in a single variable named rentalDateTime declared as **DateTime rentalDateTime**

2.20   Write the code to compute the due date for payment by adding 30 days to the invoice date declared as **DateTime invoiceDate = DateTime.Today;**

Exercises

Exercise 2-1: Cost of oil change

The service department of an auto dealership computes the cost of an oil change by adding a fixed labor charge to the cost of oil used (quarts used x cost per quart). Develop an application that allows the user to enter the labor charge, quarts used, cost per quart and sales tax rate in percents. The application should compute the total charge and display the subtotal (labor + cost of oil), sales tax and total cost (subtotal + sales tax). Develop the necessary form within a project named Exercises_Ch2. Use appropriate data types and names for variables. Use decimal type for variables that store dollar amounts.

Exercise 2-2: User name

Make necessary changes in the code in the UserNameString form (line 27 in Figure 2-5) so that the user name consists of the last 5 characters of the last name, the second character of the first name, and the first 2 digits of the id.

Exercise 2-3: Auto rental charge

"Ace Auto Rentals" is a small business that rents low-cost automobiles. Ace would like you to develop an application that lets them compute the rental charges. The application should let the user enter the date rented and date returned (use the DateTimePicker control), enter the charge/day, and compute and display the days rented and the total rental cost (days rented x charge per day). Note that this application doesn't take into account the times of rental and return.

Extra credit (30%): Modify the application so that the user also enters the time of rental and time of return. These times are used in computing the days rented to include fractional days. For fractional days, a customer is charged only for the fraction of the day the vehicle is used.

Exercise 2-4: Late fee for video game rental

Develop an application for a video game rental store to compute the late fee for videos that are returned late. Develop a form named VidioRentals that lets the store to do the following:

Enter a fixed handling charge, late charges per day per item, number of items rented and the due date. The user doesn't have to enter the date returned; use today's date as the date returned.

Compute the number of days late (today's date - due date), late charge (= No of items × No of days late × late charges per day per item), and total charge (= handling charge + late charge). Today's date is given by **DateTime.Today**."

Display the number of days, late charge and total charge in a ListBox.

Chapter 3

Decision Structures and Validation

The programs presented in Chapters 1 and 2 have simple structures, called **sequence structures,** where the statements are processed sequentially, starting with the first statement and ending with the last, processing each statement, exactly one time. In this chapter, we look at **decision (selection) structure**, which lets you execute a block of statements only under certain conditions.

Topics

3.1	Flowchart for Decision Structure	3.4	Nested if Statements
3.2	if else Statement	3.5	The switch Statement
3.3	if else if Statement	3.6	Data Validation

3.1 Flowchart for Decision Structure

As an example of decision structure, consider an updated version of Tutorial 3 from Chapter 1 to include a 20% discount for orders that are greater than or equal to 10 dollars, as shown in Figure 3-1.

Figure 3-1: The IceCreamTotalDiscounted form

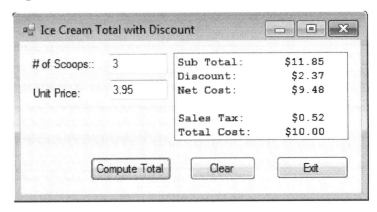

Because the discount applies only to orders greater than 10 dollars, the program has to check the value of subtotal and select alternate routes depending on the value. Such a structure that allows a program to execute one or more statements only if a certain condition is true is called a **decision (or selection) structure**.

A popular graphical tool to represent program logic is a flowchart. The flowchart that represents the logic for computing the discount percentage is shown in Figure 3-2.

Figure 3-2: Flowchart for a simple selection structure

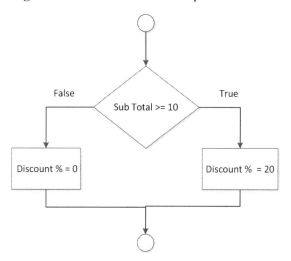

In a flowchart, the diamond is used to check whether a certain condition is true or false. The connector symbol, circle, is used to connect the selection structure to the rest of the program logic that is not shown in the diagram.

A rectangle is used to represent the action to be taken in each path. The action may consist of processing one or more statements with a sequence structure, another selection structure or other structures.

First, we will look at the use of the **if** statement to implement a decision structure.

3.2 if else Statement

This is how you compute the discount via an if statement with an else clause using the decision structure shown in Figure 3-2:

Example 1:

```
if (subTotal >= 10)
    discountPercent = 20;
else
    discountPercent = 0;
```

The above if statement uses the expression (subTotal >= 10), which evaluates to true or false. Such an expression that evaluates to true or false is called a **Boolean expression**.

The Boolean expression (subTotal >= 10) uses the **relational operator** ">=" to compare the value of the decimal type variable subTotal and the literal 10. Table 3-1 presents additional relational operators. Note that when the "< " (less than) operator is used with dates, it means "earlier than." For example, in Table 3-1, the Boolean expression (dateDue < DateTime.Today) evaluates to true if dateDue is earlier than today's date. Similarly, when the ">" (greater than) operator is used with dates, it means "later than."

How does the **if else** statement work? C# first evaluates the Boolean expression (subTotal >= 10).

If it is true, the statement within the **if** clause (discountPercent = 20) is executed, and the else clause (discount = 0) is skipped.

If expression (subTotal >= 10) evaluates to false, the statement within the if clause is skipped, and the statement within the else clause (discount = 0) is executed.

Table 3-1: Relational operators

Relational operator	What it means	Examples
>	Greater than	if (roomCapacity > enrollment)
<	Less than	if (dateDue < DateTime.Today)
>=	Greater than or equal to	if (examScore >= 90)
<=	Less than or equal to	if (dateOfBirth <= DateTime.Parse("12/31/1980")

== (combination of two "=" signs)	Equal to	if (city == "Oshkosh") // city is string type
		if (inState ==true) // inState is bool type
		// The expression if (inState ==true) may be replaced by
		// if (inState), because inState itself is true or false.

!=	Not equal to	if (middleInitial != 'C') // middleInitial is of char type.

Tutorial 1: Ice Cream Total with Discount

This tutorial develops the updated version of Tutorial 3 from Chapter 1 to include a discount, as represented in the flowchart in Figure 3-2.

Step 1-1: Open the project, Ch3_DecisionAndValidation from the Tutorial_Starts folder. Open the form IceCreamTotalDiscounted.

Edit Program.cs file to specify IceCreamTotalDiscounted as the form to be run.

Step 1-2: Modify the code for computing total cost, as shown in Figure 3-3: The changes include

Lines 40–43: Add if else statement to compute discountPercent.

Lines 46–52: Uncomment statements to compute discount, net cost and total cost.

Lines 57–60: Uncomment statements to display subtotal, discount and net cost.

Lines 62–63: Add statements to display sales tax and net cost.

Test the program with combinations of unit price and number of scoops that yield subtotals that are less than 10, equal to 10 and more than 10.

Figure 3-3: Compute total cost with discount

```csharp
private void btnComputeTotal_Click(object sender, EventArgs e)
{
    //Compute the total cost for an order of icecream:

    //Decalre variables to hold input data and computed results

    decimal unitPrice, subTotal, salesTax, totalCost;
    int scoops;
    //Declare constants to hold the discount and tax rates
    const decimal TAX_RATE = 5.5m;

    //Get input data from Text Boxes:
    unitPrice = decimal.Parse(txtUnitPrice.Text);
    scoops = int.Parse(txtScoops.Text);

    // Compute sub-total
    subTotal = unitPrice * scoops;

    // Compute discount percent
    decimal discountPercent, discount;
    if (subTotal >= 10)
        discountPercent = 20;
    else
        discountPercent = 0;

    // Compute discount
    discount = subTotal * discountPercent / 100;

    // Compute Total Cost
    decimal netCost;
    netCost = subTotal - discount;
    salesTax = netCost * TAX_RATE / 100;
    totalCost = netCost + salesTax;

    /* Display Results with decimals aligned
       Important Note: To align characters in a column, the Font type of
       the ListBox must be of fixed size, like Courier New.              */
    string formatString = "{0,-12}{1,10:C}";
    lstOutput.Items.Add(string.Format(formatString, "Sub Total:", subTotal));
    lstOutput.Items.Add(string.Format(formatString, "Discount:", discount));
    lstOutput.Items.Add(string.Format(formatString, "Net Cost:", netCost));
    lstOutput.Items.Add("");   // Add a blank line
    lstOutput.Items.Add(string.Format(formatString, "Sales Tax:", salesTax));
    lstOutput.Items.Add(string.Format(formatString, "Total Cost:", totalCost));
}
```

Mixing Different Data Types in a Boolean Expression

The rules you learned in Chapter 2 about mixing different data types in a formula (mathematical expression) involving numbers also apply to Boolean expressions—that is, you may mix different real and integer data types in a Boolean expression, but you cannot mix a decimal with a float or double without explicit conversion. For example, the expression in the following if statement is invalid:

```
decimal price = 12.5m;
if (price > 10.5)   // invalid; cannot compare a decimal to double
```

The expression should be changed to

```
if (price > 10.5m)        // valid
```

The following comparisons are valid:

```
decimal price = 12.5m;
if (price > 10)     // valid; decimal and integer types may be mixed

float price = 12.5f;
if (price > 10.5)   // valid; float and double types may be mixed
```

An if statement may use one or more of the following **relational operators** to compare data items in a Boolean expression:

if Statement with No else Clause

Example 2: An alternate way to represent the decision structure shown in Figure 3-2 is to remove the statements in the **false** path and place them before the decision structure, as shown below:

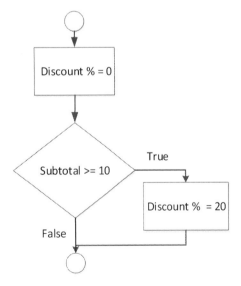

This is how you would translate the flowchart to programming statements:

```
float discountPercent = 0;
if (subTotal >= 10)
    discountPercent = 20;
```

If the value of the Boolean expression (subTotal >= 10) is true, discountPercent is assigned the value 20. What happens if subTotal is less than 10? discountPercent keeps its initial value, zero.

Caution! If there is no else clause, discountPercent must be assigned a value before the variable can be used in the program.

In computing the discountPercent, the above if statement with no else clause will have the same effect as the following code with an else clause:

```
float discountPercent;
if (subTotal >= 10)
   discountPercent = 20;
else
   discountPercent = 0;
```

The two if statements are equivalent because the value assigned in the else clause is the same as the initial value of the variable.

A Block of Statements in the if or else Clause

When the if or else clause includes multiple statements, they are enclosed in a pair of braces. Optionally, you may use braces also when the if or else clause includes only a single statement.

Example 3: if else statement with a block of statements in the if and else clauses.

Assume that the customer gets a free topping if the subtotal is $10 or above. In this case, the if clause assigns values to discountPercent and to the Boolean variable, freeTopping.

```
bool freeTopping;
if (subTotal >= 10)
{
        discountPercent = 20;
        freeTopping = true;
}
else
{
        discountPercent = 0;
        freeTopping = false;
}
```

Note that the block of statements in the if clause is enclosed in a pair of braces.

Step 1-3: Modify the if statement in the click event handler of btnComputeTotal as follows:

```
bool freeTopping;
if (subTotal >= 10)
{
        discountPercent = 20;
        freeTopping = true;
}
else
{
        discountPercent = 0;
        freeTopping = false;
}
```

Step 1-4: Add the following statements at the end of the click event handler to display the message "*** Free Topping ***" if freeTopping has the value, **true**.

```
if (freeTopping)              // same as, if (freeTopping == true)
        lstOutput.Items.Add(" *** Free Topping ***");
```

Note that it is not necessary to use the Boolean expression "(freeTopping = true)" because the value of freeTopping is true or false.

Step 1-5: Run the form to test the code.

The general syntax of the if else statement is

```
if (Boolean expression)
{
        A block of statements  (or, a single statement -- doesn't require braces).
}
else     // else clause is optional
{
        A block of statements (or, a single statement -- doesn't require braces).
}
```

Note that there is no semicolon at the end of the line starting with if (Boolean expression) and the else line.

Logical Operators

If statements may use **logical operators** to combine multiple Boolean expressions that specify the selection criterion. As stated earlier, a Boolean expression is an expression that evaluates to true or false. Table 3-2 shows a list of such logical operators and examples of their use.

Table 3-2: Logical operators

Logical operator	Name	Example	What it means
&&	Conditional And	if((ACT >=20) && (GPA>=2.5)) The second Boolean expression is evaluated only if the first one is **true**.	If **both** expressions are true, "&&" returns **true**; if not, "&&" returns **false**. That is, **true&&true** yields **true**; **true&&false** yields **false**; **false&&true** yields **false**; **false&&false** yields **false**;
&	And	if((ACT >=20) & (GPA >=2.5)) Both expressions are always evaluated.	Same as "&" (i.e., if both expressions are true, "&" returns **true**; if not, "&" returns **false**).
||	Conditional Or	if ((ACT <20) || (GPA <2.5)) The second Boolean expression is evaluated only if the first one is **false**.	If **either** expression is true, "||" returns **true**; if not, "||" returns **false**. That is, **true||false** yields **true**; **false||true** yields **true**; **true||true** yields **true**; **false||true** yields **false**;
|	Or	if ((ACT <20) | (GPA <2.5)) Both expressions are always evaluated.	Same as "|" (i.e., if either expression is true, "|" returns **true**; if not, "|" returns **false**).
!	Not	if (!(ACT >= 20))	if (ACT >= 20), "!" returns **false**; if not, "!" returns **true**.

if statement with logical operators

Table 3-2 presented different logical operators like "&&" (conditional And) and "\|\|" (conditional Or) that can be used in an expression when the decision involves multiple criteria.

Next, we will look at a modified version of the decision structure that uses logical operators. In this version, discount and free topping are given if **the subtotal is at least $10 or the number of scoops is at least 3**.

The following program shows the modified if statement to implement the new decision structure:

```
bool freeTopping;

if ((subTotal >= 10) || (scoops >=3))
{
        discountPercent = 20;         // processed if either (subTotal >= 10) or (scoops >=3),
        freeTopping = true;           // that is, if one of the two Boolean expressions is true.
}
else
{
        discountPercent = 0;          // processed if both Boolean expressions are not true
        freeTopping = false;          // that is, if both are false.
}
```

Step 1-6: Modify the if statement to include the revised criteria. Run and test the form with different combinations of subtotal and scoops so that (1) both Boolean expressions are false, (2) only the first expression is true, (3) only the second expression is true and (4) both expressions are true. Observe the result in each case.

Change the logical operator from "||" (Or) to "&&" (And), run the form and observe the difference under the four different combinations of subtotal and scoops.
Change the logical operator back to "||"

The expression **((subTotal >= 10) || (scoops >=3))** checks whether subtotal is at least 10 or scoops is at least 3. An alternative to do the same thing is to check whether the expression

((subTotal < 10) **and** (scoops < 3))

is **false**—that is, both subtotal and scoops are not below the required limit. Note that the expressions have the "<" operator in place of ">=." Further, "Or" is replaced by "And."

You can use the "!" (Not) operator to check whether the expression is false, as follows:

if (!((subTotal < 10) && (scoops < 3)))

Step 1-7: Change the if statement to: if (!((subTotal < 10) && (scoops<3))).
Run the form and observe the difference under the four different combinations of subtotal and scoops specified in Step 1-6. Observe the result in each case.
Now, change "&&" to "||" as follows:

```
bool freeTopping;

if (!((subTotal < 10) || (scoops < 3)))
{
        discountPercent = 20;
        freeTopping = true;
}
else
{
        discountPercent = 0;
        freeTopping = false;
}
```

This won't give the desired result. If any one of the two Boolean expressions is false, then the combined expression, ((subTotal < 10) || (scoops < 3)) is true, which makes (!((subTotal < 10) || (scoops < 3))) false, resulting in no discount.

Run the form and observe the difference under the four different combinations of subtotal and scoops. Observe the result in each case.

Review Questions

What would be the output displayed in the ListBox lstOutput by the following programs?

3.1   int totalAmt = 80, percentDiscount = 5;
       if (totalAmt >= 100)
           percentDiscount = 20;
       lstOutput.Items.Add(percentDiscount);

3.2   int totalAmt = 80;
       int percentDiscount = 5;
       if (totalAmt >= 100)
           percentDiscount = 20;
       else
           percentDiscount = 0;
       lstOutput.Items.Add(percentDiscount);

3.3   int ACT = 23;
       float GPA = 3.2F;
       if ((ACT >= 24) || (GPA >= 3))
           lstOutput.Items.Add("Admit");
       else
           lstOutput.Items.Add("Deny");

3.4   int ACT = 23;
       float GPA = 3.2F;
       if ((ACT >= 24) && (GPA >= 3))
           lstOutput.Items.Add("Admit");
       else
           lstOutput.Items.Add("Deny");

3.5   int ACT = 23;
       float GPA = 3.2F;
       if (!((ACT >= 24) || (GPA >= 3)))
           lstOutput.Items.Add("Admit");
       else
           lstOutput.Items.Add("Deny");

3.6   int ACT = 23;
       float GPA = 3.2F;
       if (!((ACT >= 24) && (GPA >= 3)))
           lstOutput.Items.Add("Admit");
       else
           lstOutput.Items.Add("Deny");

3.3 if else if Statement

Consider a grading scale with a minimum of 90, 80 and 70 for A, B and C, respectively (anything below 70 is an F). The flowchart that represents the decision structure would look as follows:

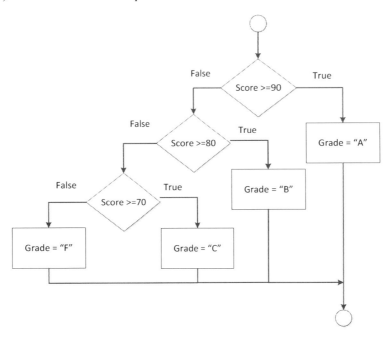

Example 4: The following program uses the **if else if** structure to compute the letter grade based on the decision structure represented by the flowchart. In addition to the grade, the program also computes the grade point.

```
string letterGrade;
int gradePoints;
if (totalScore >= 90)
{
        letterGrade = "A";
        gradePoints = 4;
}
else if (totalScore >= 80)
{
        letterGrade = "B";
        gradePoints = 3;
}
else if (totalScore >= 70)
{
        letterGrade = "C";
        gradePoints = 2;
}
else
{
        letterGrade = "F";
        gradePoints =0;
}
```

In an if else if statement, in addition to the Boolean expression(s) in the if clause, you may have additional Boolean expressions in else if clauses.

In Example 4, if the expression (totalScore >= 90) is true, the statements in the if clause,

    letterGrade = "A";
    gradePoints = 4;

are processed, and the rest of the statements are skipped.

If the expression (totalScore >= 90) is false, the statements in the if clause are ignored, and the next else if clause is processed.

If none of the Boolean expressions are true, the statements in the else clause are executed.

General syntax:
if (Boolean expression)

    one or more statements

else if (Boolean expression)

    one or more statements

...

else

    one or more statements

Review Questions

3.7   What would be the value of letterGrade after execution of the following program?

    string letterGrade = "F";
    int totalScore = 85;

    if (totalScore >= 70)
            letterGrade = "C";
    else if (totalScore >= 80)
            letterGrade = "B";
    else if (totalScore >= 90)
            letterGrade = "A";

3.8   Convert the following **if else if** statement to an **if else** statement:

    string grade;
    int totalScore = 85;

    if (totalScore >= 70)
            grade = "Pass";
    else if (totalScore < 70)
            grade = "Fail";

3.9 What would be the value of discount after execution of the following program?

    int orderQty = 120, totalAmt = 900, discount =0;

    if (totalAmt > 1000)
       discount = 20;
    else if (orderQty > 100)
       discount = 10;
    else if (totalAmt > 500)
       discount = 5;

3.4 Nested if Statements

An if statement (with or without the else clause) may be nested within another if statement's if clause or else clause.

Consider a modified version of the flowchart discussed in Example 4 to compute grade. As in the previous example, letter grades A, B and C require a minimum total score of 90, 80 and 70, respectively. In addition, an A requires a minimum of 90 for attendance, as shown in the following flowchart.

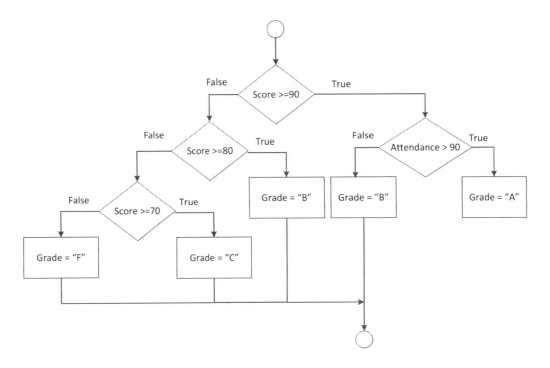

Example 5: The following program uses a nested if statement where the if statement that checks attendance is nested inside the if statement that checks total points.

```
string letterGrade;

if (totalScore >= 90)
    if (attendance >= 90)         // this statement processed only if totalScore is at least 90
        letterGrade = "A";
    else                          // else is processed only if attendance is less than 90
        letterGrade = "B";
else if (totalScore >= 80)        //else if means if totalScore is less than 90
    letterGrade = "B";
else if (totalScore >= 70)
    letterGrade = "C";
else
    letterGrade = "F";
```

If totalScore is greater than or equal to 90, the grade is assigned based on attendance, and the remaining else if clauses and the else clause are skipped.

If totalScore is less than 90, the first else if clause is processed. The process continues until a condition is satisfied.

If no Boolean expression evaluates to true, the final else clause is processed.

Tutorial 2: Compute Grade

In this tutorial, you work with different forms of if statements through an application that computes letter grade based on total score and attendance, as shown in Figure 3-4.

Figure 3-4: The ComputeGrade form

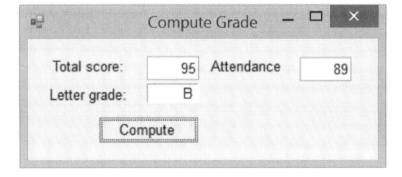

Step 2-1: Open the project, Ch3_DecisionAndValidation from the Tutorial_Starts folder. Open the form ComputeGrade that computes and displays grade based on total points and attendance.

Step 2-2: Review the user interface controls specified in Table 3-3.

Table 3-3: Controls on ComputeGrade form

Purpose	Type of control	Property	Setting
Enter total score	TextBox	Name	txtScore
Enter attendance %	TextBox	Name	txtAttendance
Display grade	Label	Name	txtGrade
Display label for total score	Label	Text	Score:
Display label for attendance	Label	Text	Attendance:
Display label for grade	Label	Text	Grade:
Compute grade	Button	Name	btnGrade
Compute grade	Button	Text	Compute Grade

First, you develop the code to compute grade based on just the total points, ignoring attendance.

Step 2-3: Add the following code to the click event handler of btnGrade to compute grade based only on score.

Figure 3-5: Code to compute grade based only on score.

```
        private void btnGrade_Click(object sender, EventArgs e)
        {
            string letterGrade;
            float totalScore = float.Parse(txtScore.Text);

            // Compute grade using if else if, based only on score:

            if (totalScore >= 90)
                letterGrade = "A";
            else if (totalScore >= 80)
                letterGrade = "B";
            else if (totalScore >= 70)
                letterGrade = "C";
            else
                letterGrade = "F";

            // Display grade:
            lblGrade.Text = letterGrade;
```

Step 2-4: Change the startup form to ComputeGrade, and run the project with values of score in the range 0 to 100, a value above 100 and one negative value. No data for attendance is needed for this version of the program.

Step 2-5: Make changes in the code, as shown in Figure 3-6, to include the requirement that an "A" grade requires an attendance of at least 90%.

The changes include:
Declare attendance in line 26.
Insert the if statement shown in lines 28-31 in place of line 28 in Figure 3-5.

Figure 3-6: Code to compute grade based on score and attendance

```csharp
20      private void btnGrade_Click(object sender, EventArgs e)
21      {
22          string letterGrade;
23          float totalScore = float.Parse(txtScore.Text);
24
25          // compute grade using if else if, based on score and attendance:
26          float attendance = float.Parse(txtAttendance.Text);
27          if (totalScore >= 90)
28              if (attendance >= 90F) // nested within the outer if
29                  letterGrade = "A";
30              else
31                  letterGrade = "B";
32          else if (totalScore >= 80)
33              letterGrade = "B";
34          else if (totalScore >= 70)
35              letterGrade = "C";
36          else
37              letterGrade = "F";
38
39          // Display grade:
40          lblGrade.Text = letterGrade;
41      }
```

Run the project with attendance below 90, and also with attendance greater than or equal to 90, while keeping the score above 90. Verify the results. Run the project when grade is less than 90, and verify that the attendance does not have any effect, in this case.

Review Questions

3.10   A restaurant offers a $10 discount if the total amount is at least $40. If the amount is less than $40, a $5 discount is given if the customer has a coupon and the amount is at least $20. If the amount is less than $20, the coupon gives a 10% discount. Draw a flowchart that represents the decision structure.

3.11   Write an if statement that implements the decision structure described in 3.10. Assume that the variables are declared as follows:

   float amount = 15.0F;
   float discount;
   bool coupon = true;

3.5 The switch Statement

The switch statement provides a convenient way to branch to one of many paths based on the value of a variable or an expression. How does it differ from an if statement with else if clause? Let's look at the code to determine the discount percent for a hotel reservation based on the discount type ("AAA," "Best Rate," "Government" or "Senior") for a hotel guest.

Here is the if statement with else if clause to determine the discount:

```
float discountPercent;
if (discountType == "AAA")
   discountPercent = 10;
else if (discountType == "Best Rate")
   discountPercent = 5;
else if (discountType == "Government")
   discountPercent = 15;
else if (discountType == "Senior")
   discountPercent = 12.5F;
else
   discountPercent = 0;
```

The corresponding switch statement is

```
float discountPercent;
switch (discountType)
{
   case "AAA":
      discountPercent = 10;
      break;
   case "Best Rate":
      discountPercent = 5;
      break;
   case "Senior":
      discountPercent = 12.5F;
      break;
   case "Government":
      discountPercent = 15;
      break;
   default:
      discountPercent = 0;
      break;
}
```

The switch statement checks the value of the **test variable**, discountType, and branches to one of the first four paths based on its value ("AAA," "Best Rate," "Government" or "Senior"). If the value does not match any one of the four values, the statement under default is processed.

Note that there is a break statement in each path, which causes the program to skip the rest of the switch statement. For example, if the discountType is "Best Rate," then the program assigns the value 5 to discountRate, and the break statement causes it to skip checking for the values "Government" and "Senior" that are specified in the rest of the case statements.

A break statement (or other statements like return) that causes the program to jump out of the switch statement is required in each case and the default statement.

The general syntax of the switch statement is

```
switch (test variable or expression)
{
    case value_1:
        one or more statements
        break;
    case value_2:
        one or more statements
        break;
    case value_3:
        one or more statements
        break;
        ...
    default:
        one or more statements
        break;
}
```

Unlike in an if statement, the value of the test variable or expression in a switch statement must be an integer, string or Boolean. Further, the expression can be compared to the different values only for equality, not for other relational operators like "<" and ">".

For example, the following code, which computes discount when discountCode is equal to 1, is valid:

```
int discountCode = int.Parse(txtDiscountCode.Text);
switch (discountCode)
{
    case 1:               // valid
        discount = 10;
        break;
}
```

The following case statement that uses the ">=" operator is invalid:

```
int score = int.Parse(txtScore.Text);
switch (score)
{
    case >= 90:   // invalid use of ">=" operator. Case = 90 also is invalid.
        letterGrade = "A";
        Break;
}
```

Similarly, the following code that uses a float type test variable is invalid:

```
float score = float.Parse(txtScore.Text);
switch (score)
{
    case 90:     // invalid; score is not int type.
        letterGrade = "A";
        Break;
}
```

Tutorial 3: switch Statement

To help understand the switch statement, you will create a form that computes the discount for a hotel room reservation based on the discount type ("AAA," "Best Rate," "Government" or "Senior"), as shown in Figure 3-7:

Figure 3-7: Reservation discount calculator

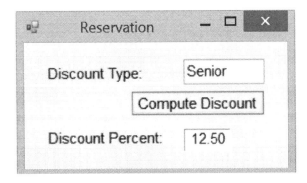

Step 3-1: Open the project Ch_DecisionAndValidation.

Step 3-2: Open the form Reservation.

Step 3-3: Add the following code to the click event of btnComputeDiscount button.

```
        private void btnComputeDiscount_Click(object sender, EventArgs e)
        {
            string discountType = txtdiscountType.Text;
            float discountPercent;

            // Compute discount percent
            switch (discountType)
            {
                case "AAA":
                    discountPercent = 10;
                    break;
                case "Best Rate":
                    discountPercent = 5;
                    break;
                case "Government":
                    discountPercent = 15;
                    break;
                case "Senior":
                    discountPercent = 12.5F;
                    break;
                default:
                    discountPercent = 0;
                    break;
            }
            // Display discount percent
            lblDiscountPercent.Text = discountPercent.ToString("N");
```

Step 3-4: Make Reservation the startup form, run the project and verify the result for the different discount types.

The following is the equivalent code to compute discount using if else if statement:

```
// Alternative to using if else if statement:
if (discountType == "AAA")
   discountPercent = 10;
else if (discountType == "Best Rate")
   discountPercent = 5;
else if (discountType == "Government")
   discountPercent = 15;
else if (discountType == "Senior")
   discountPercent = 12.5F;
else
   discountPercent = 0;
```

It's time to practice! Do Exercise 3-1.

Review Questions

3.12　In a **switch** statement, what are the valid data types for the test variable (or expression)?

3.13　A university rates its faculty members' performance on a scale of 1 to 3. The annual percent raise in pay is determined by the rating as follows:

Rating	Percent raise
3	0.25%
2	0.1%
1	-1.0%

Write a switch statement that computes the percent raise based on the above rule. Use the following declaration of variables:

```
int rating = 2;
double raise;
```

3.6 Data Validation

So far, we assumed that the user would enter valid data into TextBoxes. It is not uncommon that users inadvertently enter invalid data. Such data, if not caught properly, often result in abnormal termination of the program, frustrating the user. A good program should catch such runtime errors and let the user correct them.

Data entered into a TextBox or other controls may be validated before the user moves focus away from the control, or after the focus is moved away from the control. We refer to validation before leaving the control as **immediate validation**, and validation after leaving the control as **late validation.** You will use the **TryParse** method to of numeric data types to do both types of validation.

Consider the ComputeGrade form shown in Figure 3-8, which was presented earlier in Tutorial 2 to compute letter grade.

Figure 3-8: The ComputeGrade form from Tutorial 2

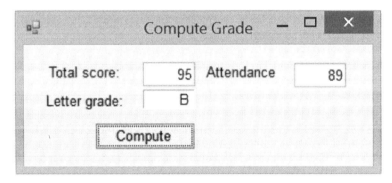

In Tutorial 2, we used the Parse method to convert the total score entered into the TextBox, txtScore, to float, as follows:

    float totalScore = float.Parse(txtScore.Text);

If you use the Parse method to convert a string that contains nonnumeric characters, it will result in an error when you run the program. You can use the TryParse method to validate the data and avoid such errors.

The TryParse method

The TryParse method tries to convert a string to a specific numeric data type. The syntax of TryPrase() method is

    numericType.TryParse(string, out resultVariable)

In general, TryParse() tries to convert the string specified as the first parameter to the specific numeric data type (represented by *numericType*). If successful, the resulting number is assigned to the variable specified in the second parameter (represented by *resultVariable*). In addition, TryParse() returns the value true. If not successful, the value zero is assigned to *resultVariable*, and TryParse() returns the value false. The key word *out* specifies that totalScore is an output parameter, which will be discussed in more detail later.

Here is an example:

    float.TryParse(txtScore.Text, out totalScore);

In this example, TryParse() tries to convert the string represented by txtScore.Text to float type. If successful, the resulting number is assigned to the variable totalScore. In addition, TryParse() returns the value true. If not successful, the value zero is assigned to totalScore, and TryParse() returns the value false.

The assignment of a numeric type to the resultVariable must follow the rules of casting discussed in Chapter 2. Thus, you cannot store a data type that has a larger range in a *resultVariable* of a smaller range.

This is how you would use TryParse to validate the total score from the TextBox, txtScore:

```
if (float.TryParse(txtScore.Text, out totalScore) == false)
{
   // Display error message and let user re-enter the data
}
```

How does it work? If conversion of the string to float type is successful, the resulting float type data will be stored in the variable totalScore, and the method will return the value **true,** causing the Boolean expression to have the value, **false.** So, the block within the if statement (to display the error message) would be skipped.

If conversion is unsuccessful, the value zero will be stored in totalScore, and the Parse method will return the value false, causing the Boolean expression to have the value, true. So, the block within the if statement (to display the error message) would be executed.

Late Validation of Data

Late validation validates the data after the user moves focus away from a control. We will use the TryParse method to validate the score and attendance when the user clicks the button, btnGrade.

Figure 3-9 shows the code to validate total score when the user clicks btnGrade.

Figure 3-9: Code to validate the score

```
20          private void btnGrade_Click(object sender, EventArgs e)
21          {
22              float totalScore;
23              // Verify that the score is a number
24              if (float.TryParse(txtScore.Text, out totalScore) == false)
25              {
26                  // Display invalid number message
27                  MessageBox.Show("Please enter a valid number for score");
28                  txtScore.Focus();
29                  txtScore.SelectAll();
30                  return;
31              }
```

In line 27, the statement,

      MessageBox.Show("Please enter a valid number for score");

uses the **Show()** method of MessageBox to display the string ("Please enter a valid number for score") specified as the parameter.

The general syntax is

> MessageBox.Show(*string*).

The second statement in line 28, **txtScore.Focus()**, uses the Focus() method of the TextBox, txtScore, to put the cursor in the TextBox.

The **SelectAll()** method in line 29 highlights the entire string in the TextBox so that the user can type over the data.

Tutorial 4: Late Validation

In this tutorial, you will create a form, **GradeWithLateValidation,** that validates the total score entered into a TextBox, as presented in Figure 3-11. The validation takes place when the user clicks the button to compute the grade. This is called late validation because the data in the TextBox is not validated before the focus leaves the TextBox.

Step 4-1: Open the form **GradeWithLateValidation** from the project Ch3_DecisionAndValidation (within Tutorial_Starts folder), shown in Figure 3-10. This form has the same user interface as the ComputeGrade form.

Figure 3-10: GradeWithLateValidation form

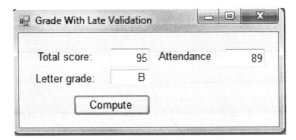

Step 4-2: Open the click event handler of btnGrade, which contains the following code to compute the letter grade. Note that the code to get the total score and attendance from TextBoxes is missing.

```
string letterGrade;

if (totalScore >= 90)
    if (attendance >= 90F) // nested within the outer if
        letterGrade = "A";
    else
        letterGrade = "B";
else if (totalScore >= 80)
    letterGrade = "B";
else if (totalScore >= 70)
    letterGrade = "C";
else
    letterGrade = "F";

// Display grade:
lblGrade.Text = letterGrade;
```

Step 4-3: Add the following code to the beginning of btnGrade click event handler to get the score and validate it:

```
float totalScore;
// Verify that the score is a number
if (float.TryParse(txtScore.Text, out totalScore) == false)
{
  // Display invalid number message
  MessageBox.Show("Please enter a valid number for score");
  txtScore.Focus();
  txtScore.SelectAll();
  return;
}

// Get attendance
float attendance = float.Parse(txtAttendance.Text);
```

Step 4-4: Run the form and test it using valid and invalid score (non-numeric score and no score).

Figure 3-11 shows the complete code for GradeWithLateValidation form.

Figure 3-11: Code for GradeWithLateValidation form

```
20    private void btnGrade_Click(object sender, EventArgs e)
21    {
22        float totalScore;
23        // Verify that the score is a number
24        if (float.TryParse(txtScore.Text, out totalScore) == false)
25        {
26            // Display invalid number message
27            MessageBox.Show("Please enter a valid number for score");
28            txtScore.Focus();
29            txtScore.SelectAll();
30            return;
31        }
32
33        // Get attendance
34        float attendance = float.Parse(txtAttendance.Text);
35
36        // Compute grade:
37        string letterGrade;
38        if (totalScore >= 90)
39            if (attendance >= 90F)
40                letterGrade = "A";
41            else
42                letterGrade = "B";
43        else if (totalScore >= 80)
44            letterGrade = "B";
45        else if (totalScore >= 70)
46            letterGrade = "C";
47        else
48            letterGrade = "F";
49
50        // Display grade:
51        lblGrade.Text = letterGrade;
52    }
```

It's time to practice!

Modify the code to verify that the attendance percent is a valid number. If invalid, display an appropriate message, highlight the current data, and allow the user to type over the existing data. If valid, verify that it is within the range 0 to 100. If not, display an appropriate message, highlight the current score and allow the user to type over the existing score.

Immediate Validation of Fields

On the GradeWithLateValidation form, the score and attendance data entered into TextBoxes are validated when the user clicks the button to compute the grade. Next, we look at how to validate data before the user moves away from a TextBox to give immediate feedback. We use the **Validating event** of the TextBox to do this.

Validating event

The Validating event of a TextBox is raised when the user tries to move the focus away from a TextBox—for example, by pressing the Tab key or clicking in another TextBox. This event is raised before another control gets the focus.

To do immediate validation, you validate the data within the Validating event handler of the TextBox and, if invalid, cancel the process of moving focus away from the TextBox.

To create the Validating event handler for txtScore TextBox, select txtScore and select the *events* tab in the Properties window, as shown in Figure 3-12. Double click Validating.

Figure 3-12: Validating event in Properties window

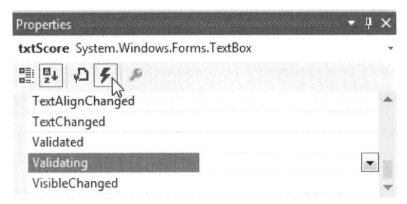

Here is the code to validate the TextBox, txtScore:

```csharp
private void txtScore_Validating(object sender, CancelEventArgs e)
{
    float totalScore;
    // Verify that the score is a number
    if (float.TryParse(txtScore.Text, out totalScore) == false)
    {
        MessageBox.Show("Please enter a valid number for score");
        e.Cancel = true; // Cancel the normal sequence of events
        txtScore.SelectAll();
        return;
    }
}
```

The code is identical to the validating code in the click event handler of the button on the previous form, except for the statement that sets focus back in the TextBox. Instead of txtScore.Focus(), here we use the statement
   e.Cancel = true;

Setting the Cancel property of the parameter e to true has the same effect as txtScore.Focus(). We use this method because setting focus from within the Validating event is not recommended because it can cause the application to stop responding.

Tutorial 5: Immediate Validation Using Validating Event

This tutorial creates the form, GradeWithImmediateValidation, that validates the input data before the focus is moved away from the TextBoxes. The user interface is the same as in GradeWithLateValidation that we discussed previously.

Step 5-1: Open the form named GradeWithImmediateValidation from Ch3_DecisionAndValidation project.

The click event handler of the btnGrade contains the code shown in Figure 3-13, which computes the letter grade with no validation of data.

Figure 3-13: Code from GradeWithImmediateValidation form

```csharp
20      private void btnGrade_Click(object sender, EventArgs e)
21      {
22          // Get totalScore and attendance
23          float totalScore = float.Parse(txtScore.Text);
24          float attendance = float.Parse(txtAttendance.Text);
25
26          // Compute grade:
27          string letterGrade;
28          if (totalScore >= 90)
29              if (attendance >= 90F)
30                  letterGrade = "A";
31              else
32                  letterGrade = "B";
33          else if (totalScore >= 80)
34              letterGrade = "B";
35          else if (totalScore >= 70)
36              letterGrade = "C";
37          else
38              letterGrade = "F";
39
40          // Display grade:
41          lblGrade.Text = letterGrade;
42      }
```

Step 5-2: Create the Validating event handler for txtScore, and add the code from lines 46–54 in Figure 3-14 to the event handler on the line following the first opening brace, "{."
Run the form, enter an invalid score and try moving away by tabbing or clicking in the second TextBox. Repeat with valid score.

Figure 3-14: Validating event handler with MessageBox

```csharp
43
44      private void txtScore_Validating(object sender, CancelEventArgs e)
45      {
46          float totalScore;
47          // Verify that the score is a number
48          if (float.TryParse(txtScore.Text, out totalScore) == false)
49          {
50              // Display invalid number message
51              MessageBox.Show("Please enter a valid number for score");
52              txtScore.Focus();
53              txtScore.SelectAll();
54              return;
55          }
56      }
```

Note that the code that validates the score, as discussed earlier, is the same as in the GradeWithLateValidation form, except that the Cancel property of the parameter **e** is set to true to cancel the process of moving away from the TextBox, thus setting focus back to the TextBox.

An alternative to using a MessageBox to alert users of data errors is to use an **ErrorProvider** control.

ErrorProvider Control

The ErrorProvider indicates an error by placing an error indicator image next to the control. Moving the mouse over the image displays the error message as a tooltip, as shown in Figure 3-15.

Figure 3-15: ErrorProvider control with error message

Unlike the MessageBox, ErrorProvider can be used to indicate errors on multiple TextBoxes simultaneously.

To use an ErrorProvider, add the ErrorProvider control from the Container group in the Toolbox. To indicate error next to a control, use the SetError method of the ErrorProvider object, which specifies the control and the error message, as in

.SetError(controlName, errorMessage);

See the code in Figure 3-16 for an example. To remove the image, set the error message to a null string.

Step 5-3: Drag and drop the ErrorProvider control from the Toolbox to the form. The ErrorProvider appears in the **component tray** at the bottom of the Designer window.
Change the name of the control from ErrorProvider1 to erpAttendance.

Create the Validating event handler for txtAttendance.
Add the code from lines 60–71 in Figure 3-16 to the Validating event handler of txtAttendance on the line following the opening brace "{."
Run the form, enter an invalid data for attendance and try moving away by tabbing or clicking in the first TextBox.
Repeat with valid data.

Figure 3-16: Validating event handler with ErrorProvider

```csharp
57
58      private void txtAttendance_Validating(object sender, CancelEventArgs e)
59      {
60          float attendance;
61          // Verify that attendance is a number
62          if (float.TryParse(txtAttendance.Text, out attendance) == false)
63          {
64              // Display invalid number message
65              erpAttendance.SetError(txtAttendance, "Please enter a valid number for attendance");
66              txtAttendance.Focus();
67              txtAttendance.SelectAll();
68              return;
69          }
70          else
71              erpAttendance.SetError(txtAttendance, "");
72      }
```

Checking for Null Values

The code in the validation event of each TextBox would validate the data entered into it, **if and only if** the user tries to move away from the TextBox. What if the user skips a TextBox and clicks the button to compute grade? That may result in an undesirable abnormal termination of the program.

So, even with the validation codes we have, it is necessary to verify that the TextBoxes are not empty. This can be done in the click event handler of the btnGrade button, as shown in Figure 3-17.

Step 5-4: Add the if statement shown in Figure 3-17 to the beginning of btnGrade_click event handler.

Step 5-5: Run the form, and test the code when attendance data is missing and when invalid attendance data is entered.

Figure 3-17: Checking for nulls in TextBoxes

```csharp
private void btnGrade_Click(object sender, EventArgs e)
{
    // Verify that the TextBoxes are not empty
    if ((txtScore.Text.Length == 0) || (txtAttendance.Text.Length == 0))
    {
        MessageBox.Show("Please enter score and attendance");
        return;
    }

    // Get totalScore and attendance
    ...
```

CauseValidation Property

When you move focus from one control to another, the Validating event for the first control is raised only if the CauseValidation property of both controls are true, which is the default.

Step 5-6: Set CauseValidation for txtAttendance to False. Run the form, enter an invalid score into txtScore and press the Tab key. No message is displayed, indicating that the Validating event for txtScore was not raised.

You can use the CauseValidation property to allow users to move away from a control without entering valid data. For example, you can let the user close the form without entering valid data by clicking the Exit button. To suppress the Validating event when the user moves to the Exit button, set the CauseValidation property of the Exit button to false.

Step 5-7: Add a new button, btnExit, and set its CauseValidation property to false. Enter the code shown in Figure 3-18 to the click event handler of btnExit. Run the form. Enter an invalid score and click the Exit button. No error message is displayed because setting CauseValidation of Exit button to false suppresses the validating event of the TextBox.

Figure 3-18: Click event handler of Exit button

```
73
74      private void btnExit_Click(object sender, EventArgs e)
75      {
76          Environment.Exit(0); // 0 is an exit code passed to the O.S.
77          // Environment.Exit closes all "threads"; but this.Close doesn't
78      }
```

Other Relevant Events

A variety of events are raised when you move from one control to another. These include

1. Enter
2. GotFocus
3. Leave
4. Validating
5. Validated
6. LostFocus

The complete code for computing grade in GradeWithImmediateValidation form is shown in Figure 3-19.

Figure 3-19: complete code for GradeWithImmediateValidation form

```csharp
20      private void btnGrade_Click(object sender, EventArgs e)
21      {
22          // Verify that the TextBoxes are not empty
23          if ((txtScore.Text.Length == 0 || (txtAttendance.Text.Length == 0)))
24          {
25              MessageBox.Show("Please enter score and attendance");
26              return;
27          }
28          // Get score and attendance
29          float totalScore = float.Parse(txtScore.Text);
30          float attendance = float.Parse(txtAttendance.Text);
31
32          // Compute grade:
33          string letterGrade;
34          if (totalScore >= 90)
35              if (attendance >= 90F)
36                  letterGrade = "A";
37              else
38                  letterGrade = "B";
39          else if (totalScore >= 80)
40              letterGrade = "B";
41          else if (totalScore >= 70)
42              letterGrade = "C";
43          else
44              letterGrade = "F";
45
46          // Display grade:
47          lblGrade.Text = letterGrade;
48      }
49
50      private void txtScore_Validating(object sender, CancelEventArgs e)
51      {
52          float totalScore;
53          // Verify that the score is a number
54          if (float.TryParse(txtScore.Text, out totalScore) == false)
55          {
56              // Display invalid number message
57              MessageBox.Show("Please enter a valid number for score");
58              txtScore.Focus();
59              txtScore.SelectAll();
60              return;
61          }
62      }
```

```csharp
63
64      private void txtAttendance_Validating(object sender, CancelEventArgs e)
65      {
66          float attendance;
67          // Verify that attendance is a number
68          if (float.TryParse(txtAttendance.Text, out attendance) == false)
69          {
70              // Display invalid number message
71              erpAttendance.SetError(txtAttendance, "Please enter a valid number for attendance");
72              txtAttendance.Focus();
73              txtAttendance.SelectAll();
74              return;
75          }
76          else
77              erpAttendance.SetError(txtAttendance, "");
78      }
79
80      private void btnExit_Click(object sender, EventArgs e)
81      {
82          Environment.Exit(0);   // 0 is an exit code passed to the O.S.
83          // Environment.Exit() closes all "threads"; but, this.close() doesn't
84      }
```

It's time to practice! Do Exercise 3-2 and Programming Assignment 1.

Review Questions

Refer to the following code to answer the first three questions:
        double salary;
        double.TryParse(txtSalary.Text, out salary);

3.14 What would be the value returned by TryParse if txtSalary contains an invalid number?

3.15 What would be the value stored in salary if txtSalary contains an invalid number?

3.16 What would be the value returned by TryParse if txtSalary contains 245?

3.17 When is the Validating event of a TextBox raised?

3.18 When you move focus from one control to a second control, what is the requirement for the validating event of the first control to be raised?

Exercises

Exercise 3-1

Create a new form named Exercise_3.1 within the project Ch3_DecisionAndValidation. Copy the controls and the code from the form IceCreamTotalDiscounted to the new form Exercise_3.1. Modify the form as shown below, so that it computes discount percent based on the discount code, not based on the subtotal: 20% discount if discount code is "1" and 10% discount if discount code is "2;" otherwise, there is no discount (if there is no discount, the user wouldn't enter any discount code). The discount code is to be entered into a TextBox as shown below. Compute the discount using an if statement and also using a switch statement.

When the user clicks the compute button, the program should validate the data entered into the TextBoxes. The unit price must be a valid number; the number of scoops must be an integer; and if the user enters a discount code, it must be a "1" or "2."

Hint: To check whether a number is an integer, you may compare the number to the truncated number, as in
    if (scoops == (int)scoops)
or,
    if (scoops == Math.Truncate(scoops))

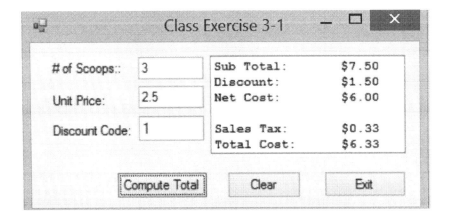

Exercise 3-2

Create a new form named Exercise_3.2 within the project Ch3_DecisionAndValidation. Copy the controls and code from the form IceCreamTotalDiscounted to the new form Exercise_3.2. Modify the form so that the user won't be able to move away from the TextBox for scoops, if the data entered into the TextBox is not an integer. Display a message using the Error Provider. Highlight the data so that the user can type over it.

Hint: To check whether a number is an integer, you may compare the number to the truncated number, as in
    if (scoops == (int)scoops)
or,
    if (scoops == Math.Truncate(scoops))

Similarly, make the necessary changes so that the user won't be able to move away from the TextBox for unit price if the data entered into the TextBox is not a valid number. Display a message using a MessageBox. Highlight the data so that the user can type over it.

Make sure the user is able to close the form by clicking the Exit button, even when the data entered into a TextBox is not valid.

Exercise 3-3

A retail clothing store offers a 10% discount if total sales is $50 or above and less than $100. If total sales is $100 or above, the store offers a 20% discount, and those who have a coupon get $10 off of total sales before applying the 20% discount.

Draw a flow chart that shows the logic for computing the subtotal (= total sales - all discounts).

Develop a new form named Exercise 3.3. You may develop the form within the existing project, Ch3_DecisionAndValidation, or a new project named Ch3_Exercises. Let the user enter the total sales into a TextBox on the form. Compute and display the total sales, coupon amount (if any), discount and subtotal, with the decimals aligned.

Hint: Use a Boolean variable for coupon. See Table 3-1 and Example 3 in this chapter. You may assign the value **true** or **false** to the Boolean variable within the program to test the code.

Exercise 3-4

A hotel chain offers special discounts for customers who have a gold or platinum status:
    Gold: 20% discount on reservations for 2 to 4 days and 30% discount for stays of 5 or more days
    Platinum: 30% discount on reservations for 2 to 4 days and 40% discount for stays of 5 or more days.

Develop a new form named Exercise 3.4. You may develop the form within the existing project, Ch3_DecisionAndValidation, or a new project named Ch3_Exercises. Let the user enter the number of days, customer status (Gold or Platinum) and room rate/day into three different TextBoxes. Compute and display the total cost (= rate/day * number of days), discount and net amount (= total cost - discount), with the decimals aligned.

Programming Assignment 1

This is an expanded version of Exercise 2-3 from Chapter 2.

"Ace Auto Rentals" is a small business that rents automobiles. Ace would like you to develop an application that lets them compute the rental charges. Specifically, the system should do the following:

Let the user enter the date rented, time rented, date returned, time returned, charge/day and a discount code. (If the study of Chapter 2 was postponed to a later time, you may enter the number of days rented instead of the date/time of rental/return.)

Compute and display the days rented, the total rental cost (days rented x charge per day), discount and net amount.

Create a project named Assignment1, and develop a form named Rental to do the above tasks.

The specifications on the data and the computation and display of results are as follows:

The discount code consists of an alphabetic letter followed by a digit. The first character of the discount code must be an "A," "B" or "C." The digit in the code is not used in this application.

Number of days = Date/time returned - date/time rented.

Any fractional day is considered as one full day (that is, if the item is returned later than the time rented, it is counted as a full day).

Total charge = Days rented x charge/day.

Discount is 20% of total charge for discount codes that start with "A," 10% for discount codes starting with "B" and 5% for codes starting with "C." There is no discount if there is no discount code entered into the field.

The decimals of dollar amounts displayed in the ListBox must be aligned, even when they are of different sizes.

Validate the data as follows:

Date Rented cannot be later than Date Returned.

Make sure that a valid number is entered in the charge/day field. If the charge/day is not a valid number, do not let the user move away from the Textbox where the number is entered. Use an Error Provider to display an appropriate message, and let the user type over.

Make sure that the first character of the discount code is an "A," "B" or "C."

Use Try Catch to catch other errors.

Other general requiements are

Use efficient code that is easy to understand.

You must use meaningful and self-explanatory names for all variables and objects like forms, TextBoxes, etc., using the naming conventions followed in the textbook. Don't use variable names like tc (for total charge). You don't have to name Labels that are not used in the code.

Chapter 4

Iteration Structure: Loops

All programs we discussed so far process each statement at most one time. A program, however, may need to process the same set of statements repeatedly. For example, searching a customer database to find a particular customer record may require reading thousands of similar records using the same set of statements.

Most programming languages, including C#, provide different types of loops to iteratively process a set of statements. In this chapter, you learn how to use different types of loops.

Topics

4.1	The while Loop	4.4	The do-while Loop
4.2	Console Applications	4.5	The for Loop
4.3	Incrementing Variables in a Loop	4.6	break and continue Statements

4.1 The while Loop

A commonly used form of loop is the *while* loop. This type of loop processes a set of statements while a Boolean expression is true. Figure 4-1 represents the structure of the while loop.

Figure 4-1: Structure of the while loop

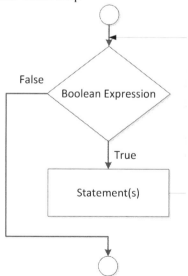

Let's look at an application that uses a while loop to check the password entered by a user when the user clicks the "Continue" button on a form, as shown in Figure 4-2. The while loop is used to check the password repeatedly and prompt the user for a valid password until the user enters the valid one. If the password is valid, a message tells the user to proceed.

Figure 4-2: The VerifyPassword form

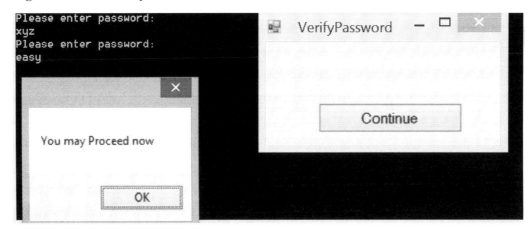

Figure 4-3 shows the structure of the while loop. If the password is not equal to "easy" (which is the correct password), then the user is prompted to enter a new password, which is checked again. The process continues as long as the password entered is invalid.

Figure 4-3: Structure of loop to verify password

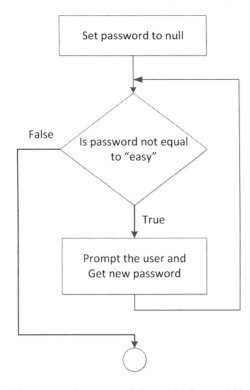

The general syntax of the code for a while statement is

```
while (Boolean expression)
{
    Statements
}
```

The statements in the body of the while loop are processed repeatedly, while the test condition represented by the Boolean expression is true. The loop terminates if the test condition is false.

Next, you create the application. To simplify development, we will run it as a Console Application so that the user enters the password on the Console window, rather than into a TextBox as it typically is done.

4.2 Console Applications

In a Console Application, the user interacts with the application through the console rather than through the controls on a form. So, the development of the form is simpler; however, the user interface is limited.

To use the console, you need to change the ***Output type*** setting of the project using the following steps:

> Open the project. We will use the project named Ch4_Loops from Tutorial_Starts folder in the next tutorial.
> Open Properties window by double clicking Properties in Solution Explorer.
> Select *Applications*. Change *Output type* to *Console Application*.

The use of console is illustrated in Tutorial 1.

Tutorial 1: while Loop to Verify Password

This tutorial builds the application to verify the password using the while loop and illustrates the use of console.

Step 1-1: Open the project named Ch4_Loops from Tutorial_Starts folder. Open the form, VerifyPassword, shown in Figure 4.2. The form has a button named btnContinue with the caption "Continue."

Step 1-2: Open the Properties window by double clicking Properties in Solution Explorer. Change *Output type* to *Console Application*.

The code for while statement

The code that translates the flow diagram in Figure 4-3 is shown in Figure 4-4.

Figure 4-4: Code to verify password using the Console

```csharp
19          private void btnContinue_Click(object sender, EventArgs e)
20          {
21              string password = null;
22              while (password != "easy")   // "!=" means not equal to
23              {
24                  // Display prompt in the Console.
25                  Console.WriteLine("Please enter password:");
26                  // Get password and store it in the variable password
27                  password = Console.ReadLine();
28              }
29              MessageBox.Show("You may proceed now");
30              // Code to open the next form goes here
31          }
```

How does the program work? The first time the while statement is processed, the password is null. Thus, the Boolean expression (password! = "easy ") is true, which causes the statements in the while clause to be processed. Let's look at the two statements in the while clause.

Line 25: Console.WriteLine ("Please enter password:") ;

The **WriteLine** method of Console class displays a line of text in the console window and positions the screen cursor at the beginning of the next line so that the user can type some input data. Thus, line 25 displays the prompt "Please enter password: " on the console window, and the application waits for the user to enter an input data. The next statement in line 27 is processed only after the user enters the data and presses the *Enter* key.

Line 27: password = Console.ReadLine ();

The **ReadLine** method of Console class reads the data typed by the user into the current line and returns it as a string. Line 27 reads the password entered by the user and assigns it to the variable, password.

After the first iteration through the loop, the Boolean expression (password != "easy") in the header of the loop is evaluated using the password entered by the user.

If the password is not valid, the statements in the while clause are processed again to get a new password. This process is repeated as long as the password is invalid.

If the password is valid ("easy"), control passes to the statement immediately following the while loop (line 29) to displays the message, "You may proceed now." A real-world application typically may have code to open another form.

Step 1-3: Add the code from Figure 4-4 to btnContinue_Click event handler, and test the code.

Step 1-4: Put a break at Line 28. Run the form after entering an invalid password, and observe the value of the password when the program breaks at line 28. Enter a valid password and note how line 28 is skipped because the Boolean expression (password != "easy") is false.

4.3 Incrementing Variables in a Loop

The ability of a loop to process statements repeatedly makes it convenient to accumulate data in a variable, like adding a set of numbers or counting the number of records read from a file. We will look at examples using the while loop.

Using a Counter Inside the Loop to Limit the Number of Attempts

Let's add a counter within the loop to limit the number of attempts to enter the valid password. Figure 4-5 shows the revised code that includes the following steps:
- declaring a counter variable, count (line 21)
- incrementing the counter (line 25), and
- checking the number of attempts (lines 26–31) and terminating the program if the number of attempts exceeds 3.

Figure 4-5: while loop with a counter

```
19      private void btnContinue_Click(object sender, EventArgs e)
20      {
21          int count = 0;
22          string password = null;
23          while (password != "easy")
24          {
25              count = count + 1;
26              if (count > 3)
27              {
28                  MessageBox.Show("Incorrect password. Only 3 attempts allowed");
29                  Environment.Exit(0);   // Close the Console window & the form
30                                          // Code 0 tells O.S. that it is a normal termination
31              }
32              Console.WriteLine("Please enter password"); // Display prompt
33              password = Console.ReadLine();   // Get password
34          }
35          MessageBox.Show(count + " Attempts. You may proceed now");
36          // Code to open the next form goes here
37      }
```

Note that the counter variable, count, is initialized to zero at line 21 **outside** the loop, and its value is increased by one **within** the loop in line 25. Let's take a more detailed look at the statements.

Incrementing and Decrementing Variables

Line 25 uses the following statement to increase the value of count by 1:
    count = count + 1;

The ++ and -- operators

In place of **count = count + 1,** you may use the shorter alternate forms of this statement using the ++ or += operators:
    count ++;   // or, ++count
    count += 1;
To decrement a variable by 1, you replace the "+" sign by the "-" sign, as in
    count = count - 1;
    count --;   // or, --count
    count -= 1;

Line 26 checks whether count exceeded 3. If it did, a message is displayed, and the form and Console windows are closed using the Exit() method of Environment class.

If count is less than or equal to 3, lines 32 and 33 ask the user to enter another password as in the original code.

If the password is valid, control jumps out of the loop to display the message from line 35 that displays the value of count to show the number of attempts.

Step 1-5: Create a new form named VerifyPasswordLimitedAttempts, and add a button named btnContinue. Add the code from Figure 4-5 to the click event handler of the button. (You may copy the code from VerifyPasssword form and modify it.) Put breaks at lines 31 and 34.

First, run the form with an invalid password two times, and then with the valid password.

When the program breaks at line 34, observe the value of password and count. Note that lines 31 and 34 are skipped when the correct password is entered.

Next, run the form with an invalid password three times, and verify that the Console and the form are closed.

Remove the breaks.

Declaring Variables: Inside versus Outside the Loop

Note that variable count is declared outside the while loop. What happens if count is declared inside the loop?

Step 1-6: Move declaration of count to inside the loop, as shown in Figure 4-6.
Put a break at the closing brace of the loop. Run the form with three or more incorrect passwords, and observe that the value of count is reset to 0 and incremented to 1 each time.

To correct the problem, move declaration of count back to the original position at line 21.

Figure 4-6: Declaring counter inside the loop—the wrong way

```
19      private void btnContinue_Click(object sender, EventArgs e)
20      {
21          // int count = 0;     //moved to line 25
22          string password = null;
23          while (password != "easy")
24          {
25              int count = 0; //Wrong place. count gets reset to 0 every time
26              count = count + 1;
```

To increment a variable inside a loop, make sure that the variable is declared outside the loop.

Incrementing a Variable: Inside versus Outside the Loop

The loop in Figure 4-5 includes the statement, **count = count +1**, in line 25, to increment count by one in each iteration through the loop. What is the effect of placing this statement before the loop? Do Step 1-7 to find out.

Step 1-7: Increment count outside the loop before the loop begins, by moving the statement,
    count = count + 1, to the line immediately following line 21, as shown in Figure 4-7.

   Put a break at the end of the loop on line 35. Run the form multiple times entering incorrect passwords, and observe that the value of count stays at one because count doesn't get incremented by the iterations through the loop.

Figure 4-7: Incrementing counter outside the loop—the wrong way.

```
19      private void btnContinue_Click(object sender, EventArgs e)
20      {
21          int count = 0;
22          count = count + 1;   // wrong place. count gets incremented only once
23          string password = null;
24          while (password != "easy")
25          {
26              // count = count + 1; moved to outside the loop
27              if (count > 3)
```

Next, we will look at the use of the while loop in a financial investment application where the value of accumulated investments determines the number of times the loop is processed.

Tutorial 2: while Loop: Financial Planning Application

This application uses a while loop in a financial planning application that helps you determine how many years it will take to reach a targeted amount of savings, as shown in Figure 4-8. This example also is used to show the application of do-while and for loops.

In this example, yearly investment and yearly growth rate are used to forecast the future value of investments for each year, while investment value is less than the target of $500,000. It is assumed that investments are made at the beginning of each year. Figure 4-9 shows the flowchart of the while statement.

Figure 4-8: The FinancialPlanner form

Year	Investment Value
1	$10,700.00
2	$22,149.00
3	$34,399.43
4	$47,507.39
5	$61,532.91
6	$76,540.21
7	$92,598.03
8	$109,779.89
9	$128,164.48
10	$147,835.99
11	$168,884.51
12	$191,406.43
13	$215,504.88
14	$241,290.22
15	$268,880.54
16	$298,402.17
17	$329,990.33
18	$363,789.65
19	$399,954.92
20	$438,651.77
21	$480,057.39
22	$524,361.41

Yearly Investment: 10000
Yearly Growth (%): 7
Target Amt: 500000

[How long?]

Figure 4-9: Flowchart of while loop in FinancialPlanner

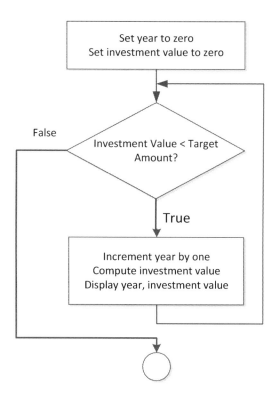

Step 2-1: Open the form named FinancialPlanner, shown in Figure 4.8, from the project Ch4_Loops in Tutorial_Starts folder. The form has the following controls:

> Three TextBoxes: txtInvestment, txtGrowthRate and txtTargetAmt
> Three Labels
> A button: btnHowLong
> A ListBox: lstDisplay

The complete code is shown in Figure 4-10.

The first three sections of the program consist of the following subtasks:

> Get input data from TextBoxes.
> Print headings.
> Set year and investment value to zero.

Figure 4-10: Code for the FinancialPlanner form

```csharp
20          private void btnHowLong_Click(object sender, EventArgs e)
21          {
22              // Get input data from Text Boxes:
23              double yearlyInvestment = double.Parse(txtInvestment.Text);
24              float YearlyGrowthPercent = float.Parse(txtGrowthRate.Text);
25              double targetAmt = double.Parse(txtTargetAmt.Text);
26
27              // Print heading
28              string formatCode = "{0,7}{1,20}";
29              lstDisplay.Items.Add(string.Format(formatCode, "years",
30                                                  "Investment Value"));
31              lstDisplay.Items.Add("");
32
33              // Set year and investment value to zero
34              int year = 0;
35              double investmentValue = 0;
36
37              // Use while loop to compute investment value for each year
38              while (investmentValue < targetAmt)
39              {
40                  year = year + 1;    // increment year by one
41                  investmentValue = (investmentValue + yearlyInvestment) *
42                                         (1+ YearlyGrowthPercent/100);
43                  lstDisplay.Items.Add(string.Format(formatCode, year,
44                                          investmentValue.ToString("C")));
45              }
```

Let's focus on the fourth section of the code that uses the while loop.

The while statement evaluates the Boolean expression (investmentValue < targetAmt). If it is true, the statements in the while clause are processed to increment the year and compute investment value; if not, control passes to the statement following the loop (the right bracket at the end of the method).

The first time the while statement is processed, investmentValue is zero, and the Boolean expression (0 < 500000) is true. So, the statements in the while clause are processed to do the following:
increment year to 1,
compute the investment value for year 1, using the expression in line 41, which yields the value 10,700, and
display year and investment value.

The formula that computes investment value is discussed later.

As the flowchart in Figure 4-9 shows, after processing the last statement in the while clause, control automatically passes to the beginning of the loop, where the Boolean expression is evaluated using the investmentValue (10,700) computed in the first iteration, to start the second iteration.

In the second iteration, the Boolean expression
(10700 < 500000)
is true. o, the statements in the while clause are processed the second time to increment year to 2, compute the new investment value (22,149) and display it.

Control again passes to the beginning of the loop for the third iteration, repeating the process.

The process continues through the twenty-second iteration because the Boolean expression in this iteration (480057 < 500000) is true. However, the investmentValue computed during the twenty-second iteration (524,361) exceeds 500,000, as shown in Table 4-1. Therefore, the Boolean expression (524361 < 500000) in the twenty-third iteration evaluates to false, causing the statements in the loop to be skipped and control to be passed to the statement following the loop (the left brace at the end of the method).

Table 4-1: Investment value in different iterations

Variable/expression	Value computed in iteration#					
	#1	#2	...	#21	#22	#23
(investmentValue<500000)	true	true	...	true	true	false
Year	1	2	...	25	26	
investmentValue	10700	22,149	...	480,057	524,361	

Step 2-2: Type in the code for the while loop from lines 38–45 in Figure 4-10.

Step 2-3: Change the project to a *Windows Application*. (Open the Properties window by double clicking Properties in Solution Explorer. Change *Output type* to *Windows Application*.) Run the form and test it.

Step 2-4: Change target amount to 50,000 (not 500,000). Put a break at the closing right brace of the while loop and another break at the closing right brace of the method.
Run the form, and observe the values of year and investmentValue for a few iterations.

After the fifth iteration, write down the values of year and investmentValue. Click Continue to complete the loop. When the program breaks at the last brace, observe that the values of year and investmentValue have not changed.

Infinite Loop

The statements in the body of a while loop are executed repeatedly while the test condition represented by the Boolean expression is true. The loop terminates when the value of the Boolean expression becomes false. However, logic errors in programming can cause the Boolean expression not to become false.

For example, if the variable investmentValue is initialized to zero within the loop, as shown below, its value may never reach the target amount, causing the test condition never to become false. This creates an "infinite loop" that continues to run until the program is interrupted.

```
int year = 0;
// double investmentValue = 0;   // moved inside the loop
while (investmentValue < targetAmt)
{
    double investmentValue = 0;   // wrong place to initialize
    year = year + 1;
    investmentValue = (investmentValue + yearlyInvestment) *
                (1+ YearlyGrowthPercent/100);
    lstDisplay.Items.Add(string.Format(formatCode, year,
                investmentValue.ToString("C")));
}
```

Pretest Loops

The while loop has a Boolean expression that is checked at the beginning of the loop. So, the while loop is called a pretest loop. Because it is a pretest loop, if the Boolean expression is false in the very first iteration, the statements in the loop will not be processed even once. For example, if investValue is initialized to a value greater than 500,000, the statements in the loop will not be processed.

Step 2-5: Change the initial value of investmentValue (in line 35) to 501,000. Run the form. Note that no output is displayed in the ListBox, indicating that the statements in the loop were not executed.

It's time to practice! Do Exercises 4-1 and 4-2.

Review Questions

4.1 Fix errors, if any, in the following program, which is intended to compute the total savings for 4 years if $750 is deposited each year, assuming that there is no interest added.

```
int yearlyDeposit = 750;
int totalSavings = 0;
while (year < 4)
{
  int year = 0;
  year = year + 1;
  totalSavings = totalSavings + yearlyDeposit;
}
```

4.2 What would be the output of the following program? (Pay special attention to the statements inside the loop.)

```
int yearlyDeposit = 750;
int totalSavings = 0;
int year = 0;
while (year < 3)
{
  year = year + 1;
  totalSavings = yearlyDeposit;
}
lstOutput.Items.Add(totalSavings);
```

4.3 What would be the output of the following program?
```
int year = 2016;
int population = 75000;
while (population < 77000)
{
year = year + 1;
    population = population + 1000;
    lstOutput.Items.Add(year + " " + population.ToString());
}
```

4.4 How many times will the loop in the following program be executed?
```
int year = 3;
while (year < 3)
{
  year = year + 1;
}
```

4.4 The do-while Loop

A variation of the while loop is the do-while loop that checks the Boolean expression at the end of the loop, as shown below. So, do-while is a post-test loop.

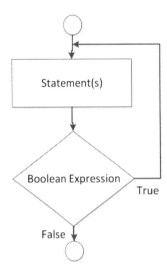

Figure 4-11 shows what the do-while version of the financial planning program looks like:

Figure 4-11: do-while version of FinancialPlanner

```
46
47              do
48              {
49                  year = year + 1;    // increment year by one
50                  investmentValue = (investmentValue + yearlyInvestment) *
51                                          (1 + YearlyGrowthPercent / 100);
52                  lstDisplay.Items.Add(string.Format(formatCode, year,
53                                          investmentValue.ToString("C")));
54              }
55              while (investmentValue < targetAmt);
```

The loop starts with the key word *do* and closes with the key word *while*, followed by the Boolean expression. It works the same way as the while loop, except that the Boolean expression is checked only at the end of the loop.

So, even if the investment value is initialized to a value greater than the target value, the statements in the loop will be processed the first time, and the iterations will continue as long as the computed investmentValue is less than the targetAmt.

Step 2-6: Comment out the while loop in lines 38-45. Add the code for the do-while loop from Figure 4-11. Make sure that investmentValue is initialized to 501,000, in line 35.
Run the form. Observe that the output is the same as when investmentValue was initialized to 0.

Because the condition is checked at the end of the loop, the statements in the loop are processed the first time when investmentValue was set to a new computed value.

Review Question

4.5 How many times will the loop in the following program be executed?
```
int year = 3;
do
{
  year = year + 1;
}
while (year < 3)
```

4.5 The for Loop

The for loop is designed to perform a fixed number of iterations, determined by the value of a counter variable. The counter variable is initialized to a certain specified value and incremented/decremented at the end of each iteration by a specified value.

To help understand the for loop, we will look at a different version of the financial planning program. Instead of finding the number of years to reach a target amount, this version specifies the number of years and displays the investment value for the specified number of years, as shown in Figure 4-12.

The form, **FinPlannerForLoop,** has a TextBox named txtYears to enter the number of years, instead of the target amount.

Figure 4-12: FinPlannerForLoop form

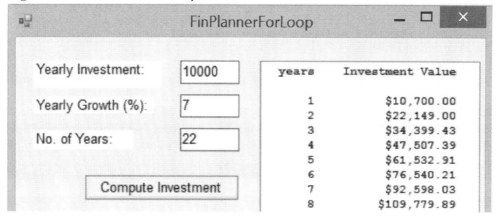

The code for the for loop version of FinancialPlanner is shown in Figure 4-13.

Figure 4-13: Code for FinPlannerForLoop form

```csharp
20          private void btnInvestment_Click(object sender, EventArgs e)
21          {
22              // Get input data from Text Boxes:
23              double yearlyInvestment = double.Parse(txtInvestment.Text);
24              float YearlyGrowthPercent = float.Parse(txtGrowthRate.Text);
25              int numberOfYears = int.Parse(txtYears.Text);
26
27              // Print heading
28              string formatCode = "{0,7}{1,20}";
29              lstDisplay.Items.Add(string.Format(formatCode, "years",
30                                                  "Investment Value"));
31              lstDisplay.Items.Add("");
32
33              // Set investment value to zero
34              double investmentValue = 0;
35
36              // Use for loop to compute investment value for each year
37              int year;
38              for (year = 1; year <= numberOfYears; year = year + 1)
39              {
40                  investmentValue = (investmentValue + yearlyInvestment) *
41                                      (1 + YearlyGrowthPercent / 100);
42                  lstDisplay.Items.Add(string.Format(formatCode, year,
43                                      investmentValue.ToString("C")));
44              }
```

The for loop header in line 38, **for (year = 1; year <= numberOfYears; year = year + 1)**, has three expressions:

1. **Initialization expression** (year = 1) that initializes the counter variable, year. You may combine the declaration and initialization of the counter variable, as in
    for (**int year = 1**; year <= numberOfYears; year = year+1)

    If you declare the variable in the loop header, the scope of the variable is limited to the loop. That is, you will not be able to access the variable outside the for loop.

2. **Boolean expression** (year <= numberOfYears). If it is true, the statements in the loop are executed; if not, control passes to the statement following the loop.

3. **Increment expression** (year = year + 1) that increments or decrements the counter variable. You may use *year++* or *year += 1* in place of *year = year + 1*.
    A variable may be incremented by any amount. For example, year = year+5 would increment year by 5 after each iteration.

Note that the three expressions are separated by semicolons, not commas.

You also may decrement a variable. For example, you may start with year 5, decrement it by one each time, and process the loop as long as year is **greater than** or equal to 1. For example,

> for (int year = 5; year >= 1; year = year - 1)

would give the following output:

```
years     Investment Value

  5          $10,700.00
  4          $22,149.00
  3          $34,399.43
  2          $47,507.39
  1          $61,532.91
```

Note that the Boolean expression has the ">=" operator.

Thus, the general syntax of the for loop is

> for (InitializationExpression, BooleanExpression, IncrementExpression)
> {
>   Statements
> }

How does the for loop shown in Figure 4-13 work? The first time the loop is processed, year is set to 1, and the Boolean expression (1 <= numberOfYears) is evaluated.

If the Boolean expression is true, the loop is processed to calculate and display the investment value. In addition, when an iteration is completed, the increment expression year=year+1 is executed to increment the value of year, and the Boolean expression is evaluated. This process continues until the Boolean expression is false.

If the Boolean expression is false, the processing of the loop ends, and control passes to the statement, if any, that follows the loop. If there are no statements after the loop, control passes to the end of the program. The equivalent flowchart is shown in Figure 4-14.

Note that the flowchart shown in Figure 4-14 is similar to that of the while loop that uses a counter variable. The for loop always has a counter variable whose value determines the termination of loop execution, whereas the while loop may use any Boolean expression to terminate the loop.

Figure 4-14: Flowchart of for loop

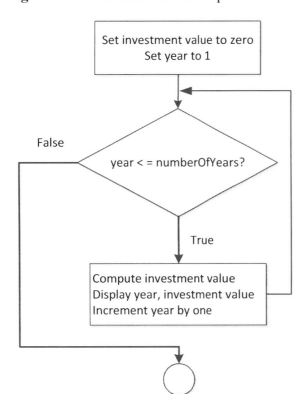

Tutorial 3: for Loop: Financial Planning Application

Let's build the FinPlannerForLoop form that uses the for loop to compute the investment value for a fixed number of years.

Step 3-1: Open the form named FinPlannerForLoop from the project, Ch4_Loops.
Add the code from lines 37-44 in Figure 4-13 to create the for loop.

Step 3-2: Put breaks at the header of the loop and at the beginning and ending braces.
Run the form, and observe the value of year the first time through the loop and how it changes between the end of the first iteration and the beginning of the second iteration.

Step 3-3: To help understand the termination of the loop, copy the statement that displays the output from lines 42 and 43 to the line right after the closing brace of the loop, as shown in Figure 4-15.
Run the form.
When the value of year become 23, the processing of the loop terminates because the Boolean expression (year<= numberOfYears) is false. So, investmentValue is not computed for year 23; it has the same value as for year 22, as shown below as shown in the following output:

```
21        $480,057.39
22        $524,361.41
23        $524,361.41
```

The variable, year, is available outside the loop because it is declared outside the loop.

Figure 4-15: Displaying output outside the loop

```
37              int year;
38              for (year = 1; year <= numberOfYears; year = year + 1)
39              {
40                  investmentValue = (investmentValue + yearlyInvestment) *
41                                    (1 + YearlyGrowthPercent / 100);
42                  lstDisplay.Items.Add(string.Format(formatCode, year,
43                                    investmentValue.ToString("C")));
44              }
45              lstDisplay.Items.Add(string.Format(formatCode, year,
46                                    investmentValue.ToString("C")));
```

Declaring Counter Variable Inside the Loop

Step 3-4: Modify the header of the loop as follows to declare the counter variable within the header:

```
37              // int year;
38              for (int year = 1; year <= numberOfYears; year = year + 1)
```

Run the form.

Error! The program would not run because the variable year is declared inside the loop, but it is used outside the loop in line 45 to display its value.

Delete the statement in lines 45 and 46 that uses the variable year outside the loop. The error disappears.

Again, the scope of a variable declared inside a loop is limited to the loop.

Incrementing Counter by a Value Other Than One

Step 3-5: Modify the header as follows to increment year by 5 in each iteration:
    for (**int** year = 1; year <= numberOfYears; year = year + 5)
Run the form. Note how the value of year changes.
Warning! The computed investment values are incorrect except for the first year, because the formula assumes one year increments.

```
    years       Investment Value

      1           $10,700.00
      6           $22,149.00
     11           $34,399.43
     16           $47,507.39
     21           $61,532.91
```

Decrementing the Counter Variable

Step 3-6: Change the header of the loop as follows to decrement the counter variable:
for (int year = 5; year >= 1; year = year - 1)
Run the form. The value of year changes from 5 to 1, as follows:

```
years      Investment Value

  5           $10,700.00
  4           $22,149.00
  3           $34,399.43
  2           $47,507.39
  1           $61,532.91
```

Review Questions

4.6   What would be the output of the following program?

```
int population = 65000;
for (int year = 2016; year <= 2018; year = year + 1)
{
      population = population + 1000;
      lstOutput.Items.Add(year + " " + population.ToString());
}
```

4.7   What would be the output of the following program?
```
int population = 65000;
for (int yr = 2018; yr > 2016; yr = yr - 1)
{
   population = population - 1000;
   lstDisplay.Items.Add(yr + " " + population.ToString());
}
```

4.6 The break and continue Statements

The break and continue statements let you deviate from the normal processing of a loop.

The break Statement

You may jump out of a loop using the break statement.

Step 3-7: Modify the code by adding an if statement and a break statement, as shown in Figure 4-16, to break out of the loop, if investmentValue exceeds 100,000. Copy the display statement from line 42 to line 47 as shown, and make sure year is declared outside the loop.

Figure 4-16: if statement with break statement

```
37      int year;
38      for (year = 1; year <= numberOfYears; year = year + 1)
39      {
40          investmentValue = (investmentValue + yearlyInvestment) *
41                            (1 + YearlyGrowthPercent / 100);
42          lstDisplay.Items.Add(string.Format(formatCode, year,
43                            investmentValue.ToString("C")));
44          if (investmentValue > 100000)
45              break;
46      }
47      lstDisplay.Items.Add(string.Format(formatCode, year,
48                            investmentValue.ToString("C")));
```

Run the code. Execution of the loop terminates at the break statement when the investmentValue exceeds 100,000 in year 8, as shown by the output in Figure 4-17, although the Boolean expression in the header of the loop is true.

A break will process the remaining statements that follow the loop within the method. So, as shown in Figure 4-17, year and investmentValue are displayed by the statement in line 47, which is outside the loop.

Delete the if statement, including the break statement.

Figure 4-17: Effect of break in a for loop

```
years       Investment Value

  1           $10,700.00
  2           $22,149.00
  3           $34,399.43
  4           $47,507.39
  5           $61,532.91
  6           $76,540.21
  7           $92,598.03
  8          $109,779.89
  8          $109,779.89
```

The continue Statement

The continue statement causes the rest of the statements in a loop to be skipped, but continues processing the loop.

Step 3-8: Modify the code, as shown in Figure 4-18, to skip displaying the results if investmentValue exceeds 100,000.

Figure 4-18: Effect of continue statement in a loop

```
37      int year;
38      for (year = 1; year <= numberOfYears; year = year + 1)
39      {
40          investmentValue = (investmentValue + yearlyInvestment) *
41                            (1 + YearlyGrowthPercent / 100);
42          lstDisplay.Items.Add(string.Format(formatCode, year,
43                                  investmentValue.ToString("C")));
44          if (investmentValue > 100000)
45              continue;
46      }
47      lstDisplay.Items.Add(string.Format(formatCode, year,
48                              investmentValue.ToString("C")));
```

Step 3-9: Run the code.

Figure 4-19 shows the output. The loop continues to be processed even after investmentValue exceeds 100,000, without displaying the year and investment value. How do we know? The final investmentValue is $524,361, and the year is 23. Note that the investment value exceeds 500,000 in year 22. But, that year is not displayed. The year is displayed only after year is incremented by 1 to 23, when control passes out of the loop.

Delete the if statement, including the continue statement.

Figure 4-19: Output with continue statement in the loop

```
years     Investment Value

  1          $10,700.00
  2          $22,149.00
  3          $34,399.43
  4          $47,507.39
  5          $61,532.91
  6          $76,540.21
  7          $92,598.03
 23         $524,361.41
```

The foreach Statement

A form of loop that is particularly useful to access each item from a collection of items is the **foreach** loop. Chapters 7, 8 and 9 discuss the foreach loop and how it is used to access items from arrays and collections.

It's time to practice! Do Exercises 4-3 and 4-4.

Review Question

4.8   What is the difference between the break and continue statements?

Exercises

Exercise 4-1
The price of laptops is expected to go down by 10% per year. Use a while loop to display the year and price when the price will be under a specified target amount. Allow the user to specify the current price and the target price using TextBoxes.

Exercise 4-2
Develop a form that allows the user to project the year when the population of your city exceeds a specified target number. Allow the user to specify the current year's population, the target number and the annual growth rate. Use a while statement to compute and display each year, starting with the current year, and the corresponding population.

Exercise 4-3
Develop a form to display the Fahrenheit temperature and corresponding Celsius temperature for Fahrenheit temperatures ranging from −40 to +40 degrees, at increments of 10. Use a for loop. The conversion formula is
$$\text{Celsius} = (\text{Fahrenheit} - 32) * 5/9$$

Exercise 4-4
Develop a form that allows the user to project the population of your city for a specified number of years and growth rate. Allow the user to specify the current year's population, the number of years and the annual growth rate. Use a for loop to compute and display each year, starting with the current year and the corresponding population.

Chapter 5

Methods

A method is a program unit that performs a particular task. The methods you wrote in previous chapters were event handlers that are connected to a certain event of a control, like the Click event of a button. The Click event handler, for example, is executed when the user clicks a button.

In this chapter, you will learn how to develop methods that are not event handlers—that is, methods not connected to any event; instead, they are called by other methods. The term method, unless qualified, would mean such methods that are not event handlers.

Topics

5.1	Introduction to Methods	5.5	Top-Down Design
5.2	Passing Values to Methods	5.6	Methods That Return a Value
5.3	Passing Arguments by Value	5.7	Enumerations: Limiting Parameter Values
5.4	Passing Arguments by Reference		

5.1 Introduction to Methods

What is the purpose of methods?

1. **Divide and conquer**: A method typically performs a subtask that is a part of a larger or more complex task. For example, processing an order might consist of the following subtasks:
   verify item numbers,
   check quantity on hand,
   compute order total.
   Rather than developing a single program to perform the entire task, breaking the task down to smaller subtasks that are performed by separate methods often helps make software development simpler and easier. This approach to software development is known as **top-down design**.

2. **Reuse**: A method, if written properly, can be reused in multiple projects/forms or called from multiple methods. A method to verify an item number, for example, could be used in multiple applications, like order processing and production scheduling.

Before we look at the process of top-down design, let's look at the structure of a method and how it is used.

Consider a form named HeartRate, shown in Figure 5-1, which computes your optimal training heart rate using your age and resting heart rate. This form has an "About" button that uses a method to display a greeting and some information about the form in a MessageBox.**Figure 5-1:** The HeartRate form with "About" MessageBox

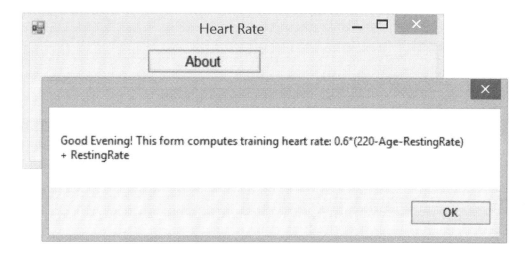

Example of a Method

Figure 5-2 shows the method named DisplayInfo() that displays the MessageBox with the greeting.

Figure 5-2: Example of a method

```
private void DisplayInfo()
{
   string greeting;
   if (DateTime.Now < DateTime.Parse("12:00 PM"))
         greeting = "Good Morning!";
   else if (DateTime.Now < DateTime.Parse("6:00 PM"))
         greeting = "Good Afternoon!";
   else
         greeting = "Good Evening!";

   MessageBox.Show(greeting + " This form computes training heart rate:" +
                  "0.6*(220-Age-RestingRate) + RestingRate");
}
```

Structure of a Method

DisplayInfo has the typical structure of a method—that is, the **header, private void DisplayInfo(),** followed by the **body** that consists of statements enclosed in curly brackets. The statements within the body of this method are not relevant in understanding the structure. For now, it is sufficient to understand that the statements display one of three greetings based on the time of the day.

The first key word in the header, **private**, specifies that this method can be called (to execute it) only from within the class where it is declared—that is, only from the form, GradeWithMethods. By default, methods are private, so the word private can be omitted.

If you want to make the method callable from other forms, you specify **public** instead of private. The key word that specifies whether the method is private or public is termed an **access modifier**.

The second key word, **void,** specifies that this method does not return a value. For a method that returns a value, you specify the data type of the value in place of void. This key word is called the **return type** because it specifies the type of value returned by the method. A method whose return type is void is called a **void method**.

The pair of empty parentheses following the **method name** DisplayInfo indicates that no data can be passed to this method when it is called.

Thus, the general syntax of a method is

> AccessModifier ReturnType MethodName([parameterList])
> {
>
>   Statements
>
> }

Naming a Method

Microsoft's recommendation is to use verbs or verb phrases to name methods, and use Pascal Casing—that is, the first letter of the method name and the first letter of subsequent words, if any, are capitalized, as in DisplayInfo.

Calling a Method

The statement to call a method is just the name of the method followed by a pair of parentheses. In Figure 5-3, the single statement, **DisplayInfo();** within the btnAbout_Click event handler calls the method DisplayInfo.

Figure 5-3: Click event handler of btnAbout that calls DisplayInfo method

```
private void btnAbout_Click(object sender, EventArgs e)
{
        DisplayInfo();
        txtRestingRate.Focus();
}
```

When a method is called, the program jumps to that method, executes the statements in the body of the method and returns to the statement that follows the call statement. So, the statement **DisplayInfo()** causes the program to branch to the method DisplayInfo, display the message and return to the statement **txtScore.Focus()**.

Tutorial 1 creates the form HeartRate.

Tutorial 1: Methods: Compute Heart Rate

To explore the use of methods with and without parameters, let's build the HeartRate form.

Step 1-1: Open the project named Ch5_Methods from the folder, Tutorial_Starts.

Step 1-2: Open the HeartRate form.

Step 1-3: Open the code window to view the code for **DisplayInfo** method.

Add the two statements to the **btnAbout_Click** event handler, as shown in Figure 5-4.

Step 1-4: Run the form and test the code.

Step 1-5: To understand the path followed by the program, put breaks on lines 21, 35 and 36 by clicking on the left margin in the code window (see Figure 5-4).

Figure 5-4: Code to create DisplayInfo and to call it

```csharp
20      private void DisplayInfo()
21      {
22          string greeting;
23          if (DateTime.Now < DateTime.Parse("12:00 PM"))
24              greeting = "Good Morning!";
25          else if (DateTime.Now < DateTime.Parse("6:00 PM"))
26              greeting = "Good Afternoon!";
27          else
28              greeting = "Good Evening!";
29          MessageBox.Show(greeting + " This form computes training heart rate:"
30                          + "0.6*(220-Age-RestingRate) + RestingRate");
31      }
32
33      private void btnAbout_Click(object sender, EventArgs e)
34      {
35          DisplayInfo();
36          txtRestingRate.Focus();
37      }
```

Step 1-6 Run the form and click the About button.

The program breaks at the statement that calls the method DisplayInfo (line 35; your line numbers may be different).

Click the Continue button. The program jumps to DisplayInfo() method and breaks at the left curly brace (line 21).

Step 1-7 Click Continue to run the method and display the message.

Click "OK" to continue. The program returns to the called program (btnAbout_Click) and breaks at line 36.

Step 1-8 Click Continue. Execution stops.

Review Questions

5.1 What is a method?

5.2 What is an event handler?

5.3 True or false: A method typically performs a subtask that is a part of a larger or more complex task.

5.4 True or false: A method can be called from multiple methods.

5.5 Write the statement to call a method named, ComputeCost.

5.2 Passing Values to Methods

The previous method, DisplayInfo, did not need any input data to be provided by the calling program to display its output. However, it often is necessary to pass input data to a method from the calling program. To help understand passing data to methods, you will develop a method, ComputeTrainingRate(), to compute your training heart rate using your age and resting rate, and display the rate in a MessageBox, as shown in Figure 5-5.

Figure 5-5: The HeartRate form with Training Rate MessageBox

The age and resting rate are passed by the calling program to ComputeTrainingRate(). Figure 5-6 depicts the **structure chart** that shows the calling and called programs and the data transferred between them.

Figure 5-6: Structure chart of program to compute training rate

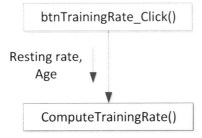

To receive the values of resting rate and age, the method ComputeTrainingRate has two **parameters**, restingRate and age, declared in the header, as shown in Figure 5-7.

Figure 5-7: ComputeTrainingRate with parameters

```
38
39          private void ComputeTrainingRate(float restingRate, float age)
40          {
41              float maximumRate, trainingRate;
42              // Compute maximum heart rate:
43              maximumRate = 220 - age;
44              // Compute training heart rate
45              trainingRate = 0.6F * (maximumRate - restingRate) + restingRate;
46
47              // Display Training heart rate:
48              MessageBox.Show("Your Training Heart Rate is: " + trainingRate.ToString("N0"));
49          }
```

The parameters restingRate and age get their values when a program calls this method. These parameters are used by the method to compute maximum heart rate and training heart rate.

The Scope of Parameters

Because parameters are declared in a method, they are available only within the method—that is, the scope of a parameter is the method in which the parameter is declared. Next, you create the method.

Step 1-9: Create the method, ComputeTrainingRate(), using the code from Figure 5-7.

Figure 5-8 shows how to call this method from the Click event handler of btnTrainingRate button to pass 72 to restingRate and 22 to age. This initial version of the event handler does not use the TextBoxes to get the resting rate and age; instead, the literals 72 and 22 are passed to the method.

Figure 5-8: Call ComputeTrainingRate

```
50
51          private void btnTrainingRate_Click(object sender, EventArgs e)
52          {
53              // Call ComputeTrainingRate. Pass 72 for restingRate and 22 for age.
54              ComputeTrainingRate(72, 22);
55          }
```

Step 1-10: Add the code from Figure 5-8 to the Click event handler of btnTrainingRate to call the method, ComputeTrainingRate().

Arguments and Parameters

The data items that are passed to a method by the calling program are called **arguments**. Thus, the values 72 and 22 in the call statement ComputeTrainingRate(72, 22) are arguments.

The variables declared in the header of a method to receive the arguments are called **parameters**. (As you will see later, parameters also may be used to pass values from the method to arguments in the calling program.) Thus, restingRate and age declared in the header of the method

        private void ComputeTrainingRate(float restingRate, float age)

are parameters.

The arguments are passed to the parameters in the order in which they appear in the argument list. That is, the first argument is passed to the first parameter, the second argument to the second parameter and so on. When ComputeTrainingRate is called, the following assignments take place automatically:

    ComputeTrainingRate(72, 22);

    private void ComputeTrainingRate(float restingRate, float age)

That is,

        restingRate = 72;
        age = 22;

The position of an argument in the argument list may be different from that of the corresponding parameter, if the name of the parameter is specified, as in

    ComputeTrainingRate(age: 22, restingRate: 72);

The data type of the parameters and those of the arguments must be compatible—that is, they must obey the rules for assigning values to variables (discussed in Chaper 2).

Let's see how it works:

Step 1-11: Insert a break inside ComputeTrainingRate at the left brace (line 40) by clicking on the left margin.

    Insert a second break inside the Click event handler at the call statement (line 54).

    Run the form. Note that the TextBoxes and the ListBox on the form are not used in this version of the form.

    Click the button to compute the training rate. The program breaks at the call statement. Click Continue. The program breaks at the left brace inside the method.

    Move the pointer over the parameters restingRate and age to observe their values, 72 and 22, respectively.

    Click Continue to display the training rate.

Step 1-12: Repeat Step 1-11 after changing the resting rate and age in the call statement.

Constants as Arguments

Instead of a literal, you may use a constant (or a variable, as discussed next) as an argument to pass a value to a method. For example, you may use a constant to represent the resting heart rate in the calling program btnTrainingRate_Click(0, as follows:

```
const float restingRate = 72;
ComputeTrainingRate(restingRate, 22);
```

Variables as Arguments

Rather than passing literals (72 and 22) or constants as arguments, variables may be used to pass data to a method to make it easy to change the values passed. Data are entered into two TextBoxes, txtRestingRate and txtAge, as shown in Figure 5-9. The data from the TextBoxes is stored in variables and passed to the method.

Figure 5-9: Variables as arguments

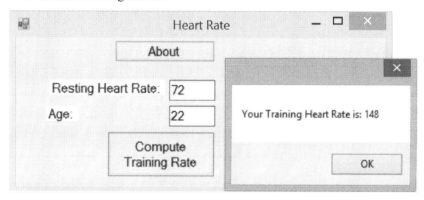

Step 1-13: Modify the code in the Click event handler, as shown in Figure 5-10, to get the resting rate and the age from TextBoxes, store them in variables and pass them as arguments to ComputeTraingRate().

Figure 5-10: Calling ComputeTrainingRate with variables as arguments

```
50
51          private void btnTrainingRate_Click(object sender, EventArgs e)
52          {
53              float restingRate, age;
54              // Get resting rate and age from Textboxes:
55              restingRate = float.Parse(txtRestingRate.Text);
56              age = float.Parse(txtAge.Text);
57
58              // Call ComputeTrainingRate
59              ComputeTrainingRate(restingRate, age);
60          }
```

Note that passing data through variables does not require any change in the called method ComputeTrainingRate(). It remains the same as in Figure 5-7, including the following header:

```
private void ComputeTrainingRate(float restingRate, float age)
```

When this method is called from line 59, the first variable in the argument list (restingRate) is assigned to the first variable in the parameter list (restingRate), and the second argument (age) to the second parameter (age).

    ComputeTrainingRate(restingRate, age);

    private void ComputeTrainingRate(float restingRate, float age)

Because the value of restingHeartRate is 72 and that of age in the calling program is 22, in effect, the following assignment takes place when ComputeTrainingRate is called:

    restingRate = 72;
    age = 22;

Again, the variables in the argument list and variables in the parameter list must be compatible.

Step 1-14: Make sure that there is a break inside ComputeTrainingRate at the left brace of the method, ComputeTrainingRate (line 40). Run the form.

Step 1-15: Enter values for resting rate and age. Click the button btnTrainingRate.

When the program breaks at the curly left bracket within the method, verify that restingRate and age have the values you entered into the TextBoxes.

Names of arguments and corresponding parameters

In the current example, the names of the two parameters (restingRate and age) in the header of the called method,

    private void ComputeTrainingRate(float restingRate, float age)

are the same as the names of the corresponding arguments in the calling statement,

    ComputeTrainingRate(restingRate, age)

However, the calling program may use any name for the arguments. Thus, it would be valid for the calling program to represent the resting rate by the variable restingHeartRate and call the method using restingHeartRate as the argument name, as in Figure 5-11.

Figure 5-11: Argument name different from the parameter name

```
50
51          private void btnTrainingRate_Click(object sender, EventArgs e)
52          {
53              float restingHeartRate, age;
54              // Get resting rate and age from Textboxes:
55              restingHeartRate = float.Parse(txtRestingRate.Text);
56              age = float.Parse(txtAge.Text);
57
58              // Call ComputeTrainingRate
59              ComputeTrainingRate(restingHeartRate, age);
60          }
```

The parameter names in the method remain the same:
    private void ComputeTrainingRate(float restingRate, float age)

Step 1-16: Change restingRate to restingHeartRate, as shown in Figure 5-11. Put a break at the statement that calls ComputeTrainingRate.

Run the form. Enter values for resting rate and age. Click the button, btnTrainingRate.

When the program breaks at the call statement, check the value of restingHeartRate.

When the program breaks at the curly left brace within the method, verify that restingRate within the method has the same value as restingHeartRate in the calling method.

Review Questions

5.6   What is an argument?

5.7   What is a parameter?

5.8   When a method has multiple parameters, how does C# determine which parameter is associated with an argument?

5.9   What is the scope of a parameter?

5.10   True or false: The name of an argument must be the same as the name of the corresponding parameter.

5.11   The following code shows a method named **Greetings**, and the Click event handler of a button, which calls the method:

```
private void btnGreeting_Click(object sender, EventArgs e)
{
    string word1 = "Hi", word2 = "There", word = "Hello";
        Greetings(word1);
        Greetings(word2);
}

private void Greetings(string word)
{
        lstDisplay.Items.Add(word);
}
```

What would be the output displayed in the List Box when you click the button?

5.3 Passing Arguments by Value

When a program calls ComputeTrainingRate(float restingRate, float age), the calling program, by default, passes the **value** of each argument to the corresponding parameter. This is called passing arguments by value, which is the default way of passing data to a method.

When an argument is passed by value, if the value of the corresponding parameter is changed within the called method, that will not affect the value of the argument in the calling program—that is, if the value of age is changed within the ComputeTrainingRate method, that will not have any effect on the value of age in the calling program, btnGrade_Click.

Step 1-17: Change the value of age to 999 (line 58) within ComputeTrainingRate, and put breaks at the left and right curly braces, as shown in Figure 5-12.

Figure 5-12: ComputeTrainaingRate that changes the value of age

```
38
39          private void ComputeTrainingRate(float restingRate, float age)
40          {
41              float maximumRate, trainingRate;
42              // Compute maximum heart rate:
43              maximumRate = 220 - age;
44              // Compute training heart rate
45              trainingRate = 0.6F * (maximumRate - restingRate) + restingRate;
46              age = 999;  // This statement is for testing purposes only
47              // Display Training heart rate:
48              MessageBox.Show("Your Training Heart Rate is: " + trainingRate.ToString("N0"));
49          }
```

Put a break at the call statement (line 59) and at the right brace in the calling program, as shown in Figure 5-13.

Figure 5-13: Break in the calling program

```
50
51          private void btnTrainingRate_Click(object sender, EventArgs e)
52          {
53              float restingHeartRate, age;
54              // Get resting rate and age from Textboxes:
55              restingHeartRate = float.Parse(txtRestingRate.Text);
56              age = float.Parse(txtAge.Text);
57
58              // Call ComputeTrainingRate
59              ComputeTrainingRate(restingHeartRate, age);
60          }
```

Step 1-18: Run the form. Enter values for resting rate and age. Click the btnTrainingRate button.

Check the value of age when the program breaks at the call statement (line 59) and at the left brace within the method (line 40) to verify that it is the value you entered. Check it when the program breaks at the end of the method (line 49) to observe that age has the value 999.

Check age again when the program returns to the calling program (line 60), to observe that the value of age has not changed from its original value. Assigning the value 999 in the called method has no impact on the argument age in the calling program, because only the value of age from the calling program is passed to the parameter age in the method.

Delete the statement "age=999" from the method.

Default Arguments

You may assign a default value (argument) for one or more parameters of a method. For example, you may assign a default value (20) to age in the header of ComputeTrainingRate method:

    private void ComputeTrainingRate(float restingRate, float age = 20)

The default value must be a literal or constant.

When you call a method that has a default value for one or more parameters, you have the option to pass no argument for age, as in the following call statement:

    ComputeTrainingRate(restingRate)

In this case, age will have the default value 20.

Or, you may pass a value for age to override the default, as in

    ComputeTrainingRate(restingRate, age)

If all parameters do not have a default value, the parameters with default value must be declared last.

Step 1-19: Modify ComputeTrainingRate to include a default argument for age, and modify the call statement so no argument is passed for age. Run the form and check the value of age within the method. Repeat by passing an argument for age with a value different from the default value.

It's time to practice! Do Exercise 5-1.

Cohesive Methods and Reusability

Consider the following method, ComputeTrainingRate, which you developed earlier:

```
private void ComputeTrainingRate(float restingRate, float age, )
{
   float maximumRate, trainingRate;
   // Compute maximum heart rate:
   maximumRate = 220 - age;
   // Compute training heart rate
   trainingRate = 0.6F * (maximumRate - restingRate) + restingRate;
   // Display Training heart rate:
   MessageBox.Show("Your Training Heart Rate is: " + trainingRate.ToString("N0"));
}
```

The above method computes the training rate and displays the result. Such a method that combines multiple tasks is not very reusable. For example, another program that needs to compute the training rate by calling this method may not want the result to be displayed in a Message Box.

Methods are more reusable when each method performs a single well-defined task, like compute training rate, and returns the results to the calling program so that the calling program can do whatever it wants with those results. Such methods are said to be **cohesive**.

Figure 5-14 shows the structure chart in which the training rate is returned by ComputeTrainingRate to the calling program.

Figure 5-14: ComputeTrainingRate that returns TrainingRate

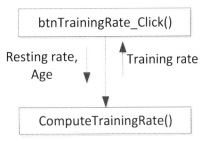

The next sections discuss how **to return values to the calling program**, which helps to create more cohesive methods.

Review Questions

5.12   What is meant by passing an argument by value?

5.13   The following code shows a method named **ComputeSum** and the Click event handler of a button, which calls the method and displays the result in a ListBox named lstDisplay:

```
private void btnExamScore_Click(object sender, EventArgs e)
{
    int totalScore = 0;
    ComputeSum(90, 85, totalScore);
    lstDisplay.Items.Add(totalScore);
}

private void ComputeSum(int score1, int score2, int totalScore)
{
    totalScore = score1 + score2;
}
```

Note that the argument, totalScore, is passed by value. What would be the output displayed in the ListBox?

5.4 Passing Arguments by Reference

Passing arguments by reference allows the called method to change the value of arguments in the calling program. This allows the results computed in a method to be made available (or **returned**) to the calling program.

It should be noted that an alternative way to return values from the called method to the calling method is to use a **value returning method** that does not use reference parameters, as discussed in section 5.6. Using value returning methods has some advantages, like ease of maintenance, over using reference parameters. However, to return multiple values from a method, you cannot use simple variables as parameters; you need to use more advanced options like a class, strut, tuple, or array. Class and arrays are discussed in later chapters.

When a variable is passed by reference, the memory location (address or reference) of the variable is passed to the parameter so that the argument and the parameter access data from the same memory location, as represented below:

So, if the value of a parameter is changed within a method, in effect, it is changing the value of the corresponding argument.

Similarly, the value assigned to an argument before calling a method becomes the value of the parameter when the method is called. That is, in effect, data can be transferred from the calling program to the called method, and also from the called method to the calling program.

Let's use passing by reference to return the computed training rate from ComputeTrainingRate to the calling program, rather than display it in a MessageBox. The modified method is shown in Figure 5-15.

It should be noted that a value returning method, discussed later in Section 5.6, is an alternative to passing an argument by reference.

Figure 5-15: Passing arguments by reference

```
38
39          private void ComputeTrainingRate(float restingRate, float age,
40                                          ref float trainingRate)
41          {
42              float maximumRate;
43              // Compute maximum heart rate:
44              maximumRate = 220 - age;
45              // Compute training heart rate
46              trainingRate = 0.6F * (maximumRate - restingRate) + restingRate;
47
48              // Training rate is not displayed; it is returned to the calling method
49          }
```

In line 40, the key word **ref** specifies that trainingRate is a **reference parameter**. Note that because trainingRate is declared as a parameter, it is removed from the declaration statement within the method. Further, the statement that displays training rate in a MessageBox also is removed.

Figure 5-16 shows the updated calling program. The call statement now includes trainingHeartRate as an argument (line 58).

Figure 5-16: Updated calling program

```
50
51          private void btnTrainingRate_Click(object sender, EventArgs e)
52          {
53              float restingHeartRate, age;
54              // Get resting rate and age from Textboxes:
55              restingHeartRate = float.Parse(txtRestingRate.Text);
56              age = float.Parse(txtAge.Text);
57
58              // Call ComputeTrainingRate
59              float trainingHeartRate = 0;
60              ComputeTrainingRate(restingHeartRate, age, ref trainingHeartRate);
61              // Display training rate
62              lstResults.Items.Add(string.Format("{0,20}{1,10}", "Training Heart Rate:",
63                                          trainingHeartRate.ToString("N0")));
64          }
```

The key word "ref" in the calling statement in line 60 specifies that the parameter is reference type. Note that the name of the argument (trainingHeartRate) is different than the name of the parameter (trainingRate).

An argument passed by reference must be declared in the calling program, and it must be set to a value. The argument trainingHeartRate is declared and it is set to zero.

What happens when ComputeTrainingRate is called?

> First, the reference to trainingHeartRate is passed to the parameter trainingRate. That means trainingRate will take the value of the argument trainingHeartRate.

> Next, the method assigns the computed value to the parameter trainingRate. Because both the argument and the parameter share the same memory location, the computed value also is available to the argument trainingHeartRate.

Thus, passing by reference makes sharing between the calling program and the called program like a two-way street, as shown, whereas passing by value is like a one-way street.

> ComputeTrainingRate(restingHeartRate, age, ref trainingHeartRate);
>
> private void ComputeTrainingRate(float restingRate, float age, ref float trainingRate)

Step 1-20: Update ComputeTrainingRate to include the parameter trainingRate, as shown in Figure 5-15.

> Put breaks at the beginning and ending braces within the called method.

> Make changes in the calling program, as shown in Figure 5-16.

> Run the form. At the first break inside ComptueTrainingRate, observe that the parameter trainingRate has the value 0 (same as the value of the argument restingHeartRate). At the second break at the end of the method, restingRate has the computed value.

When control returns to the calling program, note that the argument restingHeartRate has the value that was assigned to the parameter restingRate, because both variables share the same memory location.

Returning Multiple Values from a Method Using Parameters

The ComputeTrainingRate method returns only a single value to the calling method. A method may use multiple reference parameters, each parameter returning a single value, or a set of values using objects and different data structures like strut and tuple. The GetPersonalData method in section 5.5 provides an example of a method that returns two values.

It's time to practice! Do Exercise 5-2.

The out Parameter

When the key word "ref" is used to specify passing by reference, the parameter is called a reference parameter. Another method to pass an argument by reference is to use the key word "out" in place of "ref," as shown in the following code, to specify that the parameter is an **output parameter**.

ComputeTrainingRate(restingHeartRate, age, out trainingHeartRate);

private void ComputeTrainingRate(float restingRate, float age, out float trainingRate)

An output parameter does not require the corresponding argument to be assigned a value before calling the method. In addition, an output parameter must be assigned a value inside the method. Except for these differences, the output parameter works like a reference parameter.

You may use the out parameter when the calling program must access the value of a parameter from the called method, but the called method doesn't have to access the value of the corresponding argument from the calling program, as in the above example.

Why Pass by Value?

If pass by reference can pass data both ways, then why use pass by value, which can pass only from calling program to called method? What is wrong with using pass by reference, even when the method does not have to pass a value to the calling program?

Consider data items like resting rate and age, which are needed in a method, but their values should not be changed by the method. If such arguments are passed by reference, the called method can inadvertently change their values, which, in turn, affects other methods that might use those arguments.

For example, the developer of ComputeTrainingRate might decide to reset the values of restingRate and age to zero after using them to compute the trainingRate. If the calling program passes restingRate and age to other methods, expecting the variables to retain their original values, then those methods will not work correctly. The advantage of passing by value is that it prevents any unintended damage to the rest of the methods, due to one method changing the value of a variable.

Review Questions

5.14 What is meant by passing an argument by reference?

5.15 The following code shows a method named **ComputeSum,** and the Click event handler of a button that calls the method:

```
private void btnExamScore_Click(object sender, EventArgs e)
{
    int exam1 =150, exam2 =100, total = 0;
    ComputeSum(exam1, exam2, ref total);
    lstDisplay.Items.Add(exam1 + " " + exam2 + " " + total);
}

private void ComputeSum(int exam1, int exam2, ref int sum)
{
    sum = exam1 + exam2;
    exam1 = 0;
    exam2 = 0;
}
```

What would be the output displayed in the ListBox when you click the button?

5.16 The following code shows a method named **ComputeSum**, and the Click event handler of a button that calls the method (note that the parameters are declared differently from the previous question):

```
private void btnExamScore_Click(object sender, EventArgs e)
{
    int exam1 = 10, exam2 = 15, total = 0;
    ComputeSum(ref exam1, ref exam2, ref total);
    lstDisplay.Items.Add(exam1 + " " + exam2 + " " + total);
}

private void ComputeSum(ref int test1, ref int test2, ref int sum)
{
    sum = test1 + test2;
    test1 = 0;
    test2 = 0;
}
```

What would be the output displayed in the ListBox when you run the Click event handler?

5.5 Top-Down Design

Top-down design involves breaking down a larger problem to smaller subtasks, each performed by a separate method, to make software development simpler and easier.

Though the task of computing the training heart rate is fairly simple, for the purpose of illustrating top-down design, let's look at the subtasks of this problem. Figure 5-17 shows the code that you developed in Tutorial 1, with a couple of additional statements to display the resting rate and age, along with the training rate. This event handler calls ComputeTrainingRate method to compute the training heart rate.

Figure 5-17: Revised code to compute training rate

```
private void btnTrainingRate_Click(object sender, EventArgs e)
{
    float restingHeartRate, age;

    // Get personal data (resting rate and age) from Textboxes:
    restingHeartRate = float.Parse(txtRestingRate.Text);
    age = float.Parse(txtAge.Text);

    // Call ComputeTrainingRate
    float trainingHeartRate = 0;
    ComputeTrainingRate(restingHeartRate, age, ref trainingHeartRate);

    // Display training rate
    lstResults.Items.Add(string.Format("{0,20}{1,10}", "Resting Heart Rate:",
            restingHeartRate));
    lstResults.Items.Add(string.Format("{0,20}{1,10}", "Age:", age));
    lstResults.Items.Add(string.Format("{0,20}{1,10}", "Training Heart Rate:",
            trainingHeartRate.ToString("N0")));
}
```

The task performed in this program can be broken down to three subtasks:

1. Get personal data.
2. Compute training rate.
3. Display results.

You already developed a method for the second subtask (compute the training rate). Now, you will develop two additional methods corresponding to the subtasks (get personal data and display results) and call all three methods from btnTrainingRate_Click event handler, as represented in Figure 5-18.

Figure 5-18: Top-down design for computing training rate

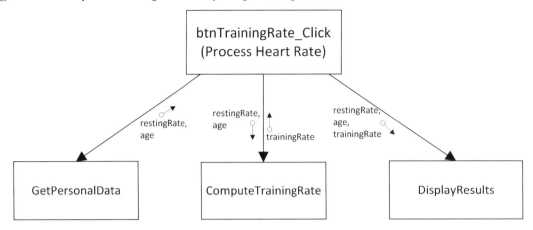

Method to Get Personal Data

The method named GetPersonalData gets restingRate and age from TextBoxes and passes them to the calling program using reference parameters, as shown in Figure 5-19.

Figure 5-19: GetPersonalData method

```
private void GetPersonalData(ref float restingRate, ref float age)
{
   restingRate = float.Parse(txtRestingRate.Text);
   age = float.Parse(txtAge.Text);
}
```

Calling GetPersonalData

To call GetPersonalData, the calling program would use reference arguments that must be assigned an initial value before calling the method. Figure 5-20 shows the Click event handler that calls GetPersonalData.

Figure 5-20: Event handler that calls all three methods

```
private void btnTrainingRate_Click(object sender, EventArgs e)
{
   // Call GetPersonalData
   float restingHeartRate = 0, age = 0;
   GetPersonalData(ref restingHeartRate, ref age);

   // Call ComputeTrainingRate
   float trainingHeartRate = 0;
   ComputeTrainingRate(restingHeartRate, age, ref trainingHeartRate);

   // Call DisplayResults
   DisplayResults(restingHeartRate, age, trainingHeartRate);
}
```

Note that the resting heart rate is represented by the variable restingHeartRate in the calling program, but the corresponding parameter in GetPersonalData is called restingRate.

Passing by reference versus class level variables

The GetPersonalData method illustrates how multiple values (restingHeartRate and age) may be passed from the called method to the calling method using refrence parameters. An alternate way to share data between the two methods is to declare the variables restingHeartRate and age at the class level, and not specify any parameters for GetPersonalData method. It should be noted that the values of such class level variables can be inadvertently changed by any method like ComputeTrainingRate.

out parameter instead of ref parameter

Because data is to be passed from GetPersonalData to the calling program, and not from the calling program to GetPersonalData, out arguments may be used in place of ref arguments, as shown in the following statement:

    private void GetPersonalData(**out** float restingRate, **out** float age)

In this case, the arguments in the calling program do not have to be initialized, and the key word **ref** is replaced by **out** in the call statement:

    float restingHeartRate, age;
    GetPersonalData(**out** restingHeartRate, **out** age);

Method to Compute Training Rate

You already developed the method to compute training rate in Tutorial 1. Figure 5-21 shows the method.

Figure 5-21: ComputeTrainingRate method

```
private void ComputeTrainingRate(float restingRate, float age, ref float trainingRate)
{
    float maximumRate; //trainingRate is not declared here; it is a parameter
    // Compute maximum heart rate:
    maximumRate = 220 - age;
    // Compute training heart rate
    trainingRate = 0.6F * (maximumRate - restingRate) + restingRate;
}
```

Figure 5-20 shows the statement that calls this method.

Method to Display Results

Data need to be passed from the calling program to DisplayResults method, but not the other way around. So, the resting rate, age, and training rate are passed by value to this method, as shown in Figure 5-22.

Figure 5-22: Method to display results

```
private void DisplayResults(float rateResting, float age, float rateTraining)
{
        // Display Results
        lstResults.Items.Add(string.Format("{0,20}{1,10}", "Resting Heart Rate:", rateResting));
        lstResults.Items.Add(string.Format("{0,20}{1,10}", "Age:", age));
        lstResults.Items.Add(string.Format("{0,20}{1,10}", "Training Heart Rate:",
            rateTraining.ToString("N0")));
}
```

The following statement calls this method, as shown in Figure 5-20, passing the three arguments by value.

```
DisplayResults(restingHeartRate, age, trainingHeartRate);
```

Note that resting heart rate and training heart rate are represented by the parameters rateResting and rateTraining, which are different from the names of the corresponding variables in the calling program. This is just to show that the name of an argument and the corresponding parameter may be different.

Getting the Big Picture

An advantage of the top-down design is that the code in the calling program, btnTrainingRate_Click, primarily consists of three statements to call the three methods, as shown in Figure 5-20. Thus, it is easier to get a high level view of the application by looking at the code in the calling program. In addition, the methods corresponding to the subtasks are more reusable, compared to having all code in a single program.

Now you can create a new form that implements the top-down design.

Step 1-21: Create a new form named HeartRateTopDown. Copy the controls from the form HeartRate. This form has the same controls as the form HeartRate. HeartRateTopDown should look as follows:

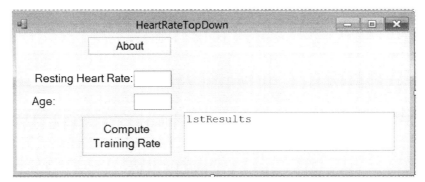

Step 1-22: Create the method GetPersonalData (see Figure 5-19).

Add the statements within btnTrainingRate_Click event handler to call GetPersonalData (see Figure 5-20).

Step 1-23: Create the method ComputeTrainingRate (see Figure 5-21).

Add the statements within btnTrainingRate_Click event handler to call ComputeTrainingRate (see Figure 5-20).

Step 1-24: Create the method DisplayResults.

Add the statements within btnTrainingRate_Click event handler to call DisplayResults (see Figure 5-20).

Step 1-25: Put breaks in the calling program at each call statement. Run the form. Verify that the values assigned to restingRate and age in GetPersonalData changes the values of the corresponding arguments in the calling program. Similarly, verify that assigning the computed value to restingRate in the ComputeHeartRate changes the value of restingHeartRate in the calling program.

The complete code for the HeartRateTopDown form looks as follows:

```csharp
private void GetPersonalData(ref float restingRate, ref float age)
{
    // Get resting rate and age from Textboxes:
    restingRate = float.Parse(txtRestingRate.Text);
    age = float.Parse(txtAge.Text);
}

private void ComputeTrainingRate(float restingRate, float age, ref float trainingRate)
{
    float maximumRate;   //trainingRate is not declared here; it is a parameter
    // Compute maximum heart rate:
    maximumRate = 220 - age;
    // Compute training heart rate
    trainingRate = 0.6F * (maximumRate - restingRate) + restingRate;
}

private void DisplayResults(float rateResting, float age, float rateTraining)
{
    // Display Results
    lstResults.Items.Add(string.Format("{0,20}{1,10}", "Resting Heart Rate:", rateResting));
    lstResults.Items.Add(string.Format("{0,20}{1,10}", "Age:", age));
    lstResults.Items.Add(string.Format("{0,20}{1,10}", "Training Heart Rate:",
                                                    rateTraining.ToString("N0")));
}

private void btnTrainingRate_Click(object sender, EventArgs e)
{
    // Call GetPersonalData
    float restingHeartRate = 0, age = 0;
    GetPersonalData(ref restingHeartRate, ref age);

    // Call ComputeTrainingRate
    float trainingHeartRate = 0;
    ComputeTrainingRate(restingHeartRate, age, ref trainingHeartRate);

    // Call DisplayResults
    DisplayResults(restingHeartRate, age, trainingHeartRate);
}
```

It's time to practice!

Modify ComputeTrainingRate, DisplayResults and the call statements as follows:

ComputeTrainingRate returns maximumRate to the calling program, in addition to trainingRate. Pass the maximumRate to DisplayResults as shown in the following structure chart. Display the maximumRate along with the trainingRate.

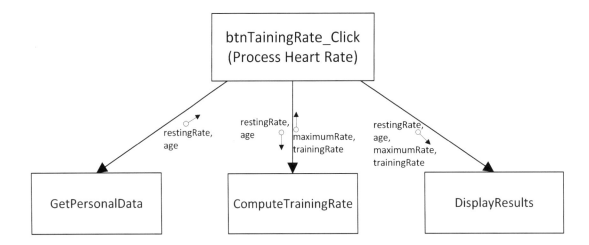

An Undesirable Design Using Noncohesive Methods

In the top-down design shown in Figure 5-18, the task of computing the training heart rate is broken down to three subtasks represented by the methods GetPersonalData, ComputeTrainingRate and DisplayResults. These methods are called in sequence by a main program (the Click event procedure, btnTrainingRate_Click). Data is shared between the calling program and the called program through parameters. As discussed earlier in this chapter, such a design makes the methods cohesive and reusable.

An alternate undesirable design that involves less reusable noncohesive methods is shown in Figure 5-23.

Figure 5-23: Alternate undesirable design involving noncohesive methods

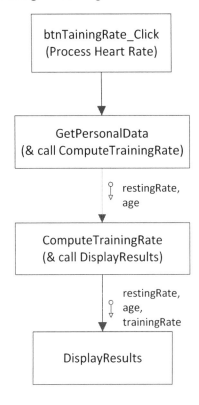

In this design, the main program (the Click event procedure, btnTrainingRate_Click) calls the method GetPersonalData as in the top-down design.

However, the method GetPersonalData gets personal data, and instead of passing the data to the main program, it calls ComputeTrainingRate, passing restingRate and age to the method.

ComputeTrainingRate computes the training rate and then calls DisplayResults passing restingRate, age and trainingRate to the method.

DisplayResults displays the results as in the top-down design.

The above design suffers from several problems:

1. The methods GetPersonalData and ComputeTrainingRate are not cohesive. For example, ComputeTrainingRate doesn't do a single well-defined task. It computes training rate and also displays results by calling DisplayResults. Similarly, GetPersonalData gets restingRate and age and also computes results by calling ComputeTrainingRate.
2. The methods GetPersonalData and ComputeTrainingRate are less reusable. For example, ComputeTrainingRate does more than computing the results, and it doesn't return the result to the calling program.
3. It might be necessary to pass certain data to a method where it is not needed. For example, if the personal data includes the person's name, the only way it can be passed to DisplayResults is by first passing it to ComputeTrainingRate, where it is not needed. Passing data to a method where is not needed makes maintenance of code difficult.

Review Questions

5.17 What is top-down design?

5.18 Describe two benefits of top-down design.

5.6 Methods That Return a Value

The methods we discussed in this chapter return values to the calling program only through parameters. Parameters allow any number of values to be returned.

The method itself can return a single value to the calling program, other than through parameters. An advantage of such a method is that the method can be used as a part of an expression. Figure 5-24 shows such a method, called TrainingRate, that computes training rate and returns it to the calling program:

Figure 5-24: Value returning method, TrainingRate

```
private float TrainingRate(float restingRate, float age)
{
  float maximumRate;
  // Compute maximum heart rate:
  maximumRate = 220 - age;
  // Compute training heart rate and return it
  return 0.6F * (maximumRate - restingRate) + restingRate;
}
```

Here are the key differences between this method and the method ComputeTrainingRate that you developed earlier:

The header,
   private float TrainingRate(float restingRate, float age)
shows no parameter to return the trainingRate. Then how is the trainingRate returned?

The method itself takes the place of a parameter and returns the value specified in the **return** statement,
   return 0.6F * (maximumRate - restingRate) + restingRate;

Because the method returns a value, the type of value returned, called **return type**, is specified in the header using the key word **float** instead of void. Further, the name of the method indicates the value returned by the method, not what the method does.

Calling a Value Returning Method

This is how you call the method:
   float trainingHeartRate = TrainingRate(restingHeartRate, age);

The value returned by TrainingRate is assigned to the float type variable, trainingHeartRate.

Next, we will develop a new form named HeartRateTopDownAlt, with an alternate version of ComputeTrainingRate that does not use a parameter to return training rate. This form has the same user interface as HeartRateTopDown, as shown in the following figure:

Chapter 5: Methods

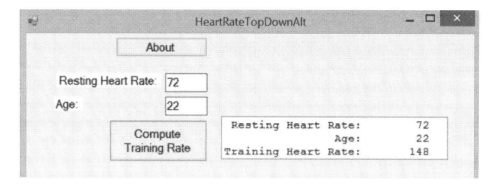

Step 1-26: Open the project, Ch5_Methods, from Tutorial_Starts folder. Open the form, HeartRateTopDownAlt.

Step 1-27: Delete the return statement from TrainingRate method, and add the code from Figure 5-24.

Step 1-28: Add the statement to call TrainingRate as follows:

trainingHeartRate = TrainingRate(restingHeartRate, age);

Step 1-29: Run the form and test it.

Figure 5-25 shows the complete code for HeartRateTopDownAlt.

Figure 5-25: Complete code for HeartRateTopDownAlt

```
20      private void btnTrainingRate_Click(object sender, EventArgs e)
21      {
22          float restingHeartRate = 0, age = 0;
23          float trainingHeartRate = 0;
24
25          // Call GetPersonalData
26          GetPersonalData(ref restingHeartRate, ref age);
27
28          // Call TrainingRate
29          trainingHeartRate = TrainingRate(restingHeartRate, age);
30
31          // Call DisplayResults
32          DisplayResults(restingHeartRate, age, trainingHeartRate);
33      }
34
35      private void GetPersonalData(ref float restingRate, ref float age)
36      {
37          // Get resting rate and age from Textboxes:
38          restingRate = float.Parse(txtRestingRate.Text);
39          age = float.Parse(txtAge.Text);
40      }
41
42      private float TrainingRate(float restingRate, float age)
43      {           // TrainingRate returns the training rate
44          float maximumRate;
45          // Compute maximum heart rate:
46          maximumRate = 220 - age;
47          // Compute training heart rate
48          return 0.6F * (maximumRate - restingRate) + restingRate;
49      }
```

```
50
51    private void DisplayResults(float rateResting, float age, float rateTraining)
52    {
53        // Display Results
54        lstResults.Items.Add(string.Format("{0,20}{1,10}", "Resting Heart Rate:", rateResting));
55        lstResults.Items.Add(string.Format("{0,20}{1,10}", "Age:", age));
56        lstResults.Items.Add(string.Format("{0,20}{1,10}", "Training Heart Rate:",
57                                                           rateTraining.ToString("N0")));
58    }
```

Additional examples of value returning methods are provided in Chapter 6.

It's time to practice! Do Exercise 5-3 and Exercise 5-4

5.7 Enumerations: Limiting Parameter Values

An enumeration is a set of named integer constants. For example, Figure 5-26 shows an enumeration named CheckState with three named constants, **Checked, Indeterminate and Unchecked**. The statement in Figure 5-26 assigns the selected value (0, 1 or 2, represented by the three named constants) to the CheckState property of the CheckBox.

Figure 5-26: Enumeration example

```
chkMember.CheckState = CheckState.
                         Checked
                         Indeterminate
                         Unchecked
```

Each name represents an integer. By default, the first name ("Checked") has the value 0; each of the following names has a value one greater than that of the previous name. So, in effect, the statement assigns a number (0, 1 or 2) to the CheckState property.

An enumeration provides a convenient way to select from a limited set of numbers. In addition, an enumeration uses meaningful names in place of numbers, which helps minimize errors in making a selection.

The syntax to create an enumeration is
enum enumerationName [: optional integer type] {list of named constants separated by commas};

For example,
```
enum CheckState { Checked, Indeterminate, Unchecked };
```
creates the enumeration CheckState, which is of **enumerated type**, where the named constants represent integer values 0, 1 and 2, respectively.

An enumeration cannot be declared inside a method. It can be declared inside the class, inside the namespace or outside the namespace.

Though the named constants represent integer values, you need to cast them to integer type before assigning to an integer variable, as in
   int checkState = (int) CheckState.UnChecked;

You may assign values other than the default set of values, which starts with zero. Here is an example:

```
enum size { Small = 10, Medium = 20, Large = 30 }
```

To better understand the use of enumeration, you will create an application that computes hotel room rates based on the discount type selected from a ComboBox, as shown in Figure 5-27.

Figure 5-27: DiscountTypeEnum form

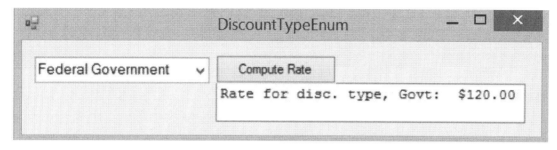

The discount types are "No Discount," "AAA," "Federal Government" and "Senior."

To make it easy to work with the above discount types within the program, we will represent them by the set of integers 0, 1, 2 and 3, respectively, and give these numbers more meaningful names: **NoDiscount, AAA, Govt and Senior**.

The following statement would create the set of named constants (an enumeration) named discountTypes that represents the set of numbers 0, 1, 2 and 3:

```
enum discountTypes { NoDiscount, AAA, Govt, Senior};
```

After creating an enumeration, you can create variables of that type. You create a variable named selectedDiscType of discountTypes type as follows:

```
discountTypes selectedDiscType;
```

The coding for the form shown in Figure 5-27 includes the following steps:

1. Declare an enumeration named discountTypes and a variable named selectedDiscType of the type discountTypes, both at the class level.

2. Add discount types ("No Discount," "AAA," "Federal Government" and "Senior") to the ComboBox, in the Load event handler of the form.

3. When the user selects a discount code from the ComboBox, assign the corresponding value (0, 1, 2 or 3, represented by NoDiscount, AAA, Govt and Senior) to the variable selectedDiscType. This is done within the SelectedIndexChanged event handler of the ComboBox.

4. Create a method that determines the room rate based on the value of selectedDiscType. This method is called to compute the room rate.

Tutorial 2: Enumeration: Enumeration Type Parameter

This tutorial illustrates the use of enumeration by building the form, DiscountTypeEnum.

Step 2.1: Open the project Ch5_Methods. Open the form named DiscountTypeEnum. Open the code window.

Step 2.2: Add the code to declare the enumeration and the variable at the class level, as shown in Figure 5-28.

Figure 5-28: Declare enumeration and a variable of enumerated type

```
19          enum discountTypes { NoDiscount, AAA, Govt, Senior};
20          discountTypes selectedDiscType;
```

Step 2.3: Insert the code to add discount types to the ComboBox, within the Load event handler of the form, as shown in Figure 5-29.

Figure 5-29: Add items to ComboBox

```
22      private void DiscountTypeEnum_Load(object sender, EventArgs e)
23      {
24          string[] discTypes = { "No Discount", "AAA", "Federal Government", "Senior" };
25          for (int index = 0; index <= 3; index++)
26              cboDiscountType.Items.Add(discTypes[index]);
27      }
```

Step 2.4: Based on the discount type selected by the user, assign a value to the variable, selectedDiscType, within the SelectedIndexChanged event of the ComboBox. See Figure 5-30.

Figure 5-30: SelectedIndexChanged event handler

```
28
29      private void cboDiscountType_SelectedIndexChanged(object sender, EventArgs e)
30      {
31          switch(cboDiscountType.SelectedItem.ToString())
32          {
33              case "No Discount":
34                  selectedDiscType = discountTypes.NoDiscount;
35                  break;
36              case "AAA":
37                  selectedDiscType = discountTypes.AAA;
38                  break;
39              case "Federal Government":
40                  selectedDiscType = discountTypes.Govt;
41                  break;
42              case "Senior":
43                  selectedDiscType = discountTypes.Senior;
44                  break;
45          }
46      }
```

Note that if the room rates are to be used just within this form, you could assign a value to the room rate within this switch statement without using an enumeration.

Here, we store the selected discount type in a class level variable of enumerated type so that it can be accessed from other classes (by making it public) where it can be used potentially for other purposes.

Next, we create a method that accepts an argument of discountTypes type and returns the room rate.

Step 2-5: Create a method that determines the room rate based on the value of selectedDiscType. See code in Figure 5-31.

Figure 5-31: Method with an enumerated type parameter

```csharp
        private void ComputeRate(discountTypes discountType, ref float rate)
        {
            switch (discountType)
            {
                case discountTypes.NoDiscount:
                    rate = 125;
                    break;
                case discountTypes.AAA:
                    rate = 110;
                    break;
                case discountTypes.Govt:
                    rate = 120;
                    break;
                case discountTypes.Senior:
                    rate = 105;
                    break;
            }
        }
```

Note that the parameter discountType is of enumerated type discountTypes, and it will not accept any value other than the enumerated constants, thus providing some data validation that is not provided by a string type parameter.

Step 2-6: Call the method ComputeRate() from the Click event handler of btnComputeRate, and display the rate in the ListBox. See Figure 5-32.

Figure 5-32: btnComputeRate Click event handler

```csharp
        private void btnComputeRate_Click(object sender, EventArgs e)
        {
            float rate = 0;
            ComputeRate(selectedDiscType, ref rate);

            lstDisplay.Items.Clear();
            lstDisplay.Items.Add("Rate for disc. type, " + selectedDiscType + ":  " + rate.ToString("C"));
        }
```

Line 70 calls ComputeRate method and passes the selectedDiscType, which is a variable of discountTypes type. Passing a value like an integer constant 1 or 2 will result in an error unless it is converted to the enumerated type.

Step 2-7: Run the form and test your code.

Review Questions

5.19   What is an enumeration?

5.20   What are the benefits of using an enumeration?

Exercises

Exercise 5-1

Create a form named GradeWithMethod_1 that lets you to do the following:

>   Enter a score (0–100) into a TextBox named txtScore, and
>   compute the letter grade by clicking a button.

Use the scale 90, 80, 70 for A, B and C, respectively. The letter grade must be computed using a separate method named ComputeGrade, which displays the letter grade using a MessageBox. Call this method from the Click event of the button passing the score as a parameter to the method.

Exercise 5-2

This is a modified version of the form specified in Exercise 5-1. You may copy the controls and code to a new form named GradeWithMethod_2 and modify the code.

Create a form named GradeWithMethod_2 that lets you do the following:
>   Enter a score (0–100) into a TextBox named txtScore, and
>   compute the letter grade by clicking a button.

Use the scale 90, 80, 70 for A, B and C, respectively. The letter grade must be computed using a separate method named ComputeGrade. Call this method from the Click event of the button passing the score as a parameter to the method. **ComputeGrade program must return the letter grade to the calling program through a parameter.** The calling program displays the data in a Label named lblLetterGrade.

Exercise 5-3

This is a modified version of the form specified in Exercise 5-1. You may copy the controls and code to a new form named GradeWithMethod_3 and modify the code.

Create a form named GradeWithMethod_3 that lets you do the following:

>   Enter a score (0–100) into a TextBox named txtScore, and
>   compute the letter grade by clicking a button.

Use the scale 90, 80, 70 for A, B and C, respectively. The letter grade must be computed using a separate method named Grade. Call this method from the Click event of the button, passing the score as a parameter to the method. Grade must return the letter grade to the calling program **without using a parameter**. The calling program displays the data in a Label named lblLetterGrade.

Exercise 5-4

Create a form named GradeTopDown that lets you to do the following:
  Enter a score (0–100) into a TextBox named txtScore, and
  compute the letter grade by clicking a button.
Use the scale 90, 80, 70 for A, B and C, respectively. The structure chart of the task to process the grade is as follows:

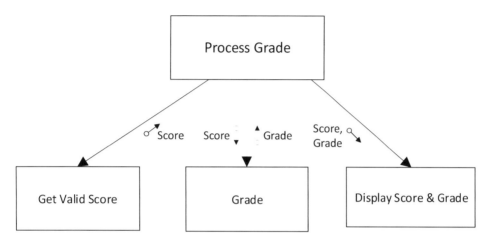

The Click event handler of the button must call three separate methods, corresponding to the three subtasks, described below:

1. A method named GetValidScore that gets the name and score from the TextBoxes, validates the score and passes them to the Click event handler.
2. A second method named Grade that computes the letter grade. Pass the score from the calling program to Grade. The method must return the letter grade **without using a parameter**. (This method is the same as in Exercise 5-3.)

A third method that displays the name and letter grade. Pass the name and letter grade as parameters to this method.

Chapter 6

Graphical User Interface Controls

This chapter introduces you to graphical user interface (GUI) controls that help minimize data entry errors and make it easier for users to interact with a program.

Topics

6.1	Working with ScrollBars	6.5	Getting User Input from Controls
6.2	Working with CheckBoxes	6.6	Working with ComboBoxes
6.3	Working with RadioButtons	6.7	Working with ListBoxes
6.4	Validating Input		

To understand the use of GUI controls, consider an application that lets you calculate the total price for tickets to a show in a theater. The user interface, shown in Figure 6-1, uses a ScrollBar to specify the number of tickets, a set of RadioButtons to select the section of the theater and a CheckBox to specify whether the customer is a member.

Figure 6-1: User interface for ticket sales application

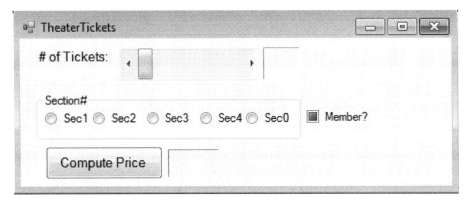

Next, we look at how to work with each of these controls.

6.1 Working with ScrollBars

The ScrollBar controls, HScrollBar and VScrollBar, let you select a number between two limits determined by the values of the **Minimum** and **Maximum** properties. The two ScrollBars are essentially the same, except that the VScrooBar lets you scroll horizontally, and the VScrollBar lets you scroll vertically. In this section, you will use the HScrollBar.

To select a number, you may move the slider by clicking the arrows at either end by clicking the inside area or by dragging the slider. The number selected by the user is given by the **Value** property. However, the selected number is not displayed by the control. Typically, a separate Label is used to display the selected value. Table 6-1 shows the important properties of HScrollBar.

Table 6-1: Properties of HScrollBar

Property name	Description
Value	The integer value of the ScrollBar, selected by moving the slider.
Minimum	Lower limit of the range of values. Default is zero. User may specify a different limit.
Maximum	Upper limit of the range of values. Default is 100. User may specify a different limit.
SmallChange	Amount by which the value changes when the user clicks on one of the two arrows. Default is 1.
LargeChange	Amount by which the value changes when the user clicks the ScrollBar area on either side of the slider. Default is 1. If LargeChange is greater than 1, you must increase the Maximum by adding (LargeChange - 1) to Maximum. For example, if you want a range of 0 to 20, and LargeChange is 4, then you must set the Maximum to 23 (20 + (4-1)).

Tutorial 1: Theater Tickets Sale Application

In this tutorial, you will create the TheaterTickets form, shown in Figure 6-1, to illustrate the use of GUI controls.

Step 1-1: Create a new project named Ch6_GUIcontrols. Rename the form to TheaterTickets.

Add an HScrollBar (found in the *All Windows Forms* group in the Toolbox) to the form, and name it **hsbNumOfTickets**.

Set Maximum property to 10. It is assumed that purchases of more than 10 tickets are processed differently at group rates in a different form.

Change the value of the Value property from 0, and observe how it affects the position of the slider. Make sure you change it back to 0.

Step 1-2: Run the form. Move the slider to change the value of the ScrollBar by clicking the arrows, clicking the ScrollBar area, or dragging the slider. But you will not see the value.

How do you view the value of the ScrollBar?

Events of ScrollBars

There are two important events of the ScrollBar that can be used to display the value in a Label: the **ValueChanged** event and the **Scroll** event.

ValueChanged event

This event is triggered when the value of a ScrollBar is changed by moving the slider manually, and also when the value is changed at runtime by executing code. The following code uses the ValueChanged event handler to display the value of the ScrollBar in a Label whenever the value changes at runtime:

```
private void hsbNumOfTickets_ValueChanged(object sender, EventArgs e)
{
    lblNumOfTickets.Text = hsbNumOfTickets.Value.ToString();
}
```

The ValueChanged event handler allows you to display the value in the Label when the user moves the slider and also when the ScrollBar is reset to zero at runtime using code by clicking the clear button. So, you will use this event handler to display the value of the ScrollBar.

Scroll event

This event is triggered only when the value of a ScrollBar is changed by moving the slider manually; not when the value is changed at runtime by code, like resetting the ScrollBar using code. You create the scroll event handler by double clicking the ScrollBar in design view.

Creating and Deleting Event Handlers

How do you create an event handler for an event like **ValueChanged**, which is **not** the default event handler that is created when you double click the ScrollBar. (**Scroll** is the default event handler.)
To create an event handler that is not the default, follow these steps:

> Select the object, which is the ScrollBar in this example.
> In the properties window, click on the **events** button to display all events for the object, as shown in Figure 6-2.
> Double click the event (ValueChanged) to open the event handler.

Figure 6-2: Events of the ScrollBar

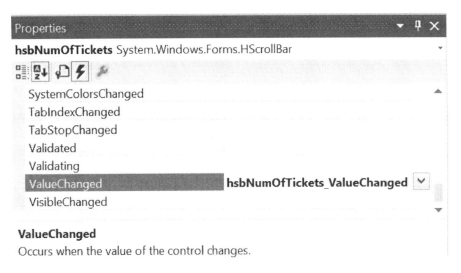

If you double clicked the ScrollBar and created the Scroll Event handler that you don't need, you may delete it using the following steps, as described in Chapter 1:

> Display the events of the object (hsbScrollBar) in the Properties window, which should look as follows:

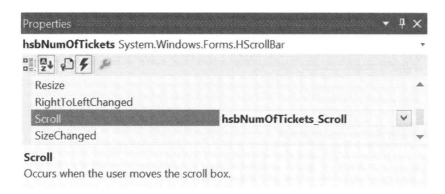

Select the event (Scroll).

Click the dropdown arrow and select the event handler (hsbNumOfTickets_Scroll) you want to delete. (You need to do it even if the event handler already is displayed.)

Right click the event handler and select **Delete**.

Step 1-3: For practice, double click the ScrollBar to create the Scroll event handler. Then, delete the event handler.

Next, create the ValueChanged event handler for the ScrollBar. Enter the following statement to display the ScrollBar's value in the Label:

lblNumOfTickets.Text = hsbNumOfTickets.Value.ToString();

Step 1-4: Put a break within the Value Changed event handler, and run the form.
Move the slider, and observe that the ValueChanged event is triggered.
Remove the break.
Move the slider from the left end to the right end, and note the range of values determined by the default values of Minimum and Maximum properties.
Change the value of LargeChange to 4.
Click inside the ScrollBar, and observe that the slider moves by 4 and the largest value is only 7, though the value of Maximum property is 10.
To get the full range of 0 to 10, set Maximum to 13 (10+4-1), and check the largest value. It should be 10.
Set LargeChange back to 1.

Step 1-5: Create the Click event handler of Clear button, and enter the statement to reset the ScrollBar value to 0, as follows:
 private void cmdClear_Click(object sender, EventArgs e)
{
    hsbNumOfTickets.Value = 0;
}

Step 1-6: Run the form. Move the slider to select a value greater than zero.
Click the Clear button. The slider moves to zero position because the ScrollBar's value is set to zero. This, in turn, triggers the ValueChanged event and also sets the Label's value to zero.

Note that if the statement,
   lblNumOfTickets.Text = hsbNumOfTickets.Value.ToString();
were in the Scroll event handler, resetting the ScrollBar value would not have reset the value displayed in the Label. You may test it by copying the above statement to the Scroll event handler and commenting out the statement inside the ValueChanged event handler.

Review Questions

6.1   A form uses a ScrollBar, hsbUnits, to let the user select whole numbers only in the range 0 to 10. What property of the ScrollBar should be changed from its default value?

6.2   Write the statement to display the value of the ScrollBar, hsbUnits, in a Label named lblUnits.

6.3   The value of a ScrollBar named hsbUnits is to be automatically displayed in a Label named lblUnits whenever the value of the ScrollBar is changed (either by clicking inside the ScrollBar, or by executing code, for example, when the ScrollBar is reset by clicking a "Clear" button). The code required to display the value should be entered in which event handler?

6.2 Working with CheckBoxes

A CheckBox is used to let the user select a true or false Boolean value. For example, we use a CheckBox in this application for the user to specify whether or not a customer is a member. Two commonly used properties of CheckBox are shown in Table 6-2.

Table 6-2: Properties of CheckBox

Property	Settings of the property	Description
CheckState	Checked	Checking the CheckBox at runtime sets CheckState to Checked.
	UnChecked	Unchecking the CheckBox at runtime sets CheckState to UnChecked. This is the default state.
	Indeterminate	The CheckState can be set to Indeterminate in the design window, or by executing code at runtime. Setting this value checks the CheckBox. When the user checks (or unchecks) the CheckBox, the CheckState changes to Checked (or UnChecked). You may set the CheckBox to Indeterminate state initially, so that the code can check whether the user forgot to make a selection.
Checked	True	Indicates that the CheckBox is in the Checked or Indeterminate state.
	False	Indicates that the CheckBox is in the UnChecked state. This is the default.

Step 1-7: Add a CheckBox named chkMember to the form. Change the value of CheckState property, and observe the effect on the CheckBox and on the value of Checked property.

How do you make sure that the user made a selection of member status (member or nonmember) at runtime? To help verify, you set the CheckState property to Indeterminate state within the Load event of the form, and within the Click event handler of Clear button using the code from Figure 6-3. If the CheckState property remains as Indeterminate when the user clicks the compute button, that means the user did not make a selection. You will later add the code to verify the CheckState status.

igure 6-3: Set CheckState property to Indeterminate

```csharp
private void cmdClear_Click(object sender, EventArgs e)
{
    hsbNumOfTickets.Value = 0;
    chkMember.CheckState = CheckState.Indeterminate;
}

private void TheaterTickets_Load(object sender, EventArgs e)
{
    chkMember.CheckState = CheckState.Indeterminate;
}
```

Step 1-8: Run the form. Note that when the form loads, the state is Indeterminate.
Click the CheckBox to change the CheckState to Checked, and then to UnChecked (you cannot set the state to Indeterminate at runtime by clicking the CheckBox).
Click the Clear button, and make sure the state changes to Indeterminate.

Events of CheckBoxes

Two important events of the CheckBox are the **CheckStateChanged** event and the **CheckedChanged** event.

The **CheckStateChanged** event is triggered when the value of CheckState property changes.
The **CheckedChanged** event is triggered when the value of Checked property changes.

You may use these event handlers to take actions when the Checked or CheckState properties change.

Review Questions

6.4 What is the value of Checked property of a CheckBox when the CheckState property is set to Indeterminate?

6.5 The CheckState property of a CheckBox is set to _____ when the Checked property is set to True?

6.3 Working with RadioButtons

RadioButtons are used as a group to let the user make a single selection from a group of limited options, like selecting a student's class in college or an employee's title. In this example, we use a set of RadioButtons to let the user select a section of the theater for seating:

The **Checked** property of the RadioButton is set to True if it is checked and False if it is not checked. At any time, only one RadioButton in a group can be checked.

You may create multiple groups by adding RadioButtons inside a **GroupBox** control on the form. If no GroupBox is used, all RadioButtons on a form are treated as part of a single group, even if they are physically in multiple groups. The GroupBox also lets the user move all RadioButtons that are within it, as a group.

RadioButtons versus CheckBoxes

RadioButtons are used when the number of options to choose from is limited (typically under 10) and also static (i.e., the number of items in the list does not change).

When not to use RadioButtons

When the selection is true/false or yes/no, the standard practice is to use a CheckBox, **not** two RadioButtons. For example, the following use of RadioButton is **not** recommended:

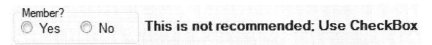

Step 1-9: Add a GroupBox to the form. Change its Text property to Section#.

Add five RadioButtons inside the GroupBox, as shown:

Change the value of Text property of the first four RadioButtons to 1, 2, 3 and 4, and that of the fifth button to zero.

Name the first four RadioButtons, **rdbSec1, rdbSec2, rdbSec3** and **rdbSec4**, respectively. Name the fifth, rdbDefault, and change its **Visible** property to False.

The first four represent four sections of the theater. The fifth button will be hidden from the user and is used as the default. It serves two purposes:

1. You can reset (uncheck) all four buttons by checking the fifth button using code.
2. If the fifth button is checked, that indicates that the user did not make a selection.

Next, change the Checked property of any one of the RadioButtons to True.

The button changes to checked status. Now, change the Checked property of another button to True. Note that its status changes to checked, and the previously checked button becomes unchecked. Change the Checked property back to False.

Step 1-10: To help verify that the user checked a RadioButton, add the following code to the Load event of the form, and to the Click event handler of Clear button:

   rdbDefault.Checked = true;

How does it help? If the Default RadioButton is checked when the user clicks the button to compute cost, that means the user did not check one of the first four buttons.

Note that the above statement assigns the Boolean value **true** to the Checked property, which has the same effect as setting the Checked property to **True** in the Desgin window of the form.

Figure 6-4 shows the modified code.

Figure 6-4: Resetting RadioButtons

```
private void cmdClear_Click(object sender, EventArgs e)
{
    hsbNumOfTickets.Value = 0;
    chkMember.CheckState = CheckState.Indeterminate;
    rdbDefault.Checked = true;
}

private void TheaterTickets_Load(object sender, EventArgs e)
{
    chkMember.CheckState = CheckState.Indeterminate;
    rdbDefault.Checked = true;
}
```

Step 1-11: Run the form. Click any RadioButton, and click Clear to test the clear button.

The code to compute the total price of tickets using the data from the controls has three major parts:

1. Verify that the user specified the number of tickets, selected a section of the theater and specified the membership status.
2. Get the number of tickets from the ScrollBar, get the unit price based on the section selected, and get the discount based on membership status.
3. Compute the total cost.

The next section discusses the code to verify user input.

Review Questions

6.6 Write the statement that would make the RadioButton, rdbDefault, selected.

6.7 A form has two GroupBoxes, each containing a set of RadioButtons. What is the maximum number of RadioButtons that can be selected by the user from the entire form at runtime?

6.4 Validating Input

The code required to verify that the user provided all input is shown in Figure 6-5.

Figure 6-5: Verify that the user provided all input data

```
41          // Verify that the user selected number of tickets:
42          if (hsbNumOfTickets.Value == 0)
43          {
44              MessageBox.Show("Please select number of tickets");
45              return;
46          }
47          // Verify that the user selected a section:
48          if (rdbDefault.Checked == true)
49          {
50              MessageBox.Show("Please select the section");
51              return;
52          }
53          // Verify that the user selected membership status:
54          if (chkMember.CheckState == CheckState.Indeterminate)
55          {
56              MessageBox.Show("plese specify membership status");
57              return;
58          }
```

Line 41 checks whether the value of the ScrollBar is 0—that is, the user did not select a value.

Line 47 checks whether the Boolean value of the Checked property of the Default button is true (as asssinged in the Load event or Click event handler of the Clear button)—that is, none of the other buttons are checked.

Line 53 checks whether the CheckState property of the CheckBox is Indeterminate—that is, the user did not specify membership status.

The return statement within each if statement would prevent processing of the remaining code, if any, in the event handler.

Step 1-12: Add the code shown in Figure 6-5 to the Click event handler of btnComputePrice. Test the code by clicking the Compute button with missing data.

6.5 Getting User Input from Controls

The previous section verified that the user provided all inputs. Next, you get the inputs from controls: number of tickets, discount and unit price.

You get the **number of tickets** from the Value property of the ScrollBar, as follows:
```
int numOfTickets = hsbNumOfTickets.Value;
```

To compute the **discount**, you create a method named Discount and invoke (call) it. Members get a $20 discount per ticket. Figure 6-6 shows the method, and the statement that invokes it.

Figure 6-6: Code to compute discount

```
float discount = Discount();

float Discount()
{
        switch (chkMember.CheckState)
        {
                case CheckState.Checked:      // or, case 0
                        return 20;
                        break;
                default:
                        return 0;
                        break;
        }
}
```

The switch statement is used to compute discount if the CheckState property of the CheckBox has the value *Checked*.

To compute **unit price**, we create a method called UnitPrice and invoke it. Unit price for sections 1, 2, 3, and 4 are $70, $60, $52, and $45, respectively. Figure 6-7 shows the method, and the statement that invokes it.

Figure 6-7: Code to compute unit price

```
float unitPrice = UnitPrice();
float UnitPrice()
{
        if (rdbSec1.Checked)
                return 70;
        else if (rdbSec2.Checked)
                return 60;
        else if (rdbSec3.Checked)
                return 52;
        else if (rdbSec4.Checked)
                return 45;
        else
                return 0;
}
```

Figure 6-8 shows the complete code that includes additional code to compute and display the amount due (lines 68, 69). Note that amtDue is declared at the class level so that it can be accessed from the Click event handler of btnComputePrice, and also from another event handler to be developed later.

Figure 6-8: Complete code to compute price

```csharp
19      float amtDue;
20      private void hsbNumOfTickets_ValueChanged(object sender, EventArgs e)
21      {
22          lblNumOfTickets.Text = hsbNumOfTickets.Value.ToString();
23      }
24
25      private void btnClear_Click(object sender, EventArgs e)
26      {
27          hsbNumOfTickets.Value = 0;
28          chkMember.CheckState = CheckState.Indeterminate;
29          rdbDefault.Checked = true;
30          lblAmtDue.Text = "";
31      }
32
33      private void TheaterTickets_Load(object sender, EventArgs e)
34      {
35          chkMember.CheckState = CheckState.Indeterminate;
36          rdbDefault.Checked = true;
37      }

39      private void btnComputePrice_Click(object sender, EventArgs e)
40      {
41          // Verify that the user selected number of tickets
42          if (hsbNumOfTickets.Value == 0)
43          {
44              MessageBox.Show("Please specify number of tickets");
45              return;
46          }
47          // Verify that the user selected a section
48          if (chkMember.CheckState == CheckState.Indeterminate)
49          {
50              MessageBox.Show("Please specify membership");
51              return;
52          }
53          // Verify that the user selected membership status
54          if (rdbDefault.Checked == true)
55          {
56              MessageBox.Show("Please specify Section");
57              return;
58          }
59          // Get number of tickets
60          int numOfTickets = hsbNumOfTickets.Value;
61
62          // Compute discount
63          float discount = Discount();
64
65          // Compute unit price
66          float unitPrice = UnitPrice();
67
68          amtDue = numOfTickets * (unitPrice - discount);
69          lblAmtDue.Text = amtDue.ToString("$#,###.00");
70      }
```

```
72      float Discount()
73      {
74          switch (chkMember.CheckState)
75          {
76              case CheckState.Checked:     // or, case 0
77                  return 20;
78                  break;
79              default:
80                  return 0;
81                  break;
82          }
83      }
84
85      float UnitPrice()
86      {
87          if (rdbSec1.Checked)
88              return 70;
89          else if (rdbSec2.Checked)
90              return 60;
91          else if (rdbSec3.Checked)
92              return 52;
93          else if (rdbSec4.Checked)
94              return 45;
95          else
96              return 0;
97      }
```

Step 1-13: Copy the rest of the code shown in Figure 6-8 from Ch6_GUIcontrols/Ch6_ Code.txt to the Click event handler of ComputePrice.

Step 1-14: Run the form. Test the code using the input data from Figure 6-9 and verify the output.
Clear the data, and run it with another set of valid selections.
Manually compute the result, and verify the output from the code.
Run it when one or more of the three items is not selected. Make sure you get the proper message.

Figure 6-9: The TheaterTickets form with sample data

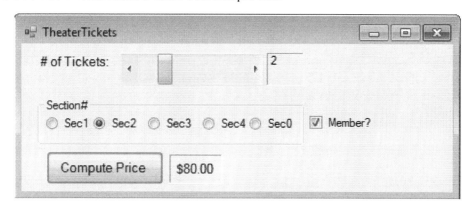

6.6 Working with ComboBoxes

A ComboBox lets the user select an item from a group of items. Consider a modified version of the TheaterTicket form that includes a ComboBox to select the name of the show, as shown in Figure 6-10. In the modified form, the name of the show and the total cost are written to a ListBox. (Typically, in a real-world application, the data would be written to a file or database.) It is assumed that the price of a ticket depends only on the section, not on the show.

Figure 6-10: TheaterTicket form with ComboBox

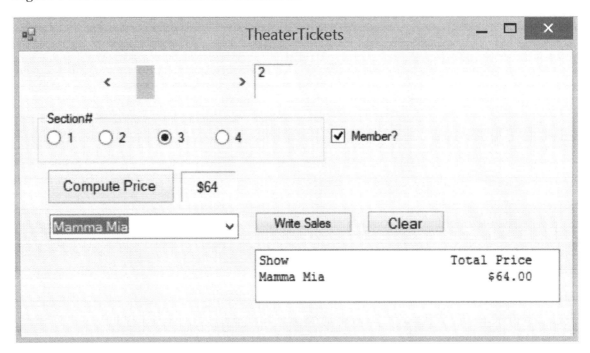

A Set of RadioButtons versus a Single ComboBox

A ComboBox, as opposed to a set of RadioButtons, is recommended when the list is large and/or the list is dynamic. The list is dynamic when the content of the list or the number of items in the list is likely to change. In the current example, we use a ComboBox because the list of shows is likely to change over time. Similarly, to let the user select a month from a list of the 12 months, for example, a ComboBox generally is recommended over a set of RadioButtons because the list is large.

Properties and Events of ComboBoxes

Table 6-3 shows commonly used properties and events of the ComboBox and ListBox.

Table 6-3: Commonly used properties and events of ComboBox and ListBox

Property	Description
DropDownStyle	Determines the style and whether user can type in an item that is not in the list
Items	The collection of all items in the list
SelectedIndex	The index of the selected item
SelectedItem	The selected item (of object type)
SelectedItems	The collection of selected items in a **ListBox**
SelectionMode	Determines whether more than one item can be selected from a **ListBox**
Sorted	If set to True, the items in the list are sorted
Text	The text displayed in the ComboBox
Events	
SelectedIndexChanged	Raised when the user selects an item from the list

The Items collection

The Items collection is an important property of the ComboBoxes and ListBoxes. It has several properties and methods, as shown in Table 6-4.

Table 6-4: Properties and methods of the Items Collection

Property	Description
Count	Number of items in the list
Item (index)	The item at the specified index
Method	
Add (object)	Adds the specified item to the list
Clear()	Removes all items from the list
Insert (index, object)	Inserts the specified item at the specified index
Remove (index)	Removes the item at the specified index
Remove (object)	Removes the specified item

Adding items to a ComboBox

You may add items to the Items collection in Design view using the GUI interface, or at runtime using code. Adding items at runtime makes future maintenance of the list easier.

To add items to the ComboBox in Design view, open the Items collection window by clicking on the ellipses on the right side of the Items property, as depicted in Figure 6-11.

Figure 6-11: Items collection of ComboBox

Next, type the items into the **String Collection Editor** window, and click OK. Figure 6-12 shows the window.

Figure 6-12: Window for adding items to a ComboBox

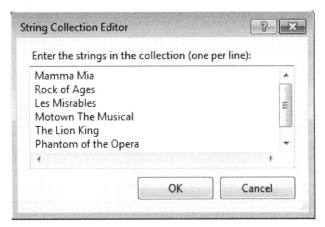

Sort property of ComboBox

You may sort the items in the collection by setting the Sort property to True. Sorting would make it easy for the user to select items from the list.

Step 1-15: Add a ComboBox named cboShows to the form (see Figure 6-10), and set its Text property to "Select a Show." Set the Sort property to True.
Open the Items collection window, type in the list of shows from Figure 6-12, and click OK.
Run the form and display the list. Note that the list is sorted. Stop the program.
Now, delete the items from the window so you can add items using code at runtime.

Adding Items to ComboBox at Runtime Using Code

To add items to the list in the ComboBox, use the Add method of the Items collection. You may add one item at a time, using a separate statement for each item, as shown in the following example:
cboShow.Items.Add("Mamma Mia");
cboShow.Items.Add("Rock of Ages");
…

However, it is more convenient to store the items in an **array** and use the **foreach** loop to add each item to the ComboBox. Unlike the simple variables you used in previous chapters, an array can store multiple data items of the same type. The foreach loop is a type of loop that can be conveniently used to access each data item stored in an array. Arrays and foreach loops are discussed in more detail in Chapter 7.

For example, the names of the shows can be stored in an array named **shows** as follows:
string[] shows = {"Mamma Mia","Rock of Ages", "Les Miserables",
"Motown The Musical","The Lion King", "The Phantom of the Opera"};

The following code gets each show from the array, stores it in the variable **show**, and adds it to the ComboBox:

```
foreach (string show in shows)
    cboShow.Items.Add(show);
```

The foreach loop loops through each data item from the array and adds the item to the ComboBox.

Step 1-16: Insert the code from Figure 6-13 to add the names of shows to the ComboBox. The code should be added to the Load event of the form so that the ComboBox will show the list when the form is displayed.

Figure 6-13: Code to add items to the ComboBox

```
34    private void TheaterTickets_Load(object sender, EventArgs e)
35    {
36        chkMember.CheckState = CheckState.Indeterminate;
37        rdbDefault.Checked = true;
38        string[] shows = { "Mamma Mia", "Rock of Ages", "Les Miserables",
39                           "Motown", "The Lion King", "The Phantom of the Opera" };
40        foreach (string show in shows)
41            cboShow.Items.Add(show);
42    }
```

Step 1-17: Run the form, and make sure the list is displayed in the sorted order when you click the arrow on the ComboBox.

ComboBox styles

You can choose one of three different styles for a ComboBox by changing the value of the DropDownStyle property shown in Figure 6-14.

Figure 6-14: DropDownStyle property

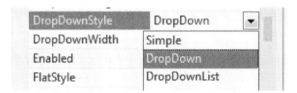

DropDown style

This is the default value of the DropDownStyle property. Setting the DropDownStyle property to DropDown style provides a dropdown list when the user clicks the arrow, as shown in Figure 6-12. The user also may type in an item that is not in the list.

Figure 6-15: The DropDown style

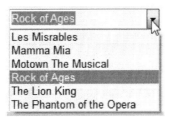

DropDownList style

This style is similar to DropDown style, except that the user cannot type in an item.

Simple style

With this style, the ComboBox does not have an arrow to provide a dropdown list. A ScrollBar is provided to view items, if any, that cannot be displayed in the available space. Figure 6-16 shows what Simple style looks like.

Figure 6-16: Simple style of ComboBox

Step 1-18: Change the DropDownStyle property of the ComboBox to DropDownList, run the form and observe the difference. Repeat for Simple style.

Getting the selected item

There are three major properties related to the item Selected in a ComboBox: SelectedIndex, SelectedItem and Text property.

SelectedIndex

SelectedIndex is the index of the selected item. If the first item is selected, the SelectedIndex would be 0; if the second item is selected, it would be 1, and so on.

If no item is selected, the SelectedIndex has the value -1. You may reset the ComboBox so that no item is selected by setting the SelectedIndex to -1, as follows:

    cboShow.SelectedIndex = -1;

It is important to note that the index of an item in the original list that was added to the ComboBox may not be the same as the index of the same item in the list displayed in a sorted ComboBox at runtime. For example, the index of Les Miserables in the original list was 2, whereas in the sorted list that is displayed in the ComboBox, it is 0. As items are added or deleted from the list, the index of the item changes. **Therefore, it is not reliable to use the SelectedIndex to identify the item selected by the user**.

SelectedItem

SelectedItem gives the item selected by the user. If no item is selected, its value is null.

Text property

The value of Text property is whatever is displayed in the TextBox part of the ComboBox, which could be an item selected by the user, the prompt specified in the Text property (e.g., "Select a Show"), or any text typed into the TextBox part of the ComboBox.

Step 1-19: Add the code from Figure 6-17 to the Click event handler of Clear button to reset the ComboBox.

Figure 6-17: Code to clear the ComboBox

```
24
25      private void btnClear_Click(object sender, EventArgs e)
26      {
27          hsbNumOfTickets.Value = 0;
28          chkMember.CheckState = CheckState.Indeterminate;
29          rdbDefault.Checked = true;
30          lblAmtDue.Text = "";
31          cboShow.SelectedIndex = -1;
32      }
```

Step 1-20: Add a button named btnWriteSales that writes the name of the selected show and the computed price to a ListBox. Add a ListBox named lstDisplay. Figure 6-18 shows the lower part of the form where button and ListBox are added.

Figure 6-18: TheTheaterTickets form with added Write button and ListBox

Step 1-21: Add the necessary code from Figure 6-19 to the Click event handler of btnWriteSales, to do the following tasks:
Verify that the user computed the price.
Verify that the user selected a Show.
Write the name of the selected show and the price to the ListBox.

Figure 6-19: Code to write the name of the show and price

```csharp
105   private void btnWriteSales_Click(object sender, EventArgs e)
106   {
107       if (cboShow.SelectedIndex == -1)
108       {
109           MessageBox.Show("Please select a show");
110           return;
111       }
112       if (amtDue == 0)
113       {
114           MessageBox.Show("Please compute amount");
115           return;
116       }
117       string fmtStr = "{0,-25}{1,12:C}";
118       lstDisplay.Items.Clear();
119       lstDisplay.Items.Add(String.Format(fmtStr, "Show", "Total Price"));
120       lstDisplay.Items.Add(String.Format(fmtStr, cboShow.SelectedItem, amtDue));
121   }
```

The two key statements related to the ComboBox are lines 107 and 120:
Line 107 checks whether the value of SelectedIndex is -1—that is, the user did not select a show.
Line 120 displays the name of the show, given by the SelectedItem property, and the price given by amtDue.

SelectedText Property

ComboBox also has a property called SelectedText. Like the SelectedItem property, the SelectedText property gives the name of the item selected by the user. However, if you select an item in a ComboBox and move focus away from the ComboBox by clicking a button, for example, the value of the SelectedText property in the Click event handler of the button will be an empty string. This is because the selection is cleared when the focus is moved away. Similarly, the SelectedText will be an empty string also within the SelectedIndexChanged and SelectedValueChanged event handlers (discussed later in this section) because at the time of these events, the selected value has not yet been set. So, the SelectedText property is not suitable to get the selected item from an event handler.

Accessing Items from the Items Collection

Any item in the collection of items in a ComboBox may be accessed using its index number. For example,
  cboShow.Items[0] gives the first item in the list, and
  cboShow.Items[1] gives the second item in the list.
To access every item in a ComboBox, you may use a loop to vary the index over the entire range, as shown below:
  for (int index=0; index < cboShow.**Items.Count**; index++)
    lstDisplay.Items.Add (cboShow.Items[index]);

In the above code, cboShow.**Items.Count** property gives the number of items in the Items collection. Because the index starts at zero, the index of the last item is one less than the count of items.

An alternate method that does not require the count of items is to use the foreach loop. The following code would display each show from the Items collection of the ComboBox, cboShow:

    foreach (string show in cboShow.Items)
        lstDisplay.Items.Add(show);

SelectedIndexChanged event

When the user selects an item from a ComboBox, the action triggers the SelectedIndexChanged event. You can use this event to take additional actions. For example, when the user selects a product from a ComboBox, the SelectedIndexChanged event handler can look up the price for that product.

Consider another revision to the TheaterTicket form where the name of the show and the price are displayed automatically in the ListBox, when the user selects a show, without having to click the Write Sales button. To do this, we add the code to the SelectedIndexChanged event handler of the ComboBox, shown in Figure 6-20.

Figure 6-20: SelectedIndexChanged event handler of ComboBox

```
123   private void cboShow_SelectedIndexChanged(object sender, EventArgs e)
124   {
125       if (amtDue == 0)
126       {
127           MessageBox.Show("Please compute amount");
128           return;
129       }
130       string fmtStr = "{0,-25}{1,-12:C}";
131       lstDisplay.Items.Clear();
132       lstDisplay.Items.Add(String.Format(fmtStr, "Show", "Total Price"));
133       lstDisplay.Items.Add(String.Format(fmtStr, cboShow.SelectedItem, amtDue));
134   }
```

This code is the same as in the Click event handler of the Write Sales button (Figure 6-19), except that this code does not check whether the user selected a show. Note that with this approach, you do not need the Write Sales button and code in the Click event handler of the button.

Step 1-22: Double click the ComboBox to create the SelectedIndexChanged event handler of the ComboBox. Add the code from Figure 6-20 to the event handler and test it.

Quite often, items are added to ComboBoxes by reading data from a text file or database. Reading from text file is discussed in Chapter 8, and getting data from a database is discussed in Chapter 12.

The following code shows the complete revised code that includes selecting a show using the ComboBox and displaying it. The code in the Click event handler of btnWrite is not shown, because the code in the SelectedIndexChanged event handler of the ComboBox does the same function.

```csharp
19         float amtDue;
20         private void hsbNumOfTickets_ValueChanged(object sender, EventArgs e)
21         {
22             lblNumOfTickets.Text = hsbNumOfTickets.Value.ToString();
23         }
24
25         private void btnClear_Click(object sender, EventArgs e)
26         {
27             hsbNumOfTickets.Value = 0;
28             chkMember.CheckState = CheckState.Indeterminate;
29             rdbDefault.Checked = true;
30             lblAmtDue.Text = "";
31             cboShow.SelectedIndex = -1;
32         }
33
34         private void TheaterTickets_Load(object sender, EventArgs e)
35         {
36             chkMember.CheckState = CheckState.Indeterminate;
37             rdbDefault.Checked = true;
38             string[] shows = { "Mamma Mia", "Rock of Ages", "Les Miserables",
39                     "Motown", "The Lion King", "The Phantom of the Opera" };
40             foreach (string show in shows)
41                 cboShow.Items.Add(show);
42         }
43
44         private void btnComputePrice_Click(object sender, EventArgs e)
45         {
46             // Verify that the user selected number of tickets
47             if (hsbNumOfTickets.Value == 0)
48             {
49                 MessageBox.Show("Please specify number of tickets");
50                 return;
51             }
52             // Verify that the user selected a section
53             if (chkMember.CheckState == CheckState.Indeterminate)
54             {
55                 MessageBox.Show("Please specify membership");
56                 return;
57             }
58             // Verify that the user selected membership status
59             if (rdbDefault.Checked == true)
60             {
61                 MessageBox.Show("Please specify Section");
62                 return;
63             }
64             // Get number of tickets
65             int numOfTickets = hsbNumOfTickets.Value;
66
67             // Compute discount
68             float discount = Discount();
69
70             // Get unit price
71             float unitPrice = UnitPrice();
72
73             amtDue = numOfTickets * (unitPrice - discount);
74             lblAmtDue.Text = amtDue.ToString("$#,###.00");
75         }
```

```csharp
 76
 77  float Discount()
 78  {
 79      switch (chkMember.CheckState)
 80      {
 81          case CheckState.Checked:    // or, case 0
 82              return 20;
 83              break;
 84          default:
 85              return 0;
 86              break;
 87      }
 88  }
 89
 90  float UnitPrice()
 91  {
 92      if (rdbSec1.Checked)
 93          return 70;
 94      else if (rdbSec2.Checked)
 95          return 60;
 96      else if (rdbSec3.Checked)
 97          return 52;
 98      else if (rdbSec4.Checked)
 99          return 45;
100      else
101          return 0;
102  }
121
122  private void cboShow_SelectedIndexChanged(object sender, EventArgs e)
123  {
124      if (amtDue == 0)
125      {
126          MessageBox.Show("Please compute amount");
127          return;
128      }
129      string fmtStr = "{0,-25}{1,-12:C}";
130      lstDisplay.Items.Clear();
131      lstDisplay.Items.Add(String.Format(fmtStr, "Show", "Total Price"));
132      lstDisplay.Items.Add(String.Format(fmtStr, cboShow.SelectedItem, amtDue));
133  }
```

Review Questions

6.8 A comboBox, cboNames, shown below, is used to display a set of student names, and let the user select a name.

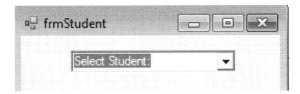

Which property of the ComboBox represents the currently selected name if the user selected an item, and would be empty if the user had not selected an item?

6.9 A comboBox, cboNames, shown below, is used to display a set of student names, and let the user select a name.

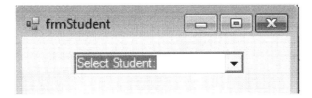

Write the statement to reset the ComboBox so that no student is selected.

6.10 A sign-up form for the activities of a club requires the user to specify whether the user is a member of the club. The prompt for the user on the form is

    Are you a member of the club?

A control (or controls) is to be placed to the right of the above prompt to let the user specify the membership status. Based on standard practice, which one of the following control(s) would you choose?

- A CheckBox
- Two CheckBoxes (one with a "Yes" label and the other with a "No" label)
- A RadioButton
- Two RadioButtons (one with a "Yes" label and the other with a "No" label)
- A TextBox

6.7 Working with ListBoxes

An important difference between a ComboBox and a ListBox is that the ListBox allows the user to select more than one item from the list.

Table 6-3 presented the commonly used properties and events of ComboBoxes and ListBoxes, which included the SelectionMode property that determines whether more than one item can be selected from a ListBox. Properties and methods of the Items collection, which is an important property of ListBoxes, are presented in Table 6-4.

A Set of CheckBoxes versus a Single ListBox

When do you use a set of ListBoxes rather than a ListBox to select multiple items from a list? Again, the guideline is to use a ListBox if the list is large and/or dynamic. A list is dynamic when the content of the list or the number of items in the list is likely to change. Thus, a ListBox is preferable to select multiple ingredients for a pizza, if the list of ingredients is dynamic.

SelectionMode Property

To enable selection of multiple items, the SelectionMode property shown in Figure 6-21 needs to be set to MultiSimple or MultiExtended. The two settings are similar, except that the MultiExtended setting makes it easier to select a set of consecutive items using the shift key.

Figure 6-21: SelectionMode property of ListBox

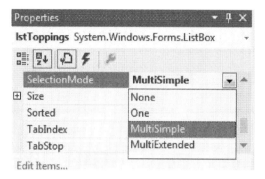

You already used ListBoxes to display output by adding text to the Items collection using the Add method, as in

    lstDisplay.Items.Add ("Total Price: " + lblAmtDue.Text);

Similar to the ComboBox, the Add method adds a new item to the ListBox.

You also may add a list of items to the ListBox by storing the items in an array and then using a loop to add the items to the ListBox. For example, the following code stores a list of pizza toppings in an array named toppings, and adds them to a ListBox named lstToppings using a foreach loop.

```
string[] toppings = {"Pepperoni","Sausage", "Onion", "Green Pepper",
                     "Mushrooms", "Xtra Cheese", "Ham", "Anchovies"};
    foreach (string topping in toppings)   // Add toppings to ListBox
         lstToppings.Items.Add(topping);
```

To better understand the use of ListBoxes, let's look at a pizza order form that lets the user select the size of pizza and the toppings, compute the total price, and display the total price and the toppings selected, as shown in Figure 6-22.

Figure 6-22: The PizzaOrder form

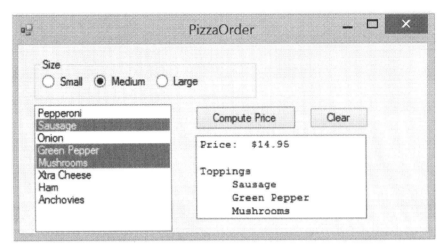

The list of toppings is added to the ListBox named lstToppings using code in the Load event of the form, shown in Figure 6-23.

Figure 6-23: Code to add toppings to the ListBox

```
19        // This form uses a ListBox to let users select multiple items from a list:
20        private void PizzaOrder_Load(object sender, EventArgs e)
21        {
22            string[] toppings = {"Pepperoni","Sausage", "Onion", "Green Pepper",
23                                 "Mushrooms", "Xtra Cheese", "Ham", "Anchovies"};
24            foreach (string topping in toppings)    // Add toppings to ListBox
25                lstToppings.Items.Add(topping);
26
27            rdbDefault.Checked = true;
28        }
```

In **line 27**, rdbDefault is a hidden RadioButton that makes it easier to reset the other RadioButtons and to check whether the user selected a size.

MultiColumn property

If there is not enough space to display the entire list of items in a ListBox, a vertical ScrollBar will be provided. If you want to avoid a vertical ScrollBar, you may set the MultiColumn property of the ListBox to True so that the items are displayed in multiple columns, as shown in Figure 6-24.

Figure 6-24: Display when MultiColumn property is set to True

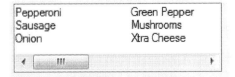

Step 1-23: Open the form named PizzaOrder. This form has the following controls. See Figure 6-22 for captions.

RadioButtons named rdbSmall, rdbMedium, rdbLarge, rdbDefault
ListBoxes named lstToppings and lstDisplay
Buttons named btnCompute, btnClear

Step 1-24: Set the **SelectionMode** property of lstToppings to **MultiSimple**.

Step 1-25: Open the load event handler of the form, which already has the code from Figure 6-23, to add the toppings to the ListBox. Review the code.

Run the form, and verify that toppings are displayed in the ListBox lstToppings and that you can select multiple items.

Getting the Selected Items

For ComboBoxes, the SelectedItem property gives the item selected by the user. How do we get the selected items from a ListBox when there are multiple items?

SelectedItems collection

The **SelectedItems** (note the "s" at the end that makes it plural) is the collection of all items selected by the user. You use a loop to get each item from the collection. The following code gets each selected item from the ListBox and displays it in another ListBox named lstDisplay:

```
foreach (string topping in lstToppings.SelectedItems)
     lstDisplay.Items.Add("   " + topping);
```

You may access any selected item from the SelectedItems collection by specifying the index. Here are some examples:

lstToppings.SelectedItems[0] gives the first selected item (Sausage)
lstToppings.SelectedItems[1] gives the second selected item (Green Pepper), etc.

Instead of using a foreach loop, you may use a for loop to vary the index using a for loop to access each selected item, and display it, as in the following code:

```
for (int index = 0; index < lstToppings.SelectedItems.Count; index++)
          lstDisplay.Items.Add("   " + lstToppings.SelectedItems[index]);
```

In the above code, **SelectedItems.Count** gives the number of items selected.

Items verses SelectedItems

Note that **Items** is the collection of all items in a ListBox, whereas **SelectedItems** is the collection of only selected items. Consider a ListBox with items selected, as follows:

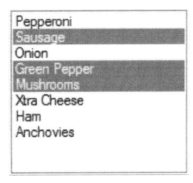

In this ListBox, Items[0] is the first item in the list (Pepperoni), but, SelectedItems[0] is the first selected item (Sausage).

Now, let's look at the code to compute the price based on the size selected and the number of toppings. Figure 6-25 shows the code, except for the statements to compute the total price and display selected items, which are left for you to complete.

Line 33 checks whether the user selected a size:
    if (rdbDefault.Checked == true),
If the default RadioButton is selected, that means none of the other RadioButtons that represent the size are selected.

Line 40, calls the GetPrice method, shown in Figure 6-25, to get the base price of the pizza with no toppings and the cost per topping:
    GetPrice(ref basePrice, ref costPerTopping);

Note that the GetPrice method returns the two values using reference parameters. You may split this method to two value returning methods named BasePrice that returns the base price, and ToppingPrice that returns the cost per topping. The statement in line 40 will have to be replaced by two statements that invoke these methods.

Step 1-26: Add the missing code in line 43 to compute the total price (base price + cost per toppings x number of toppings).
    Hint: Count property of SelectedItems collection gives the number of selected items.

Add the missing code in lines 54 and 55 to display the selected toppings in the ListBox, lstDisplay. You may use a foreach loop to access each item in the SelectedItems collection.

Figure 6-25: Code to compute price based on size and number of toppings

```csharp
            private void btnCompute_Click(object sender, EventArgs e)
            {
                // Verify that the user selected size:
                if (rdbDefault.Checked == true)
                {
                    MessageBox.Show("Please select size");
                    return;
                }
                // Get base price and cost per topping
                float basePrice = 0, costPerTopping = 0, price = 0;
                GetPrice(ref basePrice, ref costPerTopping);

                // Compute and display price:
                // (insert code to compute price: price = base price + cost of ingredients)

                lstDisplay.Items.Clear();
                lstDisplay.Items.Add ("Price:   " + price.ToString("C2"));
                // Display Toppings:
                lstDisplay.Items.Add("");
                lstDisplay.Items.Add ("Toppings");
                if (lstToppings.SelectedItems.Count == 0)  // if no item is selected
                    lstDisplay.Items.Add("    None");
                else
                {
                    // (insert code to display selected topics)

                }
            }

            private void GetPrice(ref float basePrice, ref float costPerTopping)
            {
                if (rdbSmall.Checked)
                {
                    basePrice = 6.95F;
                    costPerTopping = 1.5F;
                }
                else if (rdbMedium.Checked)
                {
                    basePrice = 8.95F;
                    costPerTopping = 2.0F;
                }
                else if (rdbLarge.Checked)
                {
                    basePrice = 11.95F;
                    costPerTopping = 2.5F;
                }
            }
```

Step 1-27: View the code in the Click event handler of Clear button, shown in Figure 6-26.

Figure 6-26: Code to reset controls

```
68
69      private void btnClear_Click(object sender, EventArgs e)
70      {
71          rdbDefault.Checked = true;
72          lstToppings.SelectedIndex = -1;
73          lstDisplay.Items.Clear();
74      }
```

Step 1-28: Run the form. Test the code by clicking the Compute button when size and/or toppings are not selected. Test it with proper selection of size and toppings.

ListView control

The ListView control is another commonly used control, which allows multiple data items to be displayed in a row. Because arrays make it easier to add data to this control, we will discuss this control in Chapter 8.

It's time to practice! Do Exercise 6-1.

6.8 Windows Presentation Foundation (WPF)

If you want to build applications with rich graphical user interfaces, Windows Presentation Foundation (WPF) is a Microsoft technology that can help you. The WPF features include extensive support for 2-D and 3-D graphics, animation, media and high quality text. Further, the layout system in WPF allows creating dynamic layouts that adjust to changes in window size and display settings. Additional components of WPF applications include

1. XAML (Extensible Application Markup Language): The user interface is created using XAML, which is a Microsoft XML-based markup language. Just like in a common Windows Form application, you can drag WPF controls from the Toolbox onto the Design widow, which will generate the corresponding XAML tags that specify the controls. You also may use XAML to create controls. XAML allows you to do extensive customization and nesting of controls. WPF controls also have properties and events that are not available in Windows Form controls.
2. Code-behind: Code-behind represents C#, Visual Basic and other .Net language programs that are attached to specific events of WPF controls, like the Click event of a button. This code is similar to the code in Windows Forms that you are familiar with.

To create a WPF project, select
   File, New, Project, C#, Windows, WPF Application

For tutorials and additional information on creating WPF applications, please see
   https://msdn.microsoft.com/en-us/library/aa970268(v=vs.110).aspx

Review Questions

6.11 You want to develop a form that allows the user to select one section of an introductory IS course from a list of all sections. The number of sections varies from semester to semester within the range 3 to 8. The prompt for the user on the form would be

   Please select a section.

   Based on the standard practice, which one of the following controls would you use to let the user select a section: a ListBox, a set of CheckBoxes, a set of RadioButtons, or a ComboBox?

6.12 It is desired to develop a form that allows the user to select one or more days from a list of all seven days of a week. The prompt for the user on the form would be

   Please select one or more days:

   Based on industry standards, which one of the following controls would you use to let the user select one or more days: a ListBox, a set of CheckBoxes, a set of RadioButtos, or a ComboBox?

6.13 A form displays a list of sports in a ListBox named lstSports. A user selects Basketball and Soccer from the list at runtime, as shown below:

   What would be the value of lstSports.SelectedItems(0)?

6.14 A form displays a list of sports in a ListBox named lstSports. A user selects Basketball and Soccer from the list at runtime, as shown below:

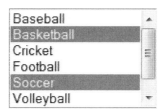

   What would be the value of lstSports.Items(2)?

Exercises

Exercise 6-1

Develop a form, Icecream.cs, as shown below:

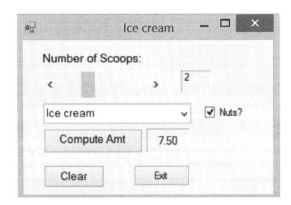

The form must let the user do the following:

Specify the number of scoops (max three) using a ScrollBar, and display the specified number in a Label named lblScoops.
Select the type of sundae from a list of sundaes (Ice Cream, Custard, Yogurt, etc.) displayed in a ComboBox sorted alphabetically. Because the items in this list may change over time, do not assume that, for example, the third item will remain as Yogurt.
Use a CheckBox to specify whether the user wants nuts for topping.
Compute the amount due.
Clear all data so that the user can make another set of selections. Exit the form.

The amount due is given by
Cost of Sundae (Ice cream - $3.50; Custard - $3:75; Yogurt - $4:00) * Number of Scoops + Cost of Nuts (0.50)

Do not use defaults for the items to be specified by the user. Defaults make it difficult to check whether the user forgot to select an item.

Follow the guidelines to make programs easy to understand, maintain, and reuse.

Exercise 6-2

A charitable organization uses volunteers to put on 10 to 15 different types of fund-raising events per year. Examples of the types of events are brat fry, walkathon, bake sale and golfing. A section of the volunteer signup form collects the following information:

 Name
 Types of events a person would like to volunteer for (a person may select one or more types)
 Number of hours available per week (The options are 5-10, 11-20, 21-30, 31-40.)
 Whether he/she is a new volunteer

Develop a form named Exercise 6.2 that collects the above information. Select appropriate controls based on the guidelines discussed in your textbook. Write the data entered by the user into a ListBox. If a volunteer is new, display the word "New"; if not, display the word "Experienced." A sample of the display is shown below:

Name: Sam Adamson

Events:
             Walkathon
             Golfing

Hours/week: 11-20

Status: New

Do not specify default values for any control. Defaults make it difficult to check whether the user forgot to select an item.

Chapter 7

Arrays

The variables we used so far can store only one value at a time. So, if you want to store 30 test scores in memory and do computations with them simultaneously, you would have to declare 30 different variables.

A single array allows you to store multiple data items of the same type. For example, a single one-dimensional array can store all 30 scores at the same time. Because one-dimensional (1-D) arrays are the simplest type of arrays, we will discuss them first.

Topics

7.1	Introduction to One-Dimensional Arrays	7.4	Accessing Elements of an Array Using foreach Loop
7.2	Assigning a Value to an Element of an Array	7.5	Copying an Array
		7.6	Looking Up Values in an Array
7.3	Accessing Elements of an Array Using the Index		

7.1 Introduction to One-Dimensional Arrays

An array can be thought of as a container of data with multiple storage locations or slots, one location for each data item. An array called testScore that stores 10 test scores may be represented as follows:

testScore

85	92	87	67	83	91	73	98	83	84
0	1	2	3	4	5	6	7	8	9

Element number:
(subscript or index)

Each storage location of the array is called an element. The elements are numbered consecutively, starting with zero. An individual element of an array is accessed using the array name, followed by the element number (subscript or index) in brackets. The subscript must be an integer. For example,

testScore[0] refers to the first element of the array, that is, the number 85,

testScore[1] refers to the second element of the array—that is, the number 92, and

testScore[9] refers to the tenth element of the array, the number 84.

Different arrays can hold data items of different types, including numbers, strings, arrays and objects. However, within an array, all elements must be of the same type. Thus, **an array is a collection of data items of the same type.**

An array, like testScore, which can store one set of data, is called a **one-dimensional** (1-D) array. A two-dimensional (2-D) array, discussed later in Chapter 8, can store multiple sets of data, all of the same type.

To help explain how to create an array, we will discuss another important property of arrays.

Value Types versus Reference Types

The integer, real and Boolean types we discussed in earlier chapters are called **value types** because each variable of these types holds the actual value assigned to that variable. Thus, the memory location allocated to a value type variable holds the actual value assigned to that variable.

By contrast, arrays (and other objects) are called **reference types** because each variable of these types holds only a reference to the data—that is, the memory location or address of the data, not the actual data. A variable that references an object, like an array, is called a **reference variable**. Let's first look at the value types through some examples.

Consider two value type variables, previousSales and currentSales, declared below:

> int previousSales, currentSales;

The effect of declaring the variables would be to store the default value zero in the memory locations for the two variables, as follows:

0		0
previousSales		currentSales

Assigning a value for currentSales using the statement, **currentSales = 245**; stores the value in the memory location for currentSales, as represented below:

0		245
previousSales		currentSales

When currentSales is assigned to previousSales using the statement

> previousSales = currentSales;

the value from currentSales is copied to the memory location for previouSales, so that **both locations store the same data,** as shown:

245		245
previousSales		currentSales

As discussed in Chapter 1, if you later change the value of currentSales using a statement like

> currentSales = 300;

it will have no effect on the value of previousSales. The locations for the two variables will appear as follows:

245		300
previousSales		currentSales

Now, let's look at how arrays differ.

Creating Arrays

Creating and using an array, which is a reference type, involves two steps:
1. Create the array object in memory to hold a set of data items, and
2. Create a variable, called reference variable, which references (holds the address of) the array, as shown in Figure 7.1

Figure 7-1: Array and reference variable

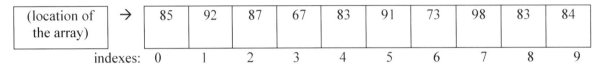

Here is a statement that creates an array called testScore:

    int[] testScore = new int[10];

The right hand side, **new int[10]**, creates an array in memory. Let's look at each part:

The key word **new** creates a new object (in memory) of the type specified by **int[10]**, and returns the memory location of the new array.

The expression **int[10]** specifies that the object is an array of int type, and that the **array size** is 10—that is, the array has 10 elements whose indexes ranges from 0 to 9. When a new array is created, each element has the default value zero.

The left-hand side, **int[] testScore**, declares a reference variable that we can use to access the array. The pair of square brackets after the variable type is what distinguishes it from a value type variable and specifies that it is a variable to reference an array.

The = operator stores a reference to the array (memory location of the array) in the reference variable testScores.

Figure 7-2 represents the effect of executing the statement, int[] testScore = new int[10], to create the array testScore and assign the reference to testScore.

Figure 7-2: Array and reference variable

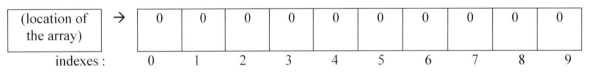

You may split the statement, int[] testScore = new int[10], to two separate statements:

    int[] testScore;
    testScore = new int[10];

The first statement, **int[] testScore,** declares testScore as a variable to reference an int type array. The second statement, **testScore = new int[10]**, creates an int type array of size 10 using the key word new. The = operator assigns the memory location of the new array to testScores.

The general syntax to create an array

The general syntax of the statement to create an array is

Type[] *VariableName* = new *Type*(*ArraySize*)

where *ArraySize* is the number of elements of the array, *Type* is the type of data to be stored in the array and *VariableName* is any valid variable name.

Alternatively, you may use two different statements:

Type[] *VariableName*

VariableName = new *Type*(*ArraySize*)

Tutorial 1: Working with 1-D Arrays: Test Scores Application

In this tutorial, you will work with simple one-dimensional arrays.

Step 1-1: Open the project, Ch7_Arrays, from the folder Tutorial_Starts.

Step 1-2: Open the form, Array_1D_Intro, shown below.

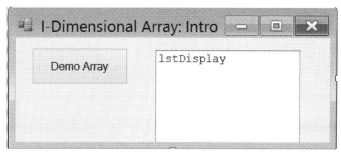

Step 1-3: Open the Click event handler of btnDemo.

Step 1-4: Add the statement to create an array testScores within the Click event handler of btnDemo, as shown in Figure 7-3, and put a break at the closing brace.

Figure 7-3: Code to create an array

```
19      private void btnDemo_Click(object sender, EventArgs e)
20      {
21          // Create array testScore with default values:
22          int[] testScore = new int[10];
```

Step 1-5: Run the project, and click the Demo button.

Step 1-6: When the program breaks, display the Locals window (Debug, Windows, Locals). Expand testScore to display the values of the 10 elements of the array corresponding to index 0 to 9, as shown in Figure 7-4.

Figure 7-4: Locals window displaying the array

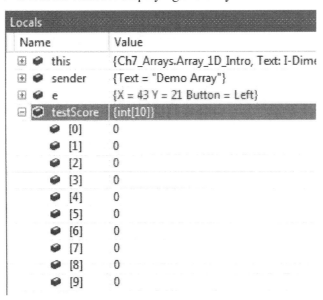

Creating Arrays with Initial Values Other Than Zero: Array Initializer

An array may be created with initial values other than the default of zeros, using **array initializer**, which is a comma-separated list of values enclosed in braces, as in

    int[] testScore = { 85, 92, 87, 67, 83, 91, 73, 98, 83, 84 };

This way of creating an array does not require the key word new, or the type and the array size specified on the right side of the = sign. The C# compiler infers the array size from the number of data items in the list and creates the array just like when the key word is used.

Alternately, you may split it to two statements as follows:

    int[] testScore;

    testScore = { 85, 92, 87, 67, 83, 91, 73, 98, 83, 84 };

Step 1-7: Comment out the statement int[] testScore = new int[10], and add the statement that creates the array using an array initializer instead of the key word new, as in Figure 7-5. Put a break at the closing brace of the event handler.

Figure 7-5: Use of the array initializer

```
19      private void btnDemo_Click(object sender, EventArgs e)
20      {
21          // Create array testScore with default values:
22          // int[] testScore = new int[10];
23
24          // Create array testScore using array initializer
25          int[] testScore = { 85, 92, 87, 67, 83, 91, 73, 98, 83, 84 };
```

Step 1-8: Run the project, and click the Demo button.

When the program breaks, display the Locals window (Debug, Windows, Locals). Expand testScore to display the values of the 10 elements of the array, as shown in Figure 7-6.

Figure 7-6: Locals window showing the effect of the array initializer

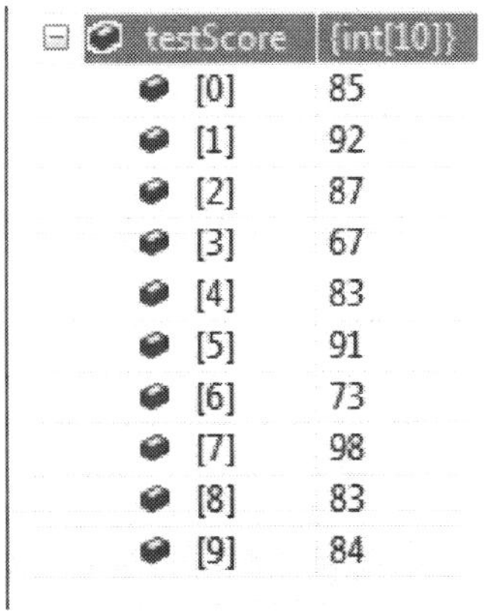

testScore	{int[10]}
[0]	85
[1]	92
[2]	87
[3]	67
[4]	83
[5]	91
[6]	73
[7]	98
[8]	83
[9]	84

Review Questions

7.1 How does a reference variable differ from a value type variable?

7.2 Consider the following code that computes and displays the total sales in a ListBox named lstDisplay:

> int sales=2, totalSales = 10;
> totalSales = totalSales + sales;
> sales = 3;
> lstDisplay.Items.Add(totalSales);

What would be the value of totalSales displayed in the ListBox?

7.3 Write a single statement to create an array that can hold a maximum of twenty student names referenced by a variable named student.

7.4 Write a statement to create a variable named student that can reference a string type array. Write a second statement to create an array that can hold a maximum of twenty student names referenced by student.7.5   Write a single statement to create an array consisting of the abbreviated names of the seven days of a week, referenced by a variable named weekDay.

7.2 Assigning a Value to an Element of an Array

Array initializer is a simple way to assign values to the elements of an array when the array is created. However, in applications where data is read from a source, like a file or database, typically data is read from its source and assigned to the elements of an array using a loop, as discussed later.

To help understand this concept, first we look at how to assign a value to an individual element of an array by specifying the index of the element. As an example, the statement

    testScore[2] = 99;

would assign the value 99 to the third element of the array testScore to replace the value 87.

You also may use a variable as the index, as in

    int index = 2;
    testScore[index] = 99;   // you may use any valid variable name to represent the index.

This method of specifying the index using a variable is used within a loop to assign values to every element of an array.

Step 1-9: Add the code from Figure 7-7 to the Click event handler to change the third test score to 99, and the first test score to 89.

Figure 7-7: Changing the values of array elements

```
19          private void btnDemo_Click(object sender, EventArgs e)
20          {
21              // Create array testScore with default values:
22              // int[] testScore = new int[10];
23
24              // Create array testScore using array initializer
25              int[] testScore = { 85, 92, 87, 67, 83, 91, 73, 98, 83, 84 };
26
27              // Change the first test score to 89, and the third test Score to 99:
28              testScore[0] = 89;
29              testScore[2] = 99;
```

Step 1-10: Run the project, and Click the Demo button.

Step 1-11: When the program breaks, display the Locals window (Debug, Windows, Locals). Expand testScore to display the values of the 10 elements of the array. The updated values of the array are shown in Figure 7-8.

Figure 7-8: Updated testScore

testScore	{int[10]}
[0]	89
[1]	92
[2]	99
[3]	67
[4]	83
[5]	91
[6]	73
[7]	98
[8]	83
[9]	84

Step 1-12: Add the necessary statement to change the last test score of the array to 79. Consider this your challenge—figure it out! Run the project, and verify that the score is changed to 79. Delete the added statement.

Chapter 8 shows you how to assign values to every element of an array by reading data items from a file using a loop.

Review Question

7.6   Write the statement to change the second element ("Mon") in the following array to "Monday":

   string[] weekDay = {"Sun", "Mon", "Tue", "Wed", "Thu", "Fri", "Sat"};

7.3 Accessing Elements of an Array Using the Index

As discussed earlier, an element of an array is accessed by specifying the corresponding index. For example, the following code displays the third element of testScore in a ListBox named lstDisplay:

   lstDisplay.Items.Add(testScore[2]);

Alternatively, you may use a variable to represent the index:

   int index = 2;
   lstDisplay.Items.Add(testScore[index]);

Step 1-13: Add the code from lines 31–33 in Figure 7-9 to display the third element of testScore.

Figure 7-9: Displaying an element of testScore

```
19      private void btnDemo_Click(object sender, EventArgs e)
20      {
21          // Create array testScore with default values:
22          // int[] testScore = new int[10];
23
24          // Create array testScore using array initializer
25          int[] testScore = { 85, 92, 87, 67, 83, 91, 73, 98, 83, 84 };
26
27          // Change the first test score to 89, and the third test Score to 99:
28          testScore[0] = 89;
29          testScore[2] = 99;
30
31          // Display 3rd element of the array
32          int index = 2;
33          lstDisplay.Items.Add(testScore[index]);
```

Step 1-14: Run the form to observe the output. You should see the score 87.

It's time to practice! Do Exercise 7-1.

Accessing Elements of an Array Using for Loops

Next, we will develop the code to display each element of the array, along with the value of its index, in the ListBox lstDisplay, as shown in Figure 7-10.

Figure 7-10: Using for Loop to display each element of an array

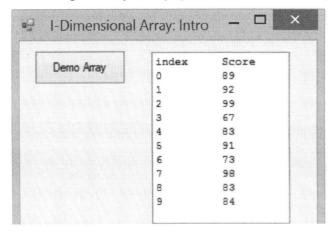

We use a **for loop** to change the value of the index of the element from 0 to 9. Note that the upper limit of the index (9) is equal to array size (10) minus 1.

Because an array is an object, it has different properties, including the **Length** property that gives the array size. Thus,

> testScore.Length

gives the array size (10). Note that the highest value of the index (9) is one less than the value of Length property (testScore.Length - 1).

The following code shows the **for** loop that changes the index from 0 to the upper limit, (testScores.Length - 1), and increments the index by 1 in each iteration.

> for (int index = 0; index <= testScore.Length-1; index = index + 1)
>   lstDisplay.Items.Add(index + "   " + testScore[index]);

Thus, first time through the loop, the index has the value 0, and the ListBox will display the value of testScore[0], and increment the index by 1.

The second time through the loop, the index has the value 1, and the ListBox will display the value of testScore[1].

The process continues while the value of the index is less than or equal to the upper limit of (testScore.Length - 1), which is equal to 9.

Step 1-15: Delete lines 31–33 in Figure 7-9, which displays a single element. Add code that displays every element of the array, as shown in Figure 7-11.

Figure 7-11: Use of for loop to display every element of an array

```
31          // Display each element of the array using a for loop
32          lstDisplay.Items.Add("index" + "   " + "Score");
33          for (int index = 0; index <= testScore.Length - 1; index = index + 1)
34              lstDisplay.Items.Add(index + "        " + testScore[index]);
```

Step 1-16: Run the form after removing the break, and observe the output.

Step 1-17: Put a break at line 34, and observe the value of the index during each iteration, and the value of testScore.Length property.

Watch Out for Out of Range Indexes!

A common source of error when using arrays is using a value for index that is outside the valid range. The statement

> lstDisplay.Items.Add(testScore[10]);   // invalid index

results in an error because 10 is outside the valid range of index for testScore, which is 0–9.

It is easy to make this error by mistakenly using the array size as the upper limit for index, as in the following statement:

> for (int index = 0; **index <= testScore.Length**; index = index + 1)
>   lstDisplay.Items.Add(index + "   " + testScore[index]);

In the above for loop, testScore.Length (10) is used as the upper limit for index, but the valid range of the index for testScore is 0–9.

Step 1-18: In line 33, change the upper limit for index to **testScore.Length** (from testScore.Length - 1). Put a break at line 34. Run the form, and iterate through the loop. Observe that when the value of index reaches 10, you get an error in line 34.

Step 1-19: Change line 33 back to the original form to specify (**testScore.Length - 1**) as the upper limit.

Using for Loop to Compute the Average Score

Let's modify the for loop that displays the scores, as shown in Figure 7-12, so that it computes the sum of all scores to compute the average.

Figure 7-12: Using for loop to compute the average

```
31          // Use a for loop to display each array element and compute the average score:
32          int totalScore = 0, currentScore;   //totalScore must be initialized to zero
33          lstDisplay.Items.Add("index" + "     " + "Score");
34          for (int index = 0; index <= testScore.Length - 1; index = index + 1)
35          {
36              currentScore = testScore[index];
37              lstDisplay.Items.Add(index + "         " + currentScore);
38              totalScore = totalScore + currentScore;
39          }
40          float averageScore = (float)totalScore / testScore.Length;
41          lstDisplay.Items.Add("Average:   " + averageScore.ToString("N"));
```

For each iteration through the loop, the first statement within the loop (line 36) stores the value of the current element in the variable currentScore so that it can be used in two different statements in lines 37 and 38 without having to access the array two times.

Step 1-20: Modify the Click event handler of the demo button so that it computes and displays the average score, as shown in Figure 7-12.

Step 1-21: Put a break on line 39, run the project, and observe the values of currentScore and totalScore in the first few iterations of the loop.

Review Questions

7.7   Write a statement to display the second element ("Mon") of the following array in a ListBox named lstDisplay:

   string[] weekDay = {"Sun", "Mon", "Tue", "Wed", "Thu", "Fri", "Sat"};

7.8   Use a **for** loop to display all elements of the following array in a ListBox named lstDisplay:

   string[] weekDay = {"Sun", "Mon", "Tue", "Wed", "Thu", "Fri", "Sat"};

7.9   Identify errors, if any, that would cause a runtime error in the second statement:

   string[] weekDay = {"Sun", "Mon", "Tue", "Wed", "Thu", "Fri", "Sat"};
   lstDisplay.Items.Add(testScore[7]);

7.4 Accessing Elements of an Array Using the foreach Loop

An alternative to using the for loop to access each element of an array is to use the foreach loop. In general, using the foreach loop is a convenient way to access individual items from any collection of items. This is how you would use a foreach loop to display each test score:

```
foreach (int score in testScore)
    lstDisplay.Items.Add(score);
```

In each iteration, the foreach loop would automatically copy the next element of the array into the **iteration variable** score, starting with the first element. Thus, the first time through the loop, score is assigned the first element (89) and it is displayed in the ListBox; the second time, score is assigned the second element (92); and the process continues until all scores are displayed.

Compare the foreach loop to the for loop introduced earlier:

```
for (int index = 0; index <= testScore.Length-1; index = index + 1)
    lstDisplay.Items.Add(index + "    " + testScore[index]);
```

Note that when using the **foreach** loop, there is no need to initialize a loop variable, increment it and specify the upper limit because it is not necessary to specify the index to access an element. Because no loop variable is used in the foreach loop, the output from the foreach loop, shown in Figure 7-13, does not include the value of the index.

Figure 7-13: Displaying each score using foreach loop

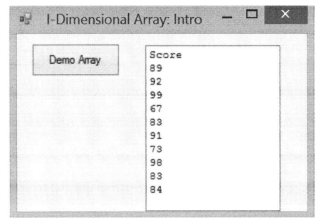

for versus foreach

Compared to the foreach loop, the for loop, in general, provides more flexibility in working with the elements. Within a for loop, you also may assign a value to an element of the array, and you can conveniently work with multiple arrays of the same size.

If the application requires accessing the elements of a single array sequentially without having to modify them, then it is mostly a matter of personal preference.

Now, let's modify the foreach loop, as shown in Figure 7-14, to compute the total score, and then compute the average.

Figure 7-14: Using foreach loop to compute the average

```
int totalScore = 0;
lstDisplay.Items.Add("Score");
foreach (int score in testScore)
{
        lstDisplay.Items.Add(score);
        totalScore = totalScore + score;
}
float averageScore = (float)totalScore / testScore.Length;
lstDisplay.Items.Add("Average:  " + averageScore.ToString("N"));
```

General syntax of foreach loop

```
Foreach (Type VariableName in ArrayName)
    {
        Statements
    }
```

Step 1-22: Comment out the code that uses the for loop to compute the average (lines 31–41 in Figure 7-12) and add the code from Figure 7-14, which uses a foreach loop.

Step 1-23: Run the project and test the code.

Should Array Names Be Singular or Plural?

Should the name of an array be singular like testScore or plural like testScores? Because an array holds multiple data items, the plural form might make more intuitive sense. When you use a foreach loop to access each element of the array, as in the following code, it is convenient to use the plural form (testScores) for the array and the singular form (testScore) for the variable representing an individual score:

```
foreach (int testScore in testScores)
    lstDisplay.Items.Add(testScore);
```

However, when you refer to an individual score using the index, the singular form of the noun makes more sense. For example, testScore[1] might be a clearer representation of a single score than testScores[1]. In addition, the singular form of the noun makes more sense in certain advanced uses of arrays. In this book, we use both forms.

Review Question

7.10   Use a **foreach** loop to display all elements of the following array, to a ListBox named lstDisplay:

```
string[] weekDay = {"Sun", "Mon", "Tue", "Wed", "Thu", "Fri", "Sat"};
```

7.5 Copying an Array

Because arrays are reference types, you cannot create a separate copy of an array by assigning the reference variable to another reference variable. For example,

    int[] temp1 = testScore;

will not create a copy of the array testScore. Instead, both temp and testScore will refer to the same array, as shown.

testScore → | 89 | 92 | 99 | 67 | 83 | 91 | 73 | 98 | 83 | 84 | ← temp1

That is, there is only one array, but two variables refer to it. That means if you change an element of testScore, in effect, you also are changing that element for temp1.

Step: 7-24: Add the following code to the end of the Click event handler of Demo button, and insert breaks at line 56 and 57:

```
53
54          // Assign testScore to another reference variable - doesn't copy the array
55          int[] temp1 = testScore;
56          testScore[9] = 69;
57      }
```

Step 1-25: Run the project. When the program breaks at line 56 (before testScore[9] is changed to 69), expand testScore and temp in Locals window (Debug, Windows, Locals) and observe that temp has the same values as testScore.

Step 1-26: Click Run to continue execution. When the program breaks at line 57, expand testScore and temp in Locals window (Debug, Windows, Locals) and observe that temp[9] has same values as testScore[9], as shown in Figure 7-15, though it looks like you changed the value of only testScore.

Figure 7-15: Changed value of temp1

temp1	{int[10]}
[0]	89
[1]	92
[2]	99
[3]	67
[4]	83
[5]	91
[6]	73
[7]	98
[8]	83
[9]	69

You may change any element of the array using temp1 or testScore. For example, temp1[8] would change the ninth element of the array that also is referenced by testScore.

Copying an Array Using Array.Copy Method

To create a separate copy of an array, you may use the **Array.Copy** method. The general syntax is

Array.Copy(source array, starting index in the source array, target array, starting index in target array, number of elements to copy)

Here is how you would copy the array referenced by testScore to another array referenced by temp2:

```
int[] temp2 = new int[10];
Array.Copy(testScore, 0, temp2, 0, 10);
```

If the entire array is to be copied, you may use the Length property in place of 10:

```
int[] temp2 = new int[testScore.Length];
Array.Copy(testScore, 0, temp2, 0, testScore.Length);
```

You may copy the last four elements of testScore (starting at index 6) to a new array, temp3, as follows:

```
int[] temp3 = new int[4];
Array.Copy(testScore, 6, temp3, 0, 4);
```

Step 1-27: Add the code from Figure 7-16 to copy the entire array testScore to another array temp2, and to copy the last four elements of testScore to temp3.

Figure 7-16: Using Array.Copy to copy an array

```
57
58          //Copy the entire array testScore using Array.Copy method to temp2:
59          int[] temp2 = new int[10];
60          Array.Copy(testScore, 0, temp2, 0, 10);
61          // copy the last four elements of testScore to temp3:
62          int[] temp3 = new int[4];
63          Array.Copy(testScore, 6, temp3, 0, 4);
64      }
```

Step 1-28: Run the project. When the program breaks at end of the method (line 64), display the Locals window and observe temp2 that should look like testScore, and temp3 that should display the values shown in Figure 7-17.

Figure 7-17: Using Array.Copy to copy a subset of the elements of an array

temp3	{int[4]}
[0]	73
[1]	98
[2]	83
[3]	69

Because the Array.Copy method makes a copy of the array referenced by testScore and references the copy by temp2, as shown in Figure 7-18, changes in one array do not affect the other array.

Figure 7-18: temp2 created by copying testScore

testScore

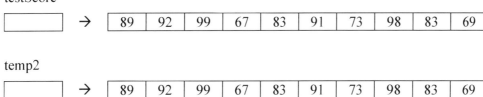

temp2

For example, executing the statement

testScore[9] = 99

changes the last element of testScore, but it does not affect temp2 or temp3.

However, temp1 is changed because temp1 was assigned the reference to testScore using the following statement; thus, both temp1 and testScore point to the same array:

int[] temp1 = testScore;

Step 1-29: Add the following code to change the last element of testScore to 99.

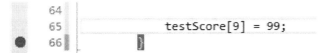

Step 1-30: Run the project. When the program breaks at line 66, display Locals window and observe that the last element of testScore and temp1 is 99, but the last element of temp2 and temp3 is unchanged.

Copying an Array Using a for Loop

You also may use a for loop to copy the elements of an array to another array. Here is the code that copies the elements of testScore to an array named temp4:

int[] temp4 = new int[10]; // or, int[] temp4 = new int[testScore.Length]
for (int index = 0; index <= testScore.Length - 1; index = index + 1)
    temp4[index] = testScore[index];

Here, the index is varied from 0 to the upper limit to access each element. So, the statement within the loop is equivalent to 10 different statements:

temp4[0] = testScore[0];
temp4[1] = testScore[1];
temp4[2] = testScore[2];
etc.

Step 1-31: Add the following statements to copy the elements of testScores to temp4:

```
66
67          // copy testScore to another array using for loop:
68          int[] temp4 = new int[10];   // or, int[] temp4 = new int[testScore.Length]
69          for (int index = 0; index <= testScore.Length - 1; index = index + 1)
70              temp4[index] = testScore[index];
71
```

Step 1-32: Run the project. When the program breaks, verify that temp4 has the same values as testScore.

Again, because testScore and temp4 are two separate arrays, the two arrays can be changed independently without affecting each other.

We will discuss one-dimensional arrays further, and introduce two-dimensional arrays, in the next chapter that discusses accessing data from files and storing in arrays.

Review Questions

7.11   The general syntax of the Array.Copy method is

Array.Copy(source array, starting index in the source array, target array,
starting index in target array, number of elements to copy)

Write a statement to copy all days, except the weekends (Sat, Sun), from the following array to another array named workDay.

string[] weekDay = {"Sun", "Mon", "Tue", "Wed", "Thu", "Fri", "Sat"};

7.12   Consider the following code that uses a ListBox named lstDisplay to display an element of an array:

string[] weekDay = {"Sun", "Mon", "Tue", "Wed", "Thu", "Fri", "Sat"};
string[] DaysOfWeek = weekDay;
weekDay[1] = "Monday";
lstDisplay.Items.Add(DaysOfWeek[1]);

What would be the value displayed in the ListBox?

7.13   Consider the following code that uses a ListBox named lstDisplay to display an element of an array:

string[] weekDay = {"Sun", "Mon", "Tue", "Wed", "Thu", "Fri", "Sat"};
string[] DaysOfWeek = new string[7];
Array.Copy(weekDay, 0, DaysOfWeek, 0, 7);
weekDay[1] = "Monday";
lstDisplay.Items.Add(DaysOfWeek [1]);

What would be the value displayed in the ListBox?

7.6 Looking Up Values in an Array

One-dimensional arrays can be used conveniently to lookup values, like finding the price for a specified item or looking up the phone number of a person.

Consider a string array of names of months and another int array that stores the units sold for a product in each month, created as shown in Figure 7-19. It should be noted that in real-world applications, such data often are read from a database table or file, rather than hard-coded.

Figure 7-19: Code to create arrays, months and unitsSold

```
string[] months = { "January", "February", "March", "April", "May", "June", "July", "August",
    "September", "October", "November", "December" };

int[] unitsSold = { 572, 645, 693, 564, 580, 756, 960, 756, 663, 678, 745, 960};
```

To find the units sold in a specified month, first find the index of the month from the months array. Then, find the data item stored in unitSold array at that index position. This method works only if the **months and units sold are stored in the same sequence** (that is, the first element of unitsSold corresponds to the first month in months array, the second element in unitsSold corresponds to second element in months array, etc.).

For example, to find the units sold in March, you find the index of March, which is 2, and then find the data item at the index position 2 in unitsSold array, which is 693. An easy way to find the index of a specific data item in an array is to use the Array.IndexOf method.

Array.IndexOf Method

The Array.IndexOf method returns the index of a specified data item in a 1-D array.

This method has the syntax

   Array.IndexOf(ArrayName, SearchValue)

For example,

   Array.IndexOf(months, "March")

ArrayName is the name of a 1-D array, and SearchValue is the value you are trying to find in an array. The method returns the index of the first element that matches the specified value. If no value is found, the method returns -1. Because months and units sold are stored in the same sequence, you can use the index of the month to find the unitsSold.

Tutorial 2: Lookup Units Sold Using 1-D Arrays

This tutorial creates a form named LookUpUsingArray that allows the user to look up the sales in a specified month, as shown in Figure 7-20.

Figure 7-20: LookUpUsingArray form

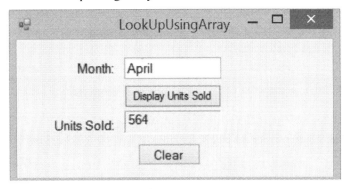

Step 2-1: Open the form named LookupUsingArray within the project Ch7_Arrays.

Open the code window. The two statements in lines 19–21 create two arrays, months and unitsSold at the class level, as shown in Figure 7-21. These arrays are created at the class level, rather than within the Click event handler of btnDisplayUnitsSold, so that they don't have to be created repeatedly each time the user clicks the button to find the sales.

Step 2-2: Add the code shown in lines 22–37 in Figure 7-21 to the Click event handler of btnDisplayUnitsSold to display units sold in a specified month.

Figure 7-21: Code to find the units sold

```
19    string[] months = { "January", "February", "March", "April", "May", "June", "July",
20                        "August", "September", "October", "November", "December" };
21    int[] unitsSold = { 572, 645, 693, 564, 580, 756, 859, 756, 663, 678, 745, 958 };
22
23    private void btnDisplayUnitsSold_Click(object sender, EventArgs e)
24    {
25        string monthName = txtMonthName.Text;
26        int monthIndex, unitsSoldInMonth;
27
28        // Find the index of the month. Use the index to find units sold
29        monthIndex = Array.IndexOf(months, monthName);
30        if (monthIndex >= 0)   // if month is not found, index would be -1
31            unitsSoldInMonth = unitsSold[monthIndex];   // find units sold
32        else
33        {
34            MessageBox.Show("Month not found");
35            return;
36        }
37        lblUnitsSold.Text = unitsSoldInMonth.ToString();
```

Line 29 finds the index of the specified month from the months array using Array.IndexOf method, described earlier.

Line 31 finds the units sold from unitsSold array, using the index of the specified month.

Step 2-3: Run the form and test the code.

Looking Up Values in an Array Using a Loop

An alternative to using the Array.IndexOf method to find the index of a data item is to use a loop to compare each element of the array to the data item whose index is to be found. For example, to find the index of "March" in the months array, compare each month in the months array to "March" until a match is found or all elements are compared. Figure 7-22 shows the code that uses a for loop.

Figure 7-22: Look up sales using a for loop

```
38
39      // Alternate method using a loop:
40      // Compare each month in the array to the specified month
41      int index = 0;
42      for (index = 0; index < months.Length; index++)
43      {
44          if (months[index] == monthName)
45          {
46              unitsSoldInMonth = unitsSold[index];
47              lblUnitsSold.Text = unitsSoldInMonth.ToString();
48              break; // if month is found jump out of the loop
49          }
50      }
51      // Diplay a message, if month was not found
52      if (index >= months.Length)
53      {
54          MessageBox.Show("Month not found");
55          return;
56      }
```

Lines 42 and 44: The for loop in line 42 varies the value of index from 0 to 11 in increments of 1, so that the statement in line 44,

    if (months[index] == monthName)

would be equivalent to 12 different statements:

    if (months[0] == monthName)
    if (months[1] == monthName)
    ...
    if (months[11] == monthName).

Line 48: If the specified month matches a month in the array, the *break* statement in line 48 transfers control out of the loop.

Line 52: If the specified month doesn't match any month in the array, the value of index would be equal to the size of the array (12), given by the Length property. Line 52 checks whether index is greater than or equal to the array size.

Step 2-4: Comment out the statements in lines 29-37, and add the code from lines 38-55 in Figure 7-22. Run the form and test the code.

It's time to practice! Do Exercises 7-2 and 7-3.

Exercises

Exercise 7-1

Open the project Ch7_Arrays from the folder Tutorial_Starts. Open the form named Exercise 7_1. Add the necessary code to the Click event handler of the btnDisplayMonth to display the name of the month for the month number typed in by the user. For example, if the user types in the number 3, the code should display "March," as follows:

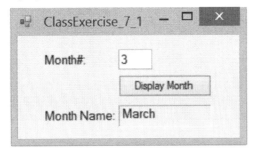

Hint: Create a string array with the twelve month names (January through December) assigned to it as the initial values. To avoid creating the array each time the button is clicked, create the array at the class level outside the Click event handler. Use the array within the Click event handler to find and display the month name.

Exercise 7-2

Part A: This exercise finds and displays the month(s) with the highest sale. Add a button named btnTopMonths to the LookUpUsingArray form that you used in Tutorial 2 . Add the necessary code to the Click event handler of the button to find and display the highest sale, and the month(s) with the highest sales, in a ListBox. You may use the following method:

First, find the highest sale by looping through the unitSold array, as follows:

> Set highest sale to the first element of unitsSold (int highestSale = unitsSold[0];)
> Set highestSalePosition to zero.
> Use a loop (that initializes and increments the index) to compare each array element to the highest sale. If the array element is greater than highest sale, update highest sale and highestSalePosition, using the following logic:
>
>> if (array element > highestSale)
>> set highestSale to the current element of unitsSold
>> set highestSalePosition to the current value *of index*

Next, display the highestSale and the month at the highestSalePosition.

Part B: This exercise is similar to Part A, except that it finds and displays the month(s) with the lowest sale. Add a button named btnBottomMonths to the LookUpUsingArray form that you used in Tutorial 2. Add the necessary code to the Click event handler of the button to find and display the lowest sale, and the month(s) with the lowest sales, in a ListBox. You may use the following method:

First, find the lowest sale by looping through the unitSold array, as follows:
> Set lowest sale to the first element of unitsSold (int lowestSale = unitsSold[0];)
> Set lowestSalePosition to zero.
> Use a loop (that initializes and increments the index) to compare each array element to the lowest sale. If the array element is less than lowest sale, update lowest sale and lowestSalePosition, using the following logic:
>> if (array element < lowestSale)
>> set lowestSale to the current element of unitsSold
>> set lowestSalePosition to the current value of index

Next, display the lowestSale and the month at the lowestSalePosition.

Exercise 7-3

The TheaterTicketsWithArray form in Tutorial_Starts/Ch7_Arrays project performs essentially the same function as TheaterTickets form from Ch6_GUIcontrols. Rather than using a button to compute the cost, TheaterTicketsWithArray form combines computing cost and displaying cost into the SelectedIndexChanged event handler of the ComboBox. In addition, this form doesn't use a separate method to compute the discount.

Modify this form so that the amount due includes a markup for the selected show. When the user selects a show from the ComboBox, add the markup for that show to the amount due and display the amount. You may use the following approach:

Add the necessary statement to the form to create a second array named markups that contains the markup amount (a positive or negative amount) for each show.

Add the code to the SelectedIndexChanged event handler of the ComboBox to find the markup amount for the selected show, and add the amount to the amount due. Display the amount in the ListBox. Make sure that the reference variable for the array, shows, is declared at the class level so that it can be used in the Load event of the form and in the SelectedIndexChanged event handler of the ComboBox.

Chapter 8

Sequential Files and Arrays

Data stored in variables and arrays are lost when the program terminates. **Sequential Access Files** provide a relatively easy way to store data for longer terms in devices like hard disks, magnetic tapes and flash drives. This chapter discusses writing data into sequential files and reading from files. Further, we look at reading data from files into one-dimensional and two-dimensional arrays.

Topics

8.1	Introduction to Text Files	8.5	Writing to Text Files
8.2	Splitting a Row: Split Method	8.6	SaveFileDialog Control
8.3	Reading Data from Files into Arrays	8.7	Passing Arrays to Methods
8.4	Additional Methods of Arrays	8.8	Two-Dimensional (2-D) Arrays

8.1 Introduction to Text Files

In a sequential access file, data are accessed sequentially, starting with the first set of data items, followed by successive sets. To find a particular data item, you need to go over all data that precedes it, like playing a song on a cassette tape. By contrast, data may be accessed directly from a **direct access** (**random access**) file, similar to playing a song on a CD.

Sequential files are of two types: **text files** and **binary files**. In a text file, all data, including numbers and dates, are stored as text. For example, the number 9.95 is stored as a combination of four characters (three digits and a decimal) that forms a string. Hence, text files can be created and edited using a text editor like Notepad. Binary files, discussed in **Appendix A**, store data in its original form without converting it to text. The number 9.95 is stored as single number using the standard internal representation of numbers.

Reading Lines from Text Files

Typically, text files store each record in a separate line, with the data items separated by delimiters like commas, tabs, or the pipe character (|). Because text files are sequential access files, data are read sequentially from the beginning to the end of the file.

Example 1: Reading and Displaying Exam Scores

Here is an example of a text file that contains the name and two exam scores of students:
Mike Burns, 75, 92
Cathy Arnold, 86, 95
Wesley Mathews, 82, 90
Jerry Luke, 70, 89

    Gina Rodriguez, 77, 88

You can create such a file in a text editor or using a program. In either case, an invisible new line character is inserted automatically at the end of each line.

Let's see how you read the above data from a text file, named **Exams.txt**. You use the .txt extension, which is used commonly to indicate that the file contains text data.

A variety of classes are available in the **System.IO** namespace of Framework Class Library to work with files and directories. You use the **StreamReader** class to read data from files. The data in the text file is viewed as a stream of data—hence the name StreamReader.

To read from a text file, you create a StreamReader object using the StreamReader class. The statement to create the StreamReader object has the following syntax:

    StreamReader VariableName = new StreamReader(@"Path and FileName");

The new key word creates the StreamReader object to read the specified file, and returns a reference to the object. The reference (address) of the object is stored in the variable specified on the left side.

Specifying the Path for a File

If no path is specified, the default path, **ProjectDirectory\Bin\DeBug**, is assumed. You do not need to use the prefix @ if no path is specified. If the file is not found, an exception is thrown.

For example, a StreamReader object to read from E:\Csharp\Exams.txt is created as follows:

    StreamReader examsReader = new StreamReader(@"E:\Csharp\Exams.txt");

If the file is in the default directory, Bin\Debug, you skip the path:

    StreamReader examsReader = new StreamReader("Exams.txt");

If the file is stored in a folder above the default folder (Bin\Debug), you may use the code "..\\" for each level. Here are examples:

    "..\\Exams.txt"—the file Exams.txt is in the Bin folder that is one level above the default folder, Debug.
    "..\\..\\Exams.txt"—the file is in the project folder that is two levels above the default folder, Debug.

Using Directive for System.IO

Because StreamReader class is in the System.IO namespace, you must add the following using directive to the form:

    using System.IO;

ReadLine method

The StreamReader provides a method called **ReadLine** to read the line at the current read position and then advance the position to the next line. This method is used within a loop to sequentially read each line of data. This is how you would use a loop to read each line and display the line in a ListBox named lstDisplay:

```
string currentLine;
while (examsReader.EndOfStream == false)  // or, !examsReader.EndOfStream
{
        // Read current line and store it in a string variable
        currentLine = examsReader.ReadLine();
        lstDisplay.Items.Add(currentLine);          // Display the line
}
examsReader.Close();
```

EndOfStream property

The EndOfStream property of StreamReader has the value true when the current read position is at the end of the file. The header of the while loop in the above code,
    while (examsReader.EndOfStream == false),
uses the **EndOfStream** property to terminate the loop when EndOfStream is true. That is, the loop continues while EndOfStream is false.

The statement
    currentLine = examsReader.ReadLine();
reads the current line and stores the entire line as a single string in the string variable, currentLine. At the end of the first iteration through the loop, the string variable currentLine would have the following value:
    "Mike Burns, 75, 92"

Thus, the loop reads each line, stores it in the string variable currentLine, and then displays it to give the output shown in Figure 8-1.

Figure 8-1: Data from the Exams.txt file

```
Mike Burns, 75,   92
Cathy Arnold, 86, 95
Wesley Mathews, 82, 90
Jerry Luke, 70, 89
Gina Rodriguez, 77, 88
```

Tutorial 1: Reading Text Files

This tutorial reads and works with the name and exam scores from the file Exams.txt.

Step 1-1: Open the project Ch8_FilesAndArrays from the Projects_Starts folder. Open the form ReadTextFiles. Figure 8.2 shows the form.

Figure 8-2: The ReadTextFiles form

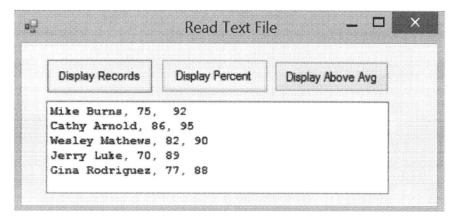

Step 1-2: Open the file Exams.txt from the Bin\Debug folder in **Ch8_FilesAndArrays** project directory, and view the records. Close the file.

Step 1-3: Make sure that the System.IO namespace is added, as shown in Figure 8-3.

Figure 8-3: The using directives with System.IO namespace

```
1  using System;
2  using System.Collections.Generic;
3  using System.ComponentModel;
4  using System.Data;
5  using System.Drawing;
6  using System.Linq;
7  using System.Text;
8  using System.Threading.Tasks;
9  using System.Windows.Forms;
10 using System.IO;    //Add this statement
11 namespace Ch8_FilesAndArrays
```

Step 1-4: Add the code from Figure 8-4 to the Click event handler of btnDisplayRecords. Please note that, in line 32, the name of the ListBox, lstDisplay, starts with the letter "l," not the number 1, though the font makes it look like number 1.

Figure 8-4: Code to read and display each line from Exams.txt

```csharp
20      private void btnDisplayRecords_Click(object sender, EventArgs e)
21      {
22          string currentLine;
23          // Create StreamReader object to read Exams.txt
24          StreamReader examsReader = new StreamReader("Exams.txt");
25
26          lstDisplay.Items.Clear();
27          while (examsReader.EndOfStream == false)
28                  // or, while (!examsReader.EndOfStream)
29          {
30              // Read the line at the current read position
31              currentLine = examsReader.ReadLine();
32              lstDisplay.Items.Add(currentLine); // Display current line
33          }
34          examsReader.Close();
35      }
```

Step 1-5: Make sure the Program.cs file specifies ReadTextFile as the startup form.

Put a break at line 33, and run the project. Click the Display Records button.

When the program breaks at the end of the loop (line 33), display the Locals window (Debug, Windows, Locals), and observe the value of currentLine for each iteration. Observe the value of EndOfStream property of examsReader after the fourth iteration (false), and after the fifth iteration, when it has the value, true, as shown in Figure 8-5.

Figure 8-5: The EndOfStream Property

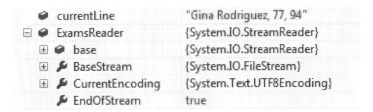

Review Questions

8.1   What namespace contains the classes that work with text files?

8.2   Describe what happens when the ReadLine method of a StreamReader is executed.

8.2 Splitting a Row: Split Method

The ReadLine method reads an entire row as a single string. How do you get the name and the two scores from the string so that you can compute the percent score for each student and display it, as shown in Figure 8-6?

Figure 8-6: The revised ReadTextFiles form with percent scores

```
Read Text File

[Display Records]  [Display Percent]  [Display Above Avg]

Name             Exam1  Exam2  Percent
Mike Burns        75     92     83.5
Cathy Arnold      86     95     90.5
Wesley Mathews    82     90     86.0
Jerry Luke        70     89     79.5
Gina Rodriguez    77     88     82.5
```

To get individual data items from a line, you use the **Split** method of the StreamReader that splits a string and **returns a string array** containing individual data items. This method uses a specified delimiter character (a character that separates the different data items in a row) to split the string. The syntax is

VariableName = StringVariableName.Split('delimiter character');

For example,

string [] fields = currentLine.Split(',');

would split the string, currentLine, to individual data items using the delimiter comma, and store the string array in the array reference variable named fields.

Note that if the delimiter is tab, it is represented by '/t,' as in currentLine.Split('/t').

The Split method requires the delimiter character to be of Char type, not string. Hence, Split(",") is invalid because "," is a string, but Split(',') is valid because ',' is a character. Note you may convert the string "," to a character, as in currentLine.Split(Char.Parse(","));.

Let's modify the code presented earlier in Figure 8-4 to compute the percent score for each student. To do this, split each line from Exams.txt, and average the two scores to get the percent. Figure 8-7 shows a modified version of the code presented earlier, within the Click event handler of a second button, btnComputePercent.

Figure 8-7: Code to get individual data items

```
37      private void btnComputPercent_Click(object sender, EventArgs e)
38      {
39          float percentScore;
40          string[] fields;
41          string fmtStr, currentLine;
42          // Create  StreamReader object to read Exams.txt
43          StreamReader examsReader = new StreamReader("Exams.txt");
44
45          lstDisplay.Items.Clear();        //Clear ListBox
46          // Print heading in the listbox
47          fmtStr = "{0,-15}{1,8}{2,8}{3,8:N1}";
48          lstDisplay.Items.Add(string.Format(fmtStr, "Name", "Exam1", "Exam2", "Percent"));
49
50          while (examsReader.EndOfStream == false)
51          {
52              currentLine = examsReader.ReadLine();
53              // Split currentLine to three string data items and store in a string array.
54              fields = currentLine.Split(',');
55
56              percentScore = (float.Parse(fields[1]) + float.Parse(fields[2]))/2;
57
58              //display the three data items and student total in the List  box
59              lstDisplay.Items.Add(string.Format(fmtStr, fields[0], fields[1], fields[2],
60                                                           percentScore));
61          }
62          examsReader.Close();
63      }
```

Let's look at some key statements:

Line 52: currentLine = examsReader.ReadLine();

As in the previous code, this line reads the current line, and stores the entire line as a single string in the string variable, currentLine. At the end of the first iteration through the loop, the string variable currentLine would have the following value:

Line 54: fields = currentLine.Split(',')

This is a key statement. Because the Split method returns a string array, the variable fields is declared as the reference to a string array, in line 40. At the end of the first iteration, the array fields would look like this:

fields	{string[3]}
[0]	"Mike Burns"
[1]	" 75"
[2]	" 92"

Line 56: percentScore = (float.Parse(fields[1]) + float.Parse(fields[2]))/2;

The two scores from the elements fields[1] and fields[2] are converted to float type and averaged to compute the percent. The percent is stored in the float variable percentScore.

Line 59: lstDisplay.Items.Add(string.Format(fmtStr, fields[0], fields[1], fields[2], percentScore));

The element, fields[0], gives the name; fields[1] and fields[2] give score1 and score2, respectively. The three items are displayed in the ListBox using the format string, fmtStr, defined in line 47 as "{0,-15}{1,8}{2,8}{3,8:N1}." The name is left justified using the "-" sign in the format code {0,-15}, and the percent is displayed with one decimal digit using the code N1 in {3,8:N1}.

Step 1-6: Open the form ReadTextFiles. Open the Click event handler of btnComputePercent.
Add the missing code from lines 51–61 in Figure 8-7.
Put a break at the end of the loop on line 61.

Make sure the Font type for the ListBox is a fixed width type, like Courier New. Otherwise, the columns may not align evenly with properly formatted codes.

Step 1-7: Run the project. Click the Compute Percent button. When the program breaks, view the values of currentLine, fields array and percentScore. View the data for the first few iterations.

Watch Out for Hidden Blank Lines at the End of Text Files!

If you accidentally hit the Enter key to insert a blank line at the end of the text file, you will get an error at line 56 (percentScore = float.Parse(fields[1]) + float.Parse(fields[2]))/2;). This is because when a blank line is split, the resulting array does not have the elements like fields[1] and fields[2].

See It for Yourself!

Step 1-8: Insert a blank line at the end of Exams.txt, as shown below, and save the file. Now run the project and click ComputePercent button. You will get the error message that "index was outside the bounds of the array." Reopen Exams.txt, delete the blank line and save it.

```
Exams.TXT - Notepad
File  Edit  Format  View  Help
Mike Burns,     75,    92
Cathy Arnold,   86,    95
Wesley Mathews, 82,    90
Jerry Luke,     70,    89
Gina Rodriguez, 77,    88

|
```

It's time to practice! Do Exercise 8-1.

Review Question

8.3 The following data on customer orders (Customer#, Order#, Order Qty, Unit Price) is stored in a text file named Orders.txt.

0155, 76533, 3, 10.95
0140, 75446, 12, 8.50
0150, 75550, 1, 89.95
0145, 75648, 25, 2.50

Consider the following code that reads the data from the above file.

```
StreamReader OrdersReader = new StreamReader("Orders.txt");
string[] fields;
while (OrdersReader.EndOfStream == false)
{
    currentLine = OrdersReader.ReadLine();
    fields = currentLine.Split(',');
}
lstDisplay.Items.Add(fields[1]);
```

What would be the output displayed in the ListBox lstDisplay?

8.3 Reading Data from Files into Arrays

Reading data from files on secondary storage devices like hard disks and flash drives is relatively slow compared to accessing data from arrays and variables, which are stored in the internal memory. So, if data that is stored in a file is to be accessed multiple times in a program, it generally would be more efficient to read the data from the file only once, store it in an array (or arrays), and access the array(s) for subsequent use of the data. For very large data sets, the consumption of internal memory by arrays should be weighed against the improvement in access speed.

A typical application that uses arrays is looking up information multiple times, like the phone number or address for different customers. Another application is computing the average for a field and then identifying all records that have above average values.

Let's consider an expanded version of the ReadTextFiles form to display a list of all students whose percent score is above the class average as shown in Figure 8-8.

Figure 8-8: The ReadTextFiles form displaying above average students

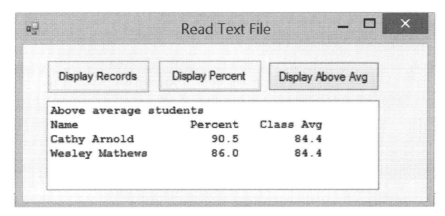

To find above average students, first you need to read all student records and compute the class average. Then, you need to access the same records again to identify who got scores above the class average.

Rather than reading the file twice, read the file once, compute the class average, and store student names in a one-dimensional array and the percent scores in another array. Then, use these arrays to identify those who got above average scores.

What array size should you use to create these arrays? The array size limits the number of data items that can be stored in an array. If a program needs to read data from a file with an increasing number of records over time, then the array size should be as large as the maximum number of records the file is likely to have. Similarly, if the program needs to read from files of different sizes, then the array size should be as large as the size of the largest file. For ease of viewing the arrays, this example limits the sizes of the two arrays to 10, as specified in the following statements that create the arrays.

```
string[] names         = new string[10];
float[] percentScores  = new float[10];
```

Storing data in the arrays involves the following process:
    Read the **first** record from Exams.txt and store the name in the first element of the names array, and store the percent score in the first element of the percentScores array, as shown in Figure 8-9. Read the **second** record from Exams.txt and store the name in the second element of the names array, and store the percent score in the second element of the percentScores array.
This process is repeated for each record in the file.

Figure 8-9: The names and percentScores arrays

	names	{string[10]}		percentScores	{float[10]}
	[0]	"Mike Burns"		[0]	83.5
	[1]	"Cathy Arnold"		[1]	90.5
	[2]	"Wesley Mathews"		[2]	86
	[3]	"Jerry Luke"		[3]	79.5
	[4]	"Gina Rodriguez"		[4]	82.5
	[5]	null		[5]	0
	[6]	null		[6]	0
	[7]	null		[7]	0
	[8]	null		[8]	0
	[9]	null		[9]	0

Figure 8-10 shows the first part of the code that reads and stores the data in the two arrays and computes the class average.

Figure 8-10: Code to compute class average using arrays

```csharp
65      private void btnAboveAvg_Click(object sender, EventArgs e)
66      {
67          //Declare two 1-D arrays
68          string[] names = new string[10];
69          float[] percentScores = new float[10];
70          // percentScores is the array of all percent scores
71          int recordCount, index = 0;
72      
73          string currentLine;
74          string[] fields;
75          float percentScore, classTotal = 0, classAvg;
76          //percentScore is the percent score for a student;
77          StreamReader examsReader = new StreamReader("Exams.txt");
78      
79          while (examsReader.EndOfStream == false)
80          {
81              currentLine = examsReader.ReadLine(); // read current line
82              fields = currentLine.Split(','); // split current line
83              percentScore = (float.Parse(fields[1]) + float.Parse(fields[2]))/2;
84      
85              names[index] = fields[0]; //Store name into current element of names array
86              percentScores[index] = percentScore;   // Store percent into percentScores
87              index = index + 1;
88              classTotal = classTotal + percentScore; // Add current percent to classTotal
89          }
90          recordCount = index;
91          classAvg = classTotal / (float) recordCount;
```

Let's examine the key statements of the code.

Lines 81-83

```csharp
81          currentLine = examsReader.ReadLine(); // read current line
82          fields = currentLine.Split(','); // split current line
83          percentScore = (float.Parse(fields[1]) + float.Parse(fields[2]))/2;
```

This is identical to the code you used earlier in Figure 8-7 to compute and display the percent scores. This code reads each record, splits the record to three string type data items that are stored in fields array, and computes the percent score. The fields array, as discussed earlier, looks as follows after the first iteration through the loop:

fields	{string[3]}
[0]	"Mike Burns"
[1]	" 75"
[2]	" 92"

Line 85: names[index] = fields[0]; //Store name into current element of names array

To access each element of the names array, this statement uses the variable index, which is initialized to zero (line 71) and incremented by one within the loop (line 87).

The first time through the loop, names[index] refers to the first element, names[0], which is assigned the value of fields[0] that always represents the name from the current record. The statement in line 85 would be equivalent to

    names[0] = fields[0];

So, after the first iteration, the names array would look as follows:

names	{string[10]}
[0]	"Mike Burns"
[1]	null
[2]	null
[3]	null
[4]	null
[5]	null
[6]	null
[7]	null
[8]	null
[9]	null

The second time through the loop, index is 1. So line 85 is equivalent to

    names[1] = fields[0];

So, the name from the second record is stored in the second element of names, as shown below:

names	{string[10]}
[0]	"Mike Burns"
[1]	"Cathy Arnold"
[2]	null

The third, fourth and fifth times, index becomes 2, 3 and 4, respectively; so line 85 would be equivalent to

    names[2] = fields[0];
    names[3] = fields[0];
    names[4] = fields[0];

After all records are read during the first five iterations of the loop, the array will have names stored in the first five elements as follows:

names	{string[10]}
[0]	"Mike Burns"
[1]	"Cathy Arnold"
[2]	"Wesley Mathews"
[3]	"Jerry Luke"
[4]	"Gina Rodriguez"
[5]	null
[6]	null
[7]	null
[8]	null
[9]	null

Line 86: percentScores[index] = percentScore;//Store percent into percentScores

This statement is similar to the previous one except that it stores the current student's percent score in the current element of percentScores array. Because the value of index increments by one each time through the loop, line 86 would be equivalent to

    percentScores[0] = percentScore;
    percentScores[1] = percentScore;
    percentScores[2] = percentScore;
    etc.

After the loop is completed, percentScores will have all five percents stored in the array as follows:

percentScores	{float[10]}
[0]	83.5
[1]	90.5
[2]	86
[3]	79.5
[4]	82.5
[5]	0
[6]	0
[7]	0
[8]	0
[9]	0

Line 88: classTotal = classTotal + percentScore;

This line calculates a running total of the percent scores for all student records that were read. The variable classTotal that accumulates the percent scores must be initialized (see line 75).

Line 90-91

```
90              recordCount = index;
91              classAvg = classTotal / (float) recordCount;
```

Because index is reinitialized to zero in the next part (not shown in Figure 8-10), its current value (that gives the number of records read) is stored in another variable, recordCount.

Step 1-9: Open the Click event handler of btnAboveAvg. Add the missing code from lines 85–91 in Figure 8-10.

Step 1-10: Put a break at the end of the loop on line 89, and run the project.
Click ComputeAvg button, and observe the values of **names, percentScores** and **fields** arrays each time through the loop.

Identifying Above Average Students

To identify students whose percent score is above the class average, the class average is compared to each element of percentScore array, using a loop. If an element is above class average, the corresponding element from the names array is displayed. For example if percentScore[2] is above average, names[2] is displayed.

This method works only if the two arrays are loaded consistently so that there is a one-to-one correspondence between the elements of names and those of percentScores (i.e., percentScores[0] is the percent for names[0], percent[1] represents the percent for names[1], etc).

Lines 93–106 in Figure 8-11 show the second part of the code, which displays the names and scores of students whose percent is above the class average. Lines 20–92 compute the class average, which is the same as Figure 8-10, discussed earlier.

Figure 8-11: Code to compute average and display above average students

```
65    private void btnAboveAvg_Click(object sender, EventArgs e)
66    {
67        //Declare two 1-D arrays
68        string[] names = new string[10];
69        float[] percentScores = new float[10];
70        // percentScores is the array of all percent scores
71        int recordCount, index = 0;
72
73        string currentLine;
74        string[] fields;
75        float percentScore, classTotal = 0, classAvg;
76        //percentScore is the percent score for a student;
77        StreamReader examsReader = new StreamReader("Exams.txt");
78
79        while (examsReader.EndOfStream == false)
80        {
81            currentLine = examsReader.ReadLine(); // read current line
82            fields = currentLine.Split(','); // split current line
83            percentScore = (float.Parse(fields[1]) + float.Parse(fields[2]))/2;
84
85            names[index] = fields[0]; //Store name into current element of names array
86            percentScores[index] = percentScore;   // Store percent into percentScores
87            index = index + 1;
88            classTotal = classTotal + percentScore; // Add current percent to classTotal
89        }
90        recordCount = index;
91        classAvg = classTotal / (float) recordCount;
92
93        string fmtStr = "{0,-15}{1,12:N1}{2,12:N1}";
94        lstDisplay.Items.Clear();
95        lstDisplay.Items.Add("Above average students");
96        lstDisplay.Items.Add(string.Format(fmtStr, "Name", "Percent", "Class Avg"));
97
98        //Find Above average students
99        for (index = 0; index < recordCount; index++)
100       {
101           if (percentScores[index] > classAvg)
102               lstDisplay.Items.Add(string.Format(fmtStr, names[index],
103                                     percentScores[index], classAvg));
104       }
105       examReader.Close();
106   }
```

Line 101: if (percentScores[index] > classAvg)

Because the loop variable assumes values 0 to 4, line 101 becomes equivalent to
> if (percentScores[0] > classAvg)
> if (percentScores[1] > classAvg)
> if (percentScores[2] > classAvg)
> etc.

Thus, the statement **if (percentScores[index] > classAvg)** compares each element of percentScores to classAvg.

Line 102: lstDisplay.Items.Add(string.Format(fmtStr, names[index], percentScores[...]...);

If any element of percentScores is greater than classAvg, then line 102 displays the name from the corresponding element of names array using the same index value.

Step 1-11: Add the missing code from lines 99–104 in Figure 8-11 to the Click event handler of btnAboveAvg to display above average students. Put breaks at lines 102 and 104.

Step 1-12: Run the project. When the program breaks, observe the values of classAvg and names and percentScore arrays.

It's time to practice! Do Exercise 8-2

ReadBlock Method of StreamReader

Instead of reading an entire line using the ReadLine method, you may read a specified number of characters, starting with a specified position in the current line, into a character array. For example, the following statement reads 10 characters starting with index 0 into a character array named studentArray:

> examsReader.ReadBlock(studentArray, 0, 10);

ReadToEnd Method of StreamReader

The ReadToEnd method of StreamReader reads the entire file, starting with the current position, and returns a string. The following statement would read the entire file Exams.txt and display it in the ListBox, lstDisplay:

> lstDisplay.Items.Add(examsReader.ReadToEnd);

Review Questions

8.4 The following is a sample of the order data (Customer#, Order#, Order Qty, Unit Price) stored in a text file named Orders.txt.

0155, 76533, 3, 10.95
0140, 75446, 12, 8.50
0150, 75550, 1, 89.95
0145, 75648, 25, 2.50

The following code is intended to read data from the above file and compute and display the extended cost (= Order Qty x Unit Price) for each order. The statement that computes the extended cost is incomplete.

```
StreamReader OrdersReader = new StreamReader("Orders.txt");
string[] fields;
float extCost;
while (OrdersReader.EndOfStream == false)
{
        currentLine = OrdersReader.ReadLine();
        fields = currentLine.Split(',');
        // insert code to compute extCost (= Order Qty x Unit Price)
        extCost =
        lstDisplay.Items.Add(extCost);
}
```

Insert the necessary code to compute the extCost.

8.5 The following is a sample of the order data (Customer#, Order#, Order Qty, Unit Price) stored in a text file named Orders.txt.

0155, 76533, 3, 10.95
0140, 75446, 12, 8.50
0150, 75550, 1, 89.95
0145, 75648, 25, 2.50

The following code is intended to read data from the above file, compute the extended cost (= Order Qty x Unit Price) for each order, and store the Order# and extended cost in two different arrays (orderNumbers and extCosts). The statements to store the Order# and extended cost into arrays are missing.

```
StreamReader OrdersReader = new StreamReader("Orders.txt");
string[] fields;
float extCost;
int index = 0;
string[] orderNumbers = new string[10];
float[] extCosts = new float[10];
while (OrdersReader.EndOfStream == false)
{
        currentLine = OrdersReader.ReadLine();
        fields = currentLine.Split(',');
        extCost = float.Parse(fields[2])*float.Parse(fields[3]);
        // insert code to store Order# into the array, orderNumbers
        ...
        // insert code to store extended cost into the array, extCosts
        ...
        index = index + 1;
}
```

Insert the missing code to store the Order# into the array, orderNumbers, and to store extended cost into the array, extCosts.

8.4 Additional Methods of Arrays

Array.Resize Method

When data are read from large files whose size may grow over time, storing data in arrays poses challenges in specifying the size of the array. An option would be to specify an initial size that is large enough to hold the current data but resize the array automatically whenever the number of records to be stored in the array exceeds the array size.

The Array.Resize() method has the following syntax:
> Array.Resize (ref *ArrayName*, *NewArraySize*)

where ref is a key word, *ArrayName* is the name of the array (must be a one-dimensional array) and *NewArraySize* is the new size of the array.

The Resize method creates a new array with the specified size, copies the data from the old array to the new array (if the new size is smaller, only a subset of the data is copied) and replaces the old array with the new.

For example, you would expand the size of the array, names, from its current size of 10 to a new size of 15, as follows:
> Array.Resize(ref names, 15);

The first 10 elements of the new array will have the data from the 10 elements of the old array. The remaining 5 elements are empty.

The statement Array.Resize (ref names, 6) copies the first six elements from the old array to the resized array. The rest of the data is discarded.

Tutorial 2: Looking Up Phone Numbers Using Arrays

This tutorial uses the **Resize** method in a new form called **LargeFiles** that looks up the telephone numbers of vendors from a large set of vendor names and phone numbers stored in two one-dimensional arrays. This form also uses the **IndexOf** method. The data are read into arrays from a file named **Vendor.txt** that contains the following seven fields:

> **Vendor#, name, street, city, state, zip, phone number.**

Here is a sample of data that are currently stored in the Vendor.txt file:

```
1,FCC,445 12th Street SW, Washington,DC,20554,888-225-5322
2,U.S.Copyright Office,101 Independence Ave, Washington,DC,20559,202-707-3000
3,CDC,1600 Clifton Road,Atlanta,GA,30333,800-232-4636
4,White House, 1600 Pennsylvania Avenue NW, Washington,DC,20500,202-456-1111
5,Labor Statistics,Postal Sq Bldg,2 Mass Ave,Washington,DC,20212,800-877-8339
6,AMTRAK,60 Massachusetts Avenue NE, Washington,DC,20002,800-523-6590
7,U.S. Air Force,1690 Air Force Pentagon,Washington,DC,20330-1670,800-423-8723
8,National Guard,111 South George Mason Dr,Arlington,VA,22204,800-864-6264
9,Office of Govt Ethics,1201 New York Ave,Washington,DC,20005,202-482-9300
10,IRS,1111 Constitution Ave NW, Washington, DC,20224,800-829-1040
```

A major feature of the code is that the size of the array is increased dynamically whenever the number of records read from the file equals the size of the array. Thus, the code is intended to read from files that contain a very large number of records and is expected to grow in size. The actual file used here (Vendor.txt), however, is very small to make it easy to track execution of the code and to test the loop.

The first part of the code that reads the seven data items from each line, and adds the names and phone numbers into two arrays, is shown in Figure 8-12.

Figure 8-12: The LargeFiles form with Array.Resize method

```csharp
20      string[] names, phoneNumbers; //Declare arrays.
21      int recordCount = 0;
22
23      private void LargeFiles_Load(object sender, EventArgs e)
24      {
25          string currentLine;
26          int arrayInitialSize = 10000, arrayIncrement = 5000;
27          string[]fields  = new string[7]; // Declare an array named fields
28          StreamReader vendorReader = new StreamReader("Vendor.txt");
29
30          names = new string[arrayInitialSize];
31          phoneNumbers = new string[arrayInitialSize];
32
33          // Read each line, split it, and store name & phone number into two arrays.
34          while (vendorReader.EndOfStream == false)
35          {
36              // If the names array is full, increase the size of arrays.
37              if (recordCount == names.Length)
38              {
39                  Array.Resize(ref names, names.Length + arrayIncrement);
40                  Array.Resize(ref phoneNumbers, phoneNumbers.Length + arrayIncrement);
41              }
42              currentLine = vendorReader.ReadLine();
43              fields = currentLine.Split(',');  // split the line
44
45              names[recordCount] = fields[1]; //Store name into the array, names
46              phoneNumbers[recordCount] = fields[6]; //Store phone number in phoneNumbers
47              recordCount = recordCount + 1;
48          }
49          vendorReader.Close();
50      }
```

The process of reading each line, splitting it into different data items and storing the seven data items in a temporary array (lines 42 and 43) is similar to the process we used earlier in the ReadTextFiles form. After a record is stored in the temporary array, fields, you store the vendor name (second element; index 1) and phone number (seventh element; index 6) into the arrays, names and phone numbers (lines 45 and 46).

Let's look at some key parts of the code:

Lines 20 and 21: The arrays, names and phoneNumbers, are declared at the class level so that they can be created and loaded in the Load event of the form and used in another method to find the phone number for a specified name. Similarly, the variable recordCount also is declared at the class level so that it can be used in two different event procedures.

Lines 30 and 31: The arrays are created with an initial size that is specified in line 26.

Lines 37–41: If the number of records loaded into arrays is equal to the size of the arrays, then the arrays are resized by adding a specified increment to the array size. This is a major difference from previous examples.

Step 2-1: Make sure that the file Vendor.txt is in the debug folder of your project folder, Ch8_FilesAndArrays. To make it easy to test the code, this file contains only 10 records.

Step 2-2: Open the form, LargeFiles, shown in Figure 8-13.

Figure 8-13: The LargeFiles form

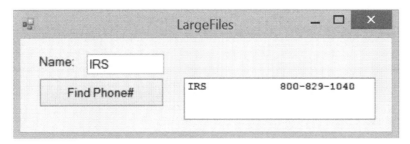

Step 2-3: Open the load event handler of the form. Fill in the missing code in lines 39 and 40 from Figure 8-12.

To make it easy to test the loop, change ArrayInitialSize to 5 (from 10,000) and ArrayIncrement to 5 (from 5,000) in line 26.

Step 2-4: Make sure that the Program.cs file specifies LargeFiles as the startup form.

Put a break at line 48, and run the form.

Go through the loop five times, and observe the values of recordCount and Names.Length by hovering the cursor over them. The length of the names array remains at 5, which is the initial value, and recordCount increments by one each time.

Click Continue, and observe that Names.Length is increased to 10.

Array.IndexOf Method

The Array.IndexOf method, introduced in Chapter 7, returns the index of a specified data item in a one-dimensional array.

This method has the syntax,
Array.IndexOf(*ArrayName, SearchValue*).

ArrayName is the name of a one-dimensional array, and *SearchValue* is the value you are trying to find in an array. The method returns the index of the first element that matches the specified value. If no value is found, the method returns -1.

For example,
   Array.IndexOf(names, "Mike Burns")
returns the index of the first element of the names array that matches the specified value, Mike Burns.

Figure 8-14 shows the code that uses this method to find a vendor's phone number by searching for a specified name.

Figure 8-14: Code to find the phone number using IndexOf method

```
53      private void btnFindPhone_Click(object sender, EventArgs e)
54      {
55          string searchName = txtVendorName.Text;
56          string fmtStr = "{0,-15}{1,8}";
57          lstDisplay.Items.Clear();
58
59          // Method 1: Use IndexOf method to find the record
60          int index = Array.IndexOf(names, searchName);
61                  //if there is no match, index would be -1.
62          if (index >= 0)
63              lstDisplay.Items.Add(string.Format(fmtStr, searchName,
64                                          phoneNumbers[index]));
65          else
66              MessageBox.Show("Sorry, name not found");
```

Line 60 uses the IndexOf method to get the index of the element in the Names array that matches the value of SearchName.

Lines 62–64 display the corresponding element from phoneNumbers array if the index is not -1. If not, a message is displayed.

Figure 8-15 shows an alternate method that uses a loop to compare the value of SearchName with each element in the Names array.

Figure 8-15: Using a loop to search a field

```
67
68          // Method 2: Use for loop to find the record
69          for (index = 0; index < recordCount; index++)
70          {
71              if (searchName == names[index])
72              {
73                  lstDisplay.Items.Add(String.Format(fmtStr, names[index],
74                                          phoneNumbers[index]));
75                  return;
76              }
77          }
78          MessageBox.Show("Sorry, there is no Vendor by that name");
79      }
```

Step 2-5: Add the code from Figure 8-15.
   Test the code by typing in the value AMTRAK, which is one of the vendor names in the Vendor.txt file.

Tutorial 3: Using a ComboBox to Select the Search Name

This tutorial creates a different version of the application where the vendor names are added to the ComboBox, and when the user selects a name, the phone number is automatically displayed in the ListBox on the form named SmallFiles, as shown in Figure 8-16. The use of ComboBox to select a vendor is appropriate only if the number of vendors is relatively small. It is time consuming to load a large data set into a ComboBox.

Figure 8-16: The SmallFiles Form with ComboBox to Select Name

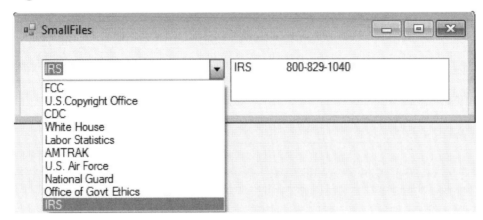

In this form called **SmallFiles**, you learn how to add items to a ComboBox from a file, and see an application of the SelectedIndexChanged event.

Step 3-1: Open the form, SmallFiles.

Two major aspects of the revised code are

1. The load event handler of the form, which reads data from Vendor.txt into two arrays, also adds vendor names to the ComboBox.
2. The SelectedIndexChanged event of the ComboBox finds the phone number for the selected name, whenever the user selects a vendor.

The complete code is shown in Figure 8-17. Because the use of ComboBox to select a vendor is appropriate only if the number of vendors is relatively small, the code does not include resizing of arrays. You use the same vendor file, **Vendor.txt**.

Figure 8-17: Code to add names to a ComboBox and find phone numbers

```csharp
19          string[] names, phoneNumbers;  //Declare arrays
20          int recordCount = 0;
21
22          private void SmallFiles_Load(object sender, EventArgs e)
23          {
24              string currentLine;
25              string[] fields = new string[7];   // Declare array named fields
26              StreamReader vendorReader = new StreamReader("Vendor.txt");
27
28              names = new string[10];
29              phoneNumbers = new string[10];
30              // Read each line, split & store name and phone number into two arrays.
31              while (vendorReader.EndOfStream == false)
32              {
33                  currentLine = vendorReader.ReadLine();
34                  fields = currentLine.Split(',');
35
36                  names[recordCount] = fields[1]; //Store name into the array, names
37                  phoneNumbers[recordCount] = fields[6];//Store phone# into phoneNumbers
38
39                  cboVendorNames.Items.Add(fields[1]);   // Add names to the ComboBox
40                  recordCount = recordCount + 1;
41              }
42              vendorReader.Close();
43          }
44
45          private void cboVendorNames_SelectedIndexChanged(object sender, EventArgs e)
46          {
47              // Get selected vendor name from the ComboBox
48              string searchName = cboVendorNames.SelectedItem.ToString();
49              string fmtStr = "{0,-15}{1,8}";
50              lstDisplay.Items.Clear();
51
52              // Use IndexOf method to find the record
53              int index = Array.IndexOf(names, searchName);
54                      //if there is no match, index would be -1.
55              if (index >= 0)
56                  lstDisplay.Items.Add(string.Format(fmtStr, searchName,
57                                              phoneNumbers[index]));
58              else
59                  MessageBox.Show("Sorry, name not found");
60          }
```

The statements related to the use of the ComboBox are as follows.

Line 39: cboVendorNames.Items.Add(fields[1])

This statement adds the vendor name to the ComboBox when the form is loaded.

Lines 45-60

These show the SelectedIndexChanged event handler of the ComboBox.

Line 48: string searchName = cboVendorNames.SelectedItem.ToString()

The SelectedItem is of object type. Therefore, it is converted to string before assigning it to the string variable searchName.

Step 3-2: Open the Load event handler of the form. Fill in the missing code in lines 36-39 from Figure 8-17.
Open the SelectedIndexChangedEvent of the ComboBox. Fill in the missing code in lines 53-59 from Figure 8-17.
Run the form and test it.

Review Questions

8.6 The following code creates an array named shows with the following initial values:
string[] shows = {"Mamma Mia","Rock of Ages","Les Miserables", "Motown The Musical", "The Lion King", "The Phantom of the Opera"};
Write the necessary code to add the names of the shows from the array to a ComboBox named cboShows. You may use a foreach loop to access each element of the array.

8.7 The following code creates two arrays named months and sales. The sales array shows the units sold for the corresponding month.
string[] months = { "Jan", "Feb", "Mar", "Apr", "May", "Jun", "Jul", "Aug", "Sep", "Oct", "Nov", "Dec" };
int[] unitsSold = { 572, 645, 693, 564, 580, 756, 960, 756, 663, 678, 745, 960};

Write the necessary code to find the units sold for a month name entered into a TextBox named txtMonthName, and display the units sold in a ListBox named lstDisplay.

8.5 Writing to Text Files

You can write data to a text file using the **StreamWriter** class available in the System.IO namespace of Framework Class Library.

Tutorial 4: Writing Scores to Text File

Let's develop a form that writes data from Text Boxes into a text file. Figure 8-18 shows the form named WriteToTextFile.

Figure 8-18: The WriteToTextFile form

The StreamWriter Object

The StreamWriter class can be used to write data into a text file. The StreamWriter object is created using the following syntax:
   StreamWriter **VariableName** = new StreamWriter(@"Path and FileName", append flag);

If **append flag** is true, data are appended to existing data, if any. If append flag is false, existing data, if any, are overwritten. In both cases, if the specified file does not exist, a new file is created.

If no path is specified, it is assumed to be Bin\Debug folder of the project directory. The "@" prefix is not needed when the path is not specified. For example, to write data to Exams.txt with the append option, the StreamWriter object is created as follows:
   StreamWriter examsWriter = new StreamWriter("Exams.txt", true);

The WriteLine Method of StreamWriter

The WriteLine method of StreamWriter writes a string as a separate line into a file. Use this method to write all three data items for each student into a separate line in Exams.txt file.

Declaring Variables at the Class Level

Because data are to be written multiple times (each time when the user clicks the Save button) to Exams.txt, you create the StreamWriter once when the form is loaded, use it each time the Save button is clicked, and close the StreamWriter when the form is closed. That means the variable (ExmasWriter) that references StreamWriter is used in three different methods:
   the **Load event** of the form,
   the Click event of the Save button, and

the **FormClosing event** of the form.

To make the variable examsWriter available in all three methods, you need to declare examsWriter at the class level (outside the three methods), and create the StreamWriter object in the **Load event** handler, as shown in the code in Figure 8-19.

Figure 8-19: Code to declare and create StreamWriter object

```
14      public partial class WriteToTextFile : Form
15      {
16          public WriteToTextFile()
17          {
18              InitializeComponent();
19          }
20          // Declare variable ExamsWriter at class level:
21          StreamWriter examsWriter;
22          private void WriteToTextFiles_Load(object sender, EventArgs e)
23          {
24              ExamsSaveDialog.ShowDialog(); // Display Save As Dialog box
25              // Get the file name selected by the user, and
26              // save it in the variable named fileName
27              string fileName = ExamsSaveDialog.FileName;
```

The Load Event of Forms

The load event of forms is triggered whenever you open a form. You create the Load event handler by double clicking the form in design view.

Step 4-1: Open the form, WriteToTextFiles, shown in Figure 8-18.

Step 4-2: Add the using directive to include System.IO namespace in the list:

```
1   using System;
2   using System.Collections.Generic;
3   using System.ComponentModel;
4   using System.Data;
5   using System.Drawing;
6   using System.Linq;
7   using System.Text;
8   using System.Threading.Tasks;
9   using System.Windows.Forms;
10  using System.IO;            // Add this statement
11
12  namespace Ch8_FilesAndArrays
```

Step 4-3: Open the code window. Fill in the code in line 21 from Figure 8-19 to declare the variable examsWriter, at the class level.

Step 4-4: Fill in the code in lines 25-26 from Figure 9-19 to create the object examsWriter. Put a break on line 25.

Step 4-5: Run the form.
   The program breaks at line 25, which creates the examsWriter object. The form is not visible at this time because the code is part of the Load event handler.
   Click Run to complete loading the form.
   Remove the break.

Because examsWriter is created in the Load event, it is available when the form is open, and it is ready to be used for writing within the Click event handler of the Save button, shown in Figure 8-20.

Figure 8-20: Code to write data to Exams.txt

```
private void btnSave_Click(object sender, EventArgs e)
{
    string name, score1, score2;
    // Store all data in string variables.
    name = txtName.Text;
    score1 = txtScore1.Text;
    score2 = txtScore2.Text;
    // Combine name & socres and store in Exams.txt
    examsWriter.WriteLine(name + "," + score1 + "," + score2);
    MessageBox.Show("Data Saved");
}
```

Note that the above code writes the data using the StreamWriter every time the user clicks the Save button without having to create the StreamWriter and close it each time.

To write the data, the name, score1 and score2 are combined into a single string,
   (name + "," + score1 + "," + score2),
with commas between each data item. The commas are inserted to make it easy to split the string if the data are read back in a program.

Step 4-6: Open the Click event handler of Save button. Fill in the missing code from Figure 8-20 to write data to Exams.txt file.

Step 4-7: Run the form. Enter data for name, score1 and score2, and click Save.
   Click OK when you get the message "Data Saved."

Step 4-8: View the Bin\Debug folder of your project directory. Open the Exams.txt file that was created earlier in the form ReadTextFile. You will not find the record that you entered and was supposed to have been written to it. What happened?

The data are written to the external device (hard disk, flash drive, etc.) only when the StreamWriter is closed. Until then, data written by StreamWriter are held in the buffer temporary memory. You have not added the code to close the Stream Writer.

FormClosing Event of Forms

The StringWriter object should be available until the form is closed. So, you should close StreamWriter only when the form is closed. This can be done by closing StreamWriter within the **FormClosing** event handler of the form, which is triggered whenever the form is closed. Figure 8-21 shows the FormClosing event handler.

Figure 8-21: FormClosing event handler

```
private void WriteToTextFiles_FormClosing(object sender,FormClosingEventArgs e)
{
    if (examsWriter != null) // Check whether ExamsWriter exits
        examsWriter.Close();
}
```

Because FormClosing is not the default event handler of forms, double clicking the form won't create this event handler. You need to select the events button in the Properties window to display form events, as shown in Figure 8-22. Then, double click the event (FormClosing) to create the event handler.

Figure 8-22: Events of the form

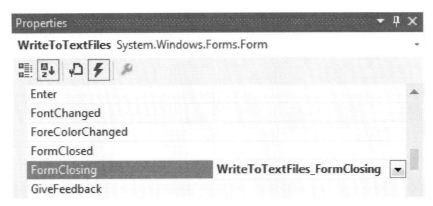

Step 4-9: Open the code window and create the FormClosing event handler.

Step 4-10: Fill in the code from Figure 8-21 to close examsWriter.

Step 4-11: Add the following code to the Click event handler of Exit button to close the form.
　　　　this.Close();　// "this" represents the current form

Step 4-12: Open Exams.txt. Make sure that the insertion point is on a new line after Gina Rodriguez's record. Close Exams.txt.
Run the project. Enter two sets of data (make up your own), and save each set.
Close the form.
Open Exams.txt. The file should have the two sets of data you entered, in addition to the five records that it already had.

Append versus Overwrite

The two sets of data that you entered were added to data that already existed in the file because the append flag was set to true in the following statement that created the StreamWriter:

```
ExamsWriter = new StreamWriter("Exams.txt", true);
```

If you set the append flag to false, all existing data will be overwritten by new data.

Step 4-13: In line 25, change the append flag to false as follows:

```
25              ExamsWriter = new StreamWriter("Exams.txt", false);
```

Add a third set of data and save it. Close the form.
Open Exams.txt and view it. You should see only the third set of data, because the previous records were overwritten.
Now, change the append flag back to true.

It's time to practice! Do Exercise 8-3

Directory Class

The Directory class in System.IO namespace provides several methods to work with directories (folders), including methods to create folders and to check whether a specified path exists. Figure 8-23 shows two commonly used methods of Directory class.

Figure 8-23: Methods of Directory class

Method	Function
Exists(path)	Returns true if the path exists and false if the path does not exist
CreateDirectory(path)	Creates the directories in the specified path

You may create Exams.txt file in the WriteToTextFiles form by specifying a path other than the one for the default directory. Use the above methods to check whether the directories in the specified path exist, and to create them if they do not exist. Figure 8-24 shows the code to create the path E:\Csharp\output if it does not exist, and to use that path to write the data.

Figure 8-24: Create a path using Directory class

```
string path = @"E:\Csharp\output\";
if (!(Directory.Exists(path)))
    Directory.CreateDirectory(path);
examsWriter = new StreamWriter(path + "Exams.txt", true);
```

If the drive E: is invalid, you will get an error. If the directory, Csharp, does not exist, it is created, and the directory, output, is created within Csharp. If Csharp folder exists but output folder does not exist, it will be created.

The last statement,
    examsWriter = new StreamWriter(path + "Exams.txt", true);
uses the newly created path to write the data.

Step 4-14: Temporarily, replace the statement in line 25 by the code in Figure 8-24 to create a directory and use it to write the data. Run the project.
Enter a new record and save it. Verify that a new directory Csharp\output is created, and the file Exams.txt is created within it.
Remove the code that you inserted, and change line 25 back to its original form with no path specified.

Review Questions

8.8 The following form, discussed earlier, lets the user save multiple sets of data into a text file by clicking the Save button after entering each set of data.

The code shown below writes the name and the two scores from TextBoxes into the file Exams.txt.

```
StreamWriter examsWriter;
private void WriteToTextFiles_Load(object sender, EventArgs e)
{
        examsWriter = new StreamWriter("Exams.txt", true);
}
private void btnSave_Click(object sender, EventArgs e)
{
        string name, score1, score2;
        name = txtName.Text;
        score1 = txtScore1.Text;
        score2 = txtScore2.Text;
        examsWriter.WriteLine(name + "," + score1 + "," + score2);
}
private void WriteToTextFiles_FormClosing(object sender, FormClosingEventArgs e)
{
        examsWriter.Close();
}
```

The streamWriter named examsWriter is declared at the class level, created in the Load event of the form, used in the Click event of the button multiple times and closed within the FormClosing event. What would be the effect of moving the declaration statement (**StreamWriter examsWriter;**) to the beginning of the Load event handler of the form, as follows?

```
// StreamWriter examsWriter;
private void WriteToTextFiles_Load(object sender, EventArgs e)
{
        StreamWriter examsWriter;
        examsWriter = new StreamWriter("Exams.txt", true);
}
```

8.9 Refer to the form described in Review Question 8.8. Consider the following two changes to the code:
1. Move the declaration statement (StreamWriter examsWriter;),to btnSave_Click event handler, as shown below.
2. Move the statement that creates examsWriter from the Load event handler to btnSave_Click event handler, as shown below.

```
// StreamWriter examsWriter;
private void WriteToTextFiles_Load(object sender, EventArgs e)
{
        // examsWriter = new StreamWriter("Exams.txt", true);
}

private void btnSave_Click(object sender, EventArgs e)

        StreamWriter examsWriter;
        examsWriter = new StreamWriter("Exams.txt", true);

        string name, score1, score2;
        name = txtName.Text;
        score1 = txtScore1.Text;
        score2 = txtScore2.Text;
        examsWriter.WriteLine(name + "," + score1 + "," + score2);
}

 private void WriteToTextFiles_FormClosing(object sender, FormClosingEventArgs e)
{
        examsWriter.Close();
}
```

What would be the combined effect of the above two changes?

8.10 Refer to the form described in Review Question 8.8. Consider the following three changes to the code:
1. Move the declaration statement (StreamWriter examsWriter;) to btnSave_Click event handler, as shown below.
2. Move the statement that creates examsWriter from the Load event handler to btnSave_Click event handler, as shown below.
3. Move the statement, examsWriter.Close(), to the end of btnSave_Click event handler, as shown below.

```
// StreamWriter examsWriter;
private void WriteToTextFiles_Load(object sender, EventArgs e)
{
        // examsWriter = new StreamWriter("Exams.txt", true);
}
```

```csharp
private void btnSave_Click(object sender, EventArgs e)
{
    StreamWriter examsWriter;
    examsWriter = new StreamWriter("Exams.txt", true);

    string name, score1, score2;
    name = txtName.Text;
    score1 = txtScore1.Text;
    score2 = txtScore2.Text;
    examsWriter.WriteLine(name + "," + score1 + "," + score2);

    examsWriter.Close();
}

private void WriteToTextFiles_FormClosing(object sender, FormClosingEventArgs e)
{
    // examsWriter.Close();
}
```

What would be the combined effect of the above three changes?

8.6 SaveFileDialog Control

The **SaveFileDialog** control allows users to select the path and file to save the data, rather than hardcoding the path/file as you did in previous examples.

To add a SaveFileDialog control to your project, double click *SaveFileDialog* under *Dialogs* group in the *Toolbox* window. The control is added to the tray at the bottom of the Designer window.

ShowDialog() method

This method displays a *Save As* dialog box to let users select the path and file.
This is how you would code to let the user specify the file:
    ExamsSaveDialog.ShowDialog();

FileName property

The FileName property stores the path and name of the file selected by the user. If the user selects Cancel from the Save As dialog box, the value of FileName property would be an empty string.

The following code gets the path and name of the file selected by the user from the FileName property and stores it in the variable filename:

    string fileName = ExamsSaveDialog.FileName;

The following code uses the selected file to create the StreamWriter object. The if statement checks whether the length of the file name is greater than zero—that is, if the user didn't cancel out:

```
            if (fileName.Length > 0)
                examsWriter = new StreamWriter(fileName, true);
        else
                MessageBox.Show("No file selected");
```

Step 4-15: Add *SaveFileDialog* control from the Toolbox to the WriteToTextFiles form. Select the control, and change its name to ExamFileDialog.

Step 4-16: To let the user select a file using the SaveFileDialog, change the Load event handler of the form, as shown in Figure 8-25.

Figure 8-25: Using SaveFileDialog to let the user select a file

```
22      private void WriteToTextFiles_Load(object sender, EventArgs e)
23      {
24          ExamsSaveDialog.ShowDialog(); // Display Save As Dialog box
25          // Get the file name selected by the user, and
26          // save it in the variable named fileName
27          string fileName = ExamsSaveDialog.FileName;
28
29          //Create StreamWriter object
30          if (fileName.Length > 0)    // if the user selected a file
31              ExamsWriter = new StreamWriter(fileName, true);
32          //true means append; false means overwrite
33          else
34          {
35              MessageBox.Show("No file selected");
36              btnSave.Enabled = false;
37          }
38      }
```

Step 4-17: Run the project. Select Exams.txt file from the Debug folder, and click Save.
When asked whether you want to replace the file, select **Yes**. (The term replace is misleading. You will not lose the existing contents of the file because the append flag is set to true.)
When the program breaks, observe the value of fileName variable that includes the path and file name. Click Continue.
Enter a new record, and Save it.
Open Exams.txt to verify that the file contains all previous records and the new record that you added.

If the user doesn't select a file, and clicks *Cancel* instead, the value of the fileName variable will be an empty string. So, line 31 will be skipped and the ExamWriter object won't be created. This will result in an error in the FormClosing event handler, shown below, when the form is closed.

```
private void WriteToTextFiles_FormClosing(object sender, FormClosingEventArgs e)
{
    ExamsWriter.Close();
}
```

Step 4-18: Run the project. Intead of selecting a file, click Cancel in the *Save As* window.
Clikc OK when the "No file selected" message is displayed.

The program breaks at line 38. Verify that the fileName variable shows an empty string.
Click the *Continue* button on the Toolbar.
Click the *Exit* button on the form.
The program breaks in the FormClosing event handler with a NullReferenceException error, as shown below, because ExamWriter object was not created in the Load event handler of the form.

```
private void WriteToTextFiles_FormClosing(object sender, FormClosingEventArgs e)
{
    //if (examsWriter != null)  // check whether examsWriter exists
        examsWriter.Close();
}
```
⚠ NullReferenceException was unhandled

To correct this problem, modify the FormClosing event handler by adding an if statement that checks whether examsWriter exits, as follows:

```
private void WriteToTextFiles_FormClosing(object sender, FormClosingEventArgs e)
{
    if (examsWriter != null)   // check wether examsWriter exists
        examsWriter.Close();
}
```

Run the project again, and verify that closing the form doesn't give an error when the user cancels out of the Save As window.

InitialDirectory Property

The InitialDirectory property allows you to specify the directory initially displayed in the *Save As* dialog box. If not specified, the Document directory or the last used directory, if any, is displayed. Here is how you would specify the initial directory:

> ExamsSaveDialog.InitialDirectory = @"E:\CSharp";

The Title Property

The title property lets you specify a title for the Save As dialog box. Here is an example:

> ExamsSaveDialog.Title = "Exams Save As dialog box";

Step 4-19: Insert the necessary statements at the beginning of the Load event handler to specify the initial directory and title. Run the project and observe the effect.

The following is the complete code to write data into a text file selected using the SaveFileDialog control:

```csharp
using System.IO;    // Add this statement

namespace Ch8_FilesAndArrays
{
    public partial class WriteToTextFile : Form
    {
        public WriteToTextFile()
        {
            InitializeComponent();
        }
        // Declare variable ExamsWriter at class level:
        StreamWriter ExamsWriter;
        private void WriteToTextFiles_Load(object sender, EventArgs e)
        {
            ExamsSaveDialog.ShowDialog(); // Display Save As Dialog box
            // Get the file name selected by the user, and
            // save it in the variable named fileName
            string fileName = ExamsSaveDialog.FileName;

            //Create StreamWriter object
            if (fileName.Length > 0)     // if the user selected a file
                ExamsWriter = new StreamWriter("Exams.txt", false);
            //true means append; false means overwrite
            else
            {
                MessageBox.Show("No file selected");
                btnSave.Enabled = false;
            }
        }

        private void btnSave_Click(object sender, EventArgs e)
        {
            string name, score1, score2;
            // Store all data in string variables.
            name = txtName.Text;
            score1 = txtScore1.Text;
            score2 = txtScore2.Text;
            // Combine name & socres and store in Exams.txt
            ExamsWriter.WriteLine(name + "," + score1 + "," + score2);
            MessageBox.Show("Data Saved");
        }

        private void WriteToTextFiles_FormClosing(object sender,
                                        FormClosingEventArgs e)
        {
            if (ExamsWriter != null) // Check whether ExamsWriter exits
                ExamsWriter.Close();
        }
```

```csharp
        private void btnExit_Click(object sender, EventArgs e)
        {
            this.Close();    // "this" represents the current form
        }

        private void btnClear_Click(object sender, EventArgs e)
        {
            txtName.Text = "";
            txtScore1.Text = "";
            txtScore2.Text = "";
        }
    }
}
```

It's time to practice. Do Exercise 8-4.

8.7 Passing Arrays to Methods

You may pass an entire array, or any other reference types, to a method just like you pass a value type like an integer or float. Tutorial 5 presents an application that passes an array to a method.

Tutorial 5: Passing Arrays to Compute Average

This tutorial creates a modified version of the ReadTextFiles form that reads the name and two exam scores from the text file Exams.txt. In the modified version, a separate method is used to read the names and score, and store the names and percent scores in arrays. These arrays are passed to two different event handlers.

Similar to ReadTextFile, the revised form named **PassingArrays** lets the user display the percent scores for each student and display the names of students with above average percent scores, as shown in Figure 8-26.

Figure 8-26: The PassingArrays form

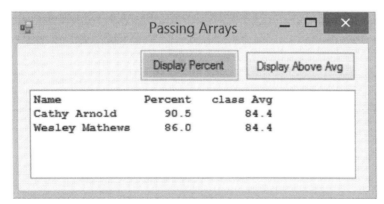

Both tasks, display percent and display above average students, require reading each student record and computing the percent score for the student. In the ReadTextFile form, you repeated the code to read and compute percents within the Click event handlers of both buttons. In the revised form, PassingArrays, you create a method called ComputePercent() and call it from both event handlers, as shown in Figure 8-27. Avoiding duplication of code makes maintenance of code easier.

Figure 8-27: Structure chart showing ComputePercent method shared by two event handlers

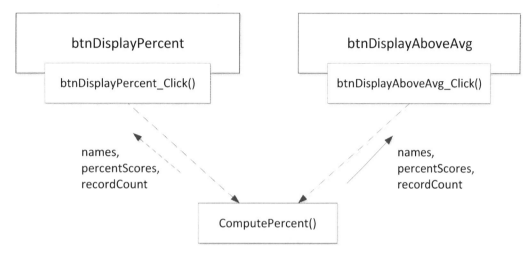

Figure 8-27 shows what happens conceptually. The names and percentScores are stored in arrays and passed from ComputePercent() method (which reads student records and computes percents) to both calling programs. But, how it is achieved in the code is almost the opposite: the reference variables for the two arrays are passed by value **to ComputePercent from the calling programs**. The ComputePercent method uses those references to access the arrays and fill the empty arrays with actual data.

For a called method to access an array, it must declare a reference variable for the array as a parameter in its header. For example, ComputePercent that updates the two arrays must declare two reference variables (with any valid names) for the arrays as shown below:

    private void ComputePercent(string[] names, float[] percentScores, ref int recordCount)

Note that the reference variables are passed by value (indicated by lack of the key word "ref") because there is no need for the method to change the reference to the arrays. ComputePercent uses the references, names and percentScores to access the arrays and update them. When control is passed back to the calling program, it can access the same arrays.

The calling program must create the arrays and pass (by value) the references to the called program, as shown below:

```
string[] names = new string[10];
float[] percentScores = new float[10];
int recordCount = 0;
```

    ComputePercent(names, percentScores, ref recordCount);

Note that recordCount is a value type variable. So, it must be passed by reference for ComputePercent to change its value.

The complete code for ComputePercent method is shown in Figure 8-28. This method reads each student's name and two exam scores, computes the percent score, stores the names in the array referenced by the parameter **names** and stores the percent scores in the array referenced by **percentScores**.

Figure 8-28: The ComputePercent method

```csharp
private void ComputePercent(string[] names, float[] percentScores, ref int recordCount )
{
    int index = 0;
    string currentLine;
    string[] fields;
    float percentScore;
    StreamReader examReader = new StreamReader("Exams.txt");

    while (examReader.EndOfStream == false)
    {
        currentLine = examReader.ReadLine();
        fields = currentLine.Split(',');
        percentScore = 100*(float.Parse(fields[1]) + float.Parse(fields[2]))/200;

        names[index] = fields[0]; //Store name into names array
        percentScores[index] = percentScore; // Store percent into percentScores
        index = index + 1;
    }
    recordCount = index;
}
```

Again, as stated earlier, the reference variables for the two arrays are passed by value because there is no need for this method to change the reference to the arrays. Passing the reference variables allows the method to access the corresponding arrays and change the values of the arrays.

The Click event handler of btnDisplayPercent that calls ComputePercent is shown in Figure 8-29.

Figure 8-29: Event handler of DisplayPercent button calling ComputePercent

```csharp
        private void btnDisplayPercent_Click(object sender, EventArgs e)
        {
            string[] names = new string[10];       //Declare names array
            float[] percentScores = new float[10]; // Declare percentScores array
            int recordCount = 0;

            ComputePercent(names, percentScores, ref recordCount); // Invoke ComputePercent

            lstDisplay.Items.Clear();
            string fmtStr = "{0,-15}{1,8:N1}";
            lstDisplay.Items.Add(string.Format(fmtStr, "Name", "Percent"));
            // Get names and percent scores from arrays and display them
            for (int index=0; index < recordCount; index ++)
                lstDisplay.Items.Add(string.Format(fmtStr, names[index], percentScores[index]));
        }
```

Line 47 calls ComputePercent method, which updates the two arrays.

Line 54 accesses the arrays to display them in the ListBox.

The Click event handler of btnDisplayAboveAvg also calls ComputePercent, as shown in Figure 8-30. This event handler uses the data from the arrays to compute the average and display names of above average students.

Figure 8-30: Event handler of DisplayAboveAvg button calling ComputePercent

```
56
57          private void btnDisplayAboveAvg_Click(object sender, EventArgs e)
58          {
59              string[] names = new string[10];
60              float[] percentScores = new float[10];
61              int index, recordCount = 0;
62
63              ComputePercent(names, percentScores, ref recordCount);
64
65              // Compute class average:
66              float classTotal = 0, classAvg;
67              foreach (float percentScore in percentScores)
68                  classTotal += percentScore;
69              classAvg = classTotal / recordCount;
70
71              lstDisplay.Items.Clear();
72              string fmtStr = "{0,-15}{1,8:N1}{2,12:N1}";
73              lstDisplay.Items.Add(string.Format(fmtStr, "Name", "Percent", "class Avg"));
74              // Display above average:
75              for (index = 0; index < recordCount; index++)
76              {
77                  if (percentScores[index] > classAvg)
78                      lstDisplay.Items.Add(string.Format(fmtStr, names[index],
                                            percentScores[index], classAvg));
79              }
80          }
```

After calling ComputePercent (line 63), this method accesses the arrays in line 67 to accumulate the total and in line 78 to display names and percents.

Step 5-1: Open the form PassingArrays, shown in Figure 8-26.

> Fill in the missing parameters for ComputePercent in line 20 of Figure 8-28.
> Add the code in lines 34 and 35 to store the name and percent into arrays.
> Uncomment line 38.
> Add the statements to call ComputePercent in the Click event procedures of btnDisplayPercent (line 47, Figure 8-29).
> Add the statements to call DisplayAboveAvg (line 60, Figure 8-30).

Step 5-2: Run the form and test the code.

8.8 Two-Dimensional (2-D) Arrays

A one-dimensional (1-D) array stores one set of data. A two-dimensional (2-D) array can store multiple sets of data, all of the same type. For example, the units sold in each quarter of a year for six different products could be stored in a 2-D array with 6 rows and 4 columns, where each cell represents an element of the array. Each row would represent a different product, and each column a different quarter, as shown in Figure 8-31.

Figure 8-31: A 2-D array

	index	0 (qtr 1)	1(qtr 2)	2(qtr 3)	3(qtr 4)
	0				
	1				
Rows	2				
(Products)	3				
	4				
	5				

Creating a 2-D Array

The syntax for creating a 2-D array is

> *type* [,] *arrayName* = new *type* [*rowCount, columnCount*]

where *rowCount* is the number of rows and *columnCount* is the number of columns. Note that the declaration of variables that represent a 2-D array requires a comma inside the pair of square brackets.

The following statement creates an int type array, called sales, that has 6 rows and 4 columns:

> int[,] sales = new int[6, 4];

Creating a 2-D Array with Initial Values

You may use the array initializer to create a 2-D array with initial values other than zero by specifying each set of values within curly braces, as shown below.

> int [,] sales = { {16, 32, 23, 40}, {144, 182, 169, 205}, {54, 49, 72, 63} }

The previous statement creates the following 2-D array, named sales, with 3 rows and 4 columns.

		Columns			
		0	1	2	3
	0	16	32	23	40
Rows	1	144	182	169	205
	2	54	49	72	63

Referring to an Element in a 2-D Array

Each element in a 2-D array is referred to by the index of the row and the index of the column. Here are some examples:

    sales[0, 0]—the element at row 0 and column 0, with the value 16
    sales[0, 1]—the element at row 0 and column 1, with the value 32
    sales[0, 2]—the element at row 0 and column 2, with the value 23
    ...

    sales[1, 0]—the element at row 1 and column 0, with the value 144
    sales[1, 1]—the element at row 1 and column 1, with the value 182
    sales[1, 2]—the element at row 1 and column 2, with the value 169

Example: Using 2-D Arrays for Sales Data

This example uses the following data that consist of the product number and quarterly sales for a set of products:

1033, 16, 32, 23, 40
1009, 144, 182, 169, 205
1014, 54, 49, 72, 63
1005, 38, 62, 55, 77
1001, 7, 9, 8, 15
1025, 98, 110, 105, 130

The application that process these data use a form that allows the user to specify any product number and display the total sales for that product, as shown in Figure 8-32.

Figure 8-32: The TwoDimArrays Form

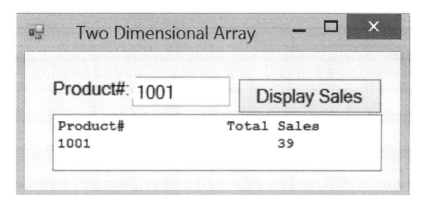

In this application, you need to store two types of data. Though the product number consists of only numeric digits, it is essentially text data because you do not do any numeric calculations with it. The sales data, however, are integers. An array has to be of a single type.

One option would be to create one string type 1D array to store product numbers, and another int type 2-D array to store the sales.

A second approach would be to store product numbers and sales in a single 2-D array. Because all data in the text file, including sales, is stored as text, we use the second approach—that is, create a string type array that stores both product numbers and sales data. You will need to convert the sales data to int type when you read them from the array so that you can add them up.

The code has two major components:
1. Load the data from the text file to the 2-D array. This is done in the **load event** of the form.
2. Search the array to find the sales for a specified production number. This is done in the **Click event** handler of a button.

Because the array is used in two different methods, it is declared outside both methods at the class level. Figure 8-33 shows part 1 of the code.

Figure 8-33: Code to create a 2-D array and load data into it

```
20      //Declare a 2-D array with 10 rows and 5 columns:
21      string[,] sales = new string[10, 5];
22      int recordCount;     // Actual number of records read
23
24      private void TwoDimArray_Load(object sender, EventArgs e)
25      {
26          string currentLine;
27          string[] fields = new string[2];
28          int row = 0, column = 0;
29          //Create StreamReader
30          StreamReader salesReader = new StreamReader("Sales.txt");
31
32          //Read each line from the file, split and store in a row of the 2-D array.
33          while (salesReader.EndOfStream == false)
34          {
35              currentLine = salesReader.ReadLine();//Read product# and 4 sales data
36              fields = currentLine.Split(','); //Store Product# & sales in a temp array
37
38              //Copy Product# and four sales to the five columns of the current row:
39              sales[row, 0] = fields[0];   // product number stored in column 0
40              sales[row, 1] = fields[1];   // Sales for qtr 1 stored in column 1
41              sales[row, 2] = fields[2];   // Sales for qtr 2 stored in column 2
42              sales[row, 3] = fields[3];   // Sales for qtr 3
43              sales[row, 4] = fields[4];   // Sales for qtr 4
44
45              // Or, use a loop to change the column of the array element:
46              for (column = 0; column <= 4; column ++)
47                  sales[row, column] = fields[column];
48
49              row = row + 1;
50          }
51          recordCount = row; // recordCount is the number of rows that has data in it.
52          salesReader.Close();
53      }
```

The code shown in Figure 8-33, which reads the data from the text file and stores them in the array, has three major steps:

    Read current line into a string variable named currentLine (line 35).
    Split it and store in a temp array named fields (line 36).
    Copy from fields to a row of the 2-D array (lines 39–43 and lines 46–47 do the same thing).

The data are copied from the current row in the text file to the 2-D array, as shown in Figure 8-34.

Figure 8-34: Copying the current line to a temp array and then to a 2-D array

Current row in the text file: 1033, 16, 32, 23, 40
string variable, currentLine: "1033, 16, 32, 23, 40"

fields (1D array):

0	1	2	3	4
"1033"	"16"	"32"	"23"	"40"

sales (2-D array):

	0	1	2	3	4
0	"1033"	"16"	"32"	"23"	"40"
1					
2					
3					
4					
5					

Rows (Products)

The code in Lines 39–43 copies the five elements of the 1D array, **fields**, to the current row of the 2-D array, **sales**, using a separate statement for each element.

Consider line 39: **sales[row, 0] = fields[0];**. The first time through the while loop, the value of row is 0. So, fields[0] (the product#) is copied to sales[0,0]—that is, row 0 and column 0.

Similarly, line 40, **sales[row, 1] = fields[1]**, copies fields[1] (sales for qtr. 1) to sales[0,1].

The five statements together copy all five fields from the current line of Sales.txt file into the five columns of the current row to fill the first row, as shown above in Figure 8-34.

The second time through the loop, the value of row is 1. So, the statements in lines 39–43 copy data from the second line of Sales.txt to sales[1,0], sales[1,1],…, sales[1,4], filling the second row of sales array, as shown in the following table.

Sales (2-D array):

	0	1	2	3	4
0	"1033"	"16"	"32"	"23"	"40"
1	"1009"	"144"	"182"	"169"	"205"
2					
3					
4					
5					

Rows (Products)

The process continues until all lines of Sales.txt are read and copied into sales array.

Using a Loop to Copy from Fields to Sales

In the following code, the five statements in lines 39–43 are replaced by a single statement within a loop in lines 46–47:

```
46                  for (column = 0; column <= 4; column ++)
47                      sales[row, column] = fields[column];
```

The first time through the loop, column is 0. So, line 47 is equivalent to line 39:
    sales[row, 0] = fields[0];
The second time, column is 1. So, line 47 is equivalent to line 40:
    sales[row, 1] = fields[1];

In effect, this loop generates the five statements in lines 39–43.

Tutorial 6: 2-D Array to Find the Sales for a Product

This tutorial reads the sales data from the file, Sales.txt, into a two-dimensional array, and uses the array to find the total sales for a product.

Part 1

Step 6-1: Open the project, Ch8_ArraysAndFiles. Open the form, TwoDimArray, shown in Figure 8-35.

Figure 8-35: The TwoDimArray form

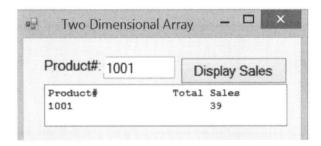

Step 6-2: Open the load event handler of the form.
   Add the code in lines 46 and 47 from Figure 8-33 to copy the data from Sales.txt to the array sales.
   Comment out lines 39–43 because lines 46 and 47 do the same thing.
   Put a break at line 50.

Step 6-3: Run the form. When the program breaks, display Locals window (Debug, Windows, Locals), and view the values of fields array. Because sales array is declared at the class level, it is not shown in Locals window.

To view sales array, display **QuickWatch** window (Debug, QuickWatch), type sales for *Expressions*, click *Add Watch*, and click *Close*. Expand sales array to view the values, displayed as follows:

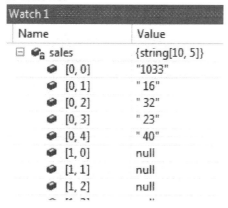

Click continue to watch the values in row 1 of sales array.
Click continue until all lines from Sales.txt are copied to the array.

Part 2

Search 2-D array to find the sales for a specified product number. This is done in the Click event handler of btnDisplaySales. The code is shown in Figure 8-36.

Figure 8-36: Code to search the 2-D array

```
54
55        private void btnDisplaySales_Click(object sender, EventArgs e)
56        {
57            int row = 0, column, rowSum=0;
58            string fmtStr = "{0,-20}{1,8}";
59            lstDisplay.Items.Clear();
60            lstDisplay.Items.Add(string.Format(fmtStr, "Product#", "Total Sales"));
61
62            for (row = 0; row < recordCount; row++)
63                // recordCount is the number of rows with data in it
64            {
65                // Check whether column 0 of current row matches Prodduct#
66                if (sales[row, 0] == txtProdNumber.Text)
67                {
68                    // Add columns 1 - 4
69                    rowSum = int.Parse(sales[row, 1]) + int.Parse(sales[row, 2]) +
70                            int.Parse(sales[row, 3]) + int.Parse(sales[row, 4]);
71                    // Alternate method using a loop:
72                    for (column = 1; column <= 4; column++)
73                        rowSum = rowSum + int.Parse(sales[row, column]);
74
75                    lstDisplay.Items.Add(string.Format(fmtStr, txtProdNumber.Text, rowSum));
76                    return;    // Exit the method, if a matching product is found
77                }
78        }
```

Let's look at the key statements:

Line 62: for (row = 0; row < recordCount; row++)

Here, recordCount is the actual number of rows with data in it. The "<" sign is used in the expression because the largest row number that has data is one less than the number of rows.

Line 66: if (sales[row, 0] == txtProdNumber.Text)

For each value of row, this statement compares Column 0 of the array (product#) with the product# that you entered into the TextBox.

Line 69: rowSum = int.Parse(sales[row, 1]) + int.Parse(sales[row, 2]) +

    int.Parse(sales[row, 3]) + int.Parse(sales[row, 4]);

This statement adds columns 1–4 (sales for quarters 1–4) of the current row.

Lines 72-73: for (column = 1; column <= 4; column++)

    rowSum = rowSum + int.Parse(sales[row, column]);

By varying the value of column from 1 to 4, the statements add columns 1–4 of the current row to rowSum.

Line 76: return

The return statement, in general, transfers control to the calling method. Here, the effect is to stop processing the current method so that the rest of the rows are not checked for a match.

Step 6-4: Open the Click event handler of btnDisplaySales. Fill in the code from lines 72 and 73 from Figure 8-36 to compute the total for each row. Comment out line 69. Put breaks at lines 73 and 78.

Step 6-5: Run the form. Enter 1014 for Product Number, and click the Display Sales button.

When the program breaks at line 78 for the first time, observe the values of row and rowSum in the Locals window.

View the data stored in sales array by displaying the QuickWatch window (Debug, Windows, Watch, Watch1). Click Continue, and observe the data for the second iteration in the Locals window. To switch to Locals window, click the Locals tab at the bottom left of the window.

Click Continue to go through the loop the third time when row 2 that contains product# 1014 is accessed. Because the expression (sales[2,0] == txtProdNumber.Text) is true, line 72 is processed, and the program breaks at line 73.

Observe the values of column and rowSum for each iteration of the loop by clicking Continue. The final value of rowSum (238) is displayed in the ListBox.

It's time to practice! Do Exercise 8-5.

Jagged Arrays

Jagged arrays are similar to two-dimensional arrays, except that the rows can have different lengths. It is a one-dimensional array where each element is an array that does not have a fixed size. Figure 8-37 shows an example of a jagged array with three rows.

Figure 8-37: Jagged array

	0	1	2	3
0	16	32	23	40
1	144	182		
2	54	49	72	

Rows

This is how you would create a jagged int type array with three rows:

   int[][] jaggedSales = new int[3][];

Note that the declaration of the variable includes two sets of square brackets, indicating that it is an array of arrays. Further, the number of columns is not specified.

You specify the number of columns for each row and the initial values (optional) as follows:

   jaggedSales[0] = new int[4] {16, 32, 23, 40};
   jaggedSales [1] = new int[2] {144, 182};
   jaggedSales [2] = new int[3] {54, 49, 72};

You access an element of the array by specifying the row and column index. For example, jaggedSales [2][1] has the value 49.

Review Question

8-11   The following code creates a 2-dimensional array, and then displays one element of the array in a ListBox named lstDisplay.

   int[,] sales = {{ 16, 32, 23, 40 }, { 144, 182, 1679, 205 },{ 54, 49, 72, 63 } };
   lstDisplay.Items.Add(sales[1, 3]);

What would be the output displayed in the ListBox, lstDisplay?

Exercises

Exercise 8-1

The following is a sample of the order data (Customer#, Order#, OrderQty, UnitPrice) to be processed. Store the data in a text file named Order.txt. You may use Notepad to enter the data.

0155, 76533, 3, 10.95
0140, 75446, 12, 8.50
0150, 75550, 1, 89.95
0145, 75648, 25, 2.50
0130, 76360, 5, 17.90

Create a form, ProcessOrder, that reads each order record, computes the extended cost, and displays Cust#, Order#, OrderQty, UnitPrice and ExtendedCost in a ListBox, with the columns aligned.

Exercise 8-2

Add a new button named btnAboveAvgForExam1. Add the code to the Click event handler of the button to display students with above average scores for the first exam. List the student's name, exam 1 score, and class average.

Hint: When the file is read, store the score for exam 1 in an array, and compute the average for exam 1. Next, use a loop to list all students whose exam 1 scores were above the class average.

Exercise 8-3

Modify the form named ReadTextFile within the project Ch8_FilesAndArrays so that when the user presses the ComputePercent button, the Click event handler will write the name, score1, score2 and percentScore to a text file named PercentScores.txt. Add the necessary code to the end of the Click event handler of ComputePercent button.

Exercise 8-4

Modify Exercise 8-3 so that the user can select the file to write the name, score1, score2 and percentScore.

Exercise 8-5

Open the project Ch8_ArraysAndFiles from the Tutorial_Starts folder. Open the form named TwoDimArray. Add a button, btnQtr1Total, that computes the total sales for a specified quarter and displays it in the ListBox. Let the user enter the required quarter (a number 1–4) into a TextBox.

Programming Assignment 2

The main goals of this assignment are to give you experience in
1. Designing and creating a user interface that includes different types of controls like RadioButtons and ComboBoxes
2. Reading from and writing to text files
3. Working with arrays

You are to develop an enhanced version of the system specified in Programming Assignment 1 in Chapter 3. "Ace Auto Rentals" is a small business that rents automobiles. Ace would like you to develop an application that allows them to process the rental charges as specified below:

Create a form, EnterRentals, that allows the user to do the following for each car rental:

1. Enter a five-digit customer#, customer name, rental date and number of days of rental.
2. Select the type of auto. The current options are

    SUV, minivan, luxury, full size, midsize and compact.

    This list is likely to change.

    To make it easy for the user to select an auto type and view the daily rental rate, first you must store each auto type and corresponding rate in a text file named AutoTypes.txt (SUV = $ 80, minivan = $70, luxury = $50, full size = $40, midsize = $35 and compact = $25).

    Every time you open the form, read the auto types and rates, and add the auto types to a ComboBox or ListBox, sorted by auto type, so that the user can select one from the list.

    In addition, store the auto types in an array named autoTypes, and store the rates in another array named dailyRates.

    When the user selects an autotype, find the rate for the selected auto type from the arrays and display the rate in a Label.

3. Select one of the discount categories. The current options are

    Normal, State Employees, Business and Favorite Customer.

    This list is **not** likely to change.

    Choose the appropriate type of GUI to let the user select a category.

4. Compute the rental charge, discount and total cost.
    Rental charge: Number of days x Daily rental rate
    Discount: Normal = 0%, State emps = 15%, Business = 10%, Favorite = 20%
    Total cost: Rental charge + Discount

5. Display the number of days, rental charge, discount and total cost, nicely formatted with proper labels.

6. Write the customer#, customer name, type of auto, number of days, rental charge, discount and total cost into a text file, Rental.txt. The user should be able to clear the data and enter data on rentals for additional customers without having to exit and re-run the form.

Validate the data to make sure that no input data are missing, date and number of days are valid and the user made a selection for auto type and discount.

Use the form to enter and save information for several rentals. Make sure that at least a few customers have more than one rental.

Store the files Autotypes.txt and Rental.txt in the default path for files (typically, the Bin/Debug folder). **Don't specify any path for files** (so that the program can be tested on a different computer by copying the project directory).

Create a second form, **DisplayRentals**, that allows the user to display all previous rentals (customer#, type of auto and rental date) for any specified customer#. Allow the user to type a customer# into a TextBox, and click a button to display the data. Format and align the columns nicely.

To keep it simple, you may read the data from the Rental.txt file each time to find the rentals. (A better way would be to read the data in the Form Load event handler, store the data in arrays one time and search the arrays to find the rentals. You can get 5% extra points if you use this method).

Create a third form, **Rentals**, that serves as a switchboard with buttons that allow the user to go to EnterRentals and DisplayRentals. To open EnterRentals, for example, from the form Rentals, do the following:

> First, declare a variable named EnterRentals to represent EnterRentals form:
> EnterRentals EnterRentals;
> Next, create an object of the type EnterRentals, and make it visible using the Show method (we will discuss this in more detail later):
> EnterRentals = new EnterRentals();
> EnterRentals.Show();

Make Rentals the startup form.

The factors that determine your grade include the following:

Use meaningful and easily understood names for forms, controls and variables used in the code. Use standard naming conventions. Do not use names like form1 and button1.

Make the code easy to understand. Use comments to specify what each part of the code does. Avoid unnecessary duplication of code.

Use the C# features you learned so far, effectively.

Chapter 9

Collections

A collection, like an array, can hold a group of elements (items). An element could be of any data type, like a simple number, a string, an array or an object. How do collections differ from arrays? This chapter describes the unique features of collections and presents their practical applications.

Topics

9.1	Introduction to Collections		9.3	Dictionary<TKey, TValue>Collection
9.2	List<T>Collection			

9.1 Introduction to Collections

You already used collections when you worked with GUI controls like ComboBox and ListBox. To create a list of items in these controls, you added items to the **Items** collection of these controls using the Items.Add() method, without having to specify the index of the element, as in a C# array. Collections provide several such methods that make it easy to work with them without having to know the underlying structure. Further, the size of a collection is increased automatically when you insert new items.

The classes provided by the .Net Framework to create collections can be grouped broadly into **generic (typed)** and **nongeneric (untyped)** collections. The generic collections are strongly typed—that is, they allow you to specify a single type that can be stored in the collection. Nongeneric collections allow elements of different types to be stored in a single collection. That means a nongeneric collection intended to be of a certain type like float may contain, for example, a string that was accidently added to it. Therefore, use of nongenerics is generally not recommended, unless there is a special reason for using them. Both types of collections may include strings, arrays, objects and value types like int and float.

Some generic types have corresponding nongenerics, and others do not. Table 9-1 shows some commonly used generic collections and their corresponding nongeneric collections, if any.

Additional features of the different collections listed in Table 9-1 include the following:

> **List** does not require a unique key, but **Dictionary, SortedList** and **SortedDictionary** require a unique key.
> **SortedList** and **SortedDictionary** maintain the data in sorted order, making insertion and deletion of items generally less efficient than unsorted collections.

Table 9-1: Commonly used collections

Generic (typed) collections	Corresponding nongeneric (untyped) collections	Implementation and other features
List<T> **T** represents the type	ArrayList	Implemented as an array; allows accessing items by index. Performance could be slightly lower compared to arrays, but easier to use.
Dictionary<TKey, TValue> **TKey** represents the key, and **TValue** represents the value	Hashtable	Implemented as a hash table. Values accessed by key.
SortedList<TKey, TValue>	SortedList	Implemented as two arrays, one for the key and another for values; allows access by index, and by key.
SortedDictionary<TKey, TValue>		Implemented as a tree. Generally, better than SortedList, when • adding and removing items into large sets • number of items unknown • populating from unsorted data
LinkedList<T>		A group of nodes, each node consisting of a data item and a link to the next node. Efficient to insert items, but inefficient to search.
Queue<T>	Queue	A FIFO (first in, first out) list
Stack<T>	Stack	A LIFO (last in, first out) list

We will look at two popular unsorted generic collections that have overall good performance:

1. List<T>, which does not require a unique key, and
2. Dictionary<TKey, TValue>, which requires a unique key.

Review Questions

9.1 List two features of C# collections that distinguish them from arrays.
9.2 What is a generic collection?

9.2 List<T> Collection

List is a generic (typed) collection of items, similar to an array. Elements, in general, could be of any data type like a simple number, a string, an array or an object; but within any one List, all items must be of the same type. For example, a List of arrays, where each item is an array that contains a product number and four quarterly sales, might look as follows:

Index

0	1033	16	32	23	40
1	1009	144	182	169	205
2	1014	54	49	72	63
3	1005	38	62	55	77
4	1001	7	9	8	15
5	1025	98	110	105	130

Methods and Properties of List

Common properties and methods of List are shown in Table 9-2.

Table 9-2: Common properties and methods of List

Property	Description
Capacity	The number of elements that the List can hold
Count	The number of elements in the List
Method	
Add(item)	Adds an element to the end of the List and returns the index of the element added
Clear()	Removes all elements from the List
Contains(item)	Returns true if the item is found in the List; if not, returns false
BinarySearch(item)	Searches a List for the item. Returns the index of the item, if found. If not, a negative integer is returned.
Insert(index, item)	Inserts the item at the specified index
Remove(item)	Removes the first occurrence of the specified item
RemoveAt(item)	Removes the item at the specified index
Sort()	Sorts the elements in the List

Creating a List

The statement to create a List has the following syntax:

    List<*Type*> *listName* = new List<*Type*>();

where *listName* is the name of the variable that references the List object, and *Type* is the type of items in the List. Note that a pair of parentheses is required in the *new* expression after the type.

Examples

List<int> unitsSold = new List<int>();   // Creates a List of int type named unitsSold

List<string> movies = new List<string>();   // Creates a List of string type named movies

List<int[]> classGrades = new List<int[]>();   // Creates a List of int arrays named classGrades

List<Array> examScores = new List<Array>();   // Creates a List of arrays named examScores

List<Customer> customers = new List<Customer>(); // Creates a List of Customer type

Note that names like unitsSold and movies in the above examples are the names of variables that reference the List, although, for the sake of brevity, we describe them as the names of Lists.

In general, the term "List" is used in place of "variable that references the List object."

Adding Items to a List

The Add() method adds a specified item to the end of the List. When the number of items in a List exceeds its initial capacity, the capacity is doubled automatically.

Examples Using the Lists Created Above

    unitsSold.Add(205);   // adds 205 to the end of the unitsSold List

    movies.Add("Pocahontas");

    int[] studentGrades = {85,91,80};   //creates the array to be inserted into a List
    classGrades.Add(studentGrades);   //Adds the array studentGrades to the List, classGrades

Accessing an Item

One way to access each element is by specifying the index of the element.

Examples

    unitsSold[0] returns the int type number stored in the first element of unitsSold

    classGrades[0] returns the array stored in the first element of classGrades

You also may use the foreach loop to access every element of the List.

Tutorial 1: Looking Up Sales Data Using a List

Let's look at an example, similar to the one discussed in Chapter 8, to look up sales data for a specified product number. The product number and sales for quarters 1–4 are read from the text file, Sales.txt, shown in Figure 9-1. The first data item in each row is the product number, which is followed by sales for quarters 1–4.

Figure 9-1: Sales.txt File

1033, 16, 32, 23, 40
1009, 144, 182, 169, 205
1014, 54, 49, 72, 63
1005, 38, 62, 55, 77
1001, 7, 9, 8, 15
1025, 98, 110, 105, 130

In Chapter 8, we used a 2-D array to store the data. In this application, you will use a generic List to store the data and to help find the total annual sales for a specified product. The product number is selected from a ComboBox, and the total sales displayed in a ListBox, as shown in Figure 9-2.

Figure 9-2: The GenericList form

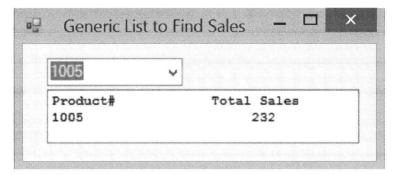

The coding for this form consists of three parts:
1. Create a List named sales.
2. Read each row from Sales.txt, split it and store it in a temporary array, and add the array to the List. Add the product number to the ComboBox. Both tasks are done in the Load event handler of the form.
3. When the user selects a product number from the ComboBox, find the corresponding item from the List. This is done in the SelectedIndexChanged event of the ComboBox.

Creating the List

Step 1-1: Open the project Ch9_Collections from Tutorial_Starts folder. Open the form GenericList, shown in Figure 9-2.

Tutorial 1: Looking Up Sales Data Using a List

Step 1-2: Create a List of string arrays, referenced by the variable, sales, as shown in line 21 of Figure 9-3. This is declared at the class level so that the List can be used in the Load event and in the SelectedIndexChanged event. The List is of string[] (string array) type, though the product number and units sold are integers, because the data read from the Text file are string type.

Figure 9-3: Code to create the List, referenced by the variable sales

```
17          public GenericList()
18          {
19              InitializeComponent();
20          }
21          private List<string[]> sales = new List<string[]>();
```

Figure 9-4 shows part 2 of the code that reads each row from Sales.txt, splits it and stores it in a temporary array, and adds the array to the List. Except for line 36, the code is the same as what you used in other examples to read each line from a text file and store it in an array called fields. The code also adds the product number to the ComboBox.

Figure 9-4: Code to add items to the List and ComboBox

```
22
23          private void GenericList_Load(object sender, EventArgs e)
24          {
25              string currentLine;
26              string[] fields = new string[2];
27              //Create StreamReader
28              StreamReader SalesReader = new StreamReader("Sales.txt");
29
30              //Read each line from the file, split and store in an array
31              while (SalesReader.EndOfStream == false)
32              {
33                  currentLine = SalesReader.ReadLine(); // Read a product# and four sales data
34                  fields = currentLine.Split(','); // Store Product# and Sales in a temp array
35
36                  sales.Add(fields);                  // Add the entire array to the List, sales.
37                  cboProdNumber.Items.Add(fields[0]); // Add product# to the ComboBox
38              }
39              SalesReader.Close();
40          }
```

Line 36 adds the array fields to the List. The array consists of the product number and sales for quarters 1-4. Figure 9-5 shows the List with the first row of the file added to it.

Figure 9-5: sales List after the first item is added

Index

| 0 | 1033 | 16 | 32 | 23 | 40 |

1

2

Step 1-3: Insert the code from line 36 of Figure 9-4 into the Load event handler of the form.

Step 1-4: Make sure that the file Sales.txt is in the Debug folder of the project.
Open Program.cs file and specify GenericList as the startup form.
Put a break at line 38 at the end of the loop, and run the form.
When the program breaks at line 38, click Continue to add a second item.
Display Watch window (Debug, Windows, Watch, Watch1) and type **sales** for Name.
Expand sales List to view the two arrays that were added to sales, as shown in Figure 9-6.

Figure 9-6: sales List after adding two items

Name	Value
[0]	{string[5]}
[0]	"1033"
[1]	" 16"
[2]	" 32"
[3]	" 23"
[4]	" 40"
[1]	{string[5]}
[0]	"1009"
[1]	" 144"
[2]	" 182"
[3]	" 169"
[4]	" 205 "
Raw View	
Capacity	4
Count	2

Click *Continue* two more times. After the fourth time, you should see the Capacity doubled to 8 from the initial capacity of 4. Figure 9-7 shows the List with six rows of the file added to it.

Figure 9-7: sales List after adding all six rows

Index

0	1033	16	32	23	40
1	1009	144	182	169	205
2	1014	54	49	72	63
3	1005	38	62	55	77
4	1001	7	9	8	15
5	1025	98	110	105	130

Finding an Item in the List Using the Index

When both the ComboBox and the List are not sorted (or when both are sorted), both have the product numbers stored in the same order because they were added in the same order from the Sales.txt file. That means the index of the selected product in the ComboBox is the same as the index of the corresponding item in the List. So, the SelectedIndex from the ComboBox can be used as the index for the List to find the item.

For example, if you select product number 1014 from the ComboBox, the SelectedIndex would be 2, and sales[2] will give you the corresponding array of four quarterly sales.

Figure 9-8 shows the code to find the quarterly sales using the SelectedIndex from the ComboBox, and find the sum of sales for the four quarters.

Figure 9-8: Code to find an item using the index of the selected item

```
41
42        private void cboProdNumber_SelectedIndexChanged(object sender, EventArgs e)
43        {
44            string fmtStr = "{0,-20}{1,8}";
45            lstDisplay.Items.Clear();
46            lstDisplay.Items.Add(string.Format(fmtStr, "Product#", "Total Sales"));
47
48            // If both the List and ComboBox are not sorted, or if both are sorted, then
49            // you can use the SelectedIndex from the ComboBox to find the item from the List:
50            int selectedIndex = cboProdNumber.SelectedIndex;
51            string[] productSale = sales[selectedIndex];
52
53            // compute total sales for quarters 1 - 4
54             int annualSales = 0;
55            for (int quarter = 1; quarter <= 4; quarter++)
56                annualSales = annualSales + int.Parse(productSale[quarter]);
57
58            lstDisplay.Items.Add(string.Format(fmtStr, cboProdNumber.SelectedItem, annualSales));
```

Line 50: int selectedIndex = cboProdNumber.SelectedIndex;

The above statement gets the index of the selected product from the ComboBox and stores it in selectedIndex.

Line 51: string[] productSale = sales[selectedIndex];

sales[selectedIndex] gets the item (array) from the position specified by the selectedIndex. This string array is stored in the reference variable productSale of string array type. The string array, productSales, would have five elements with indices 0 - 4 representing the product number and sales for quarters 1 through 4.

Lines 55-56 use a loop to add up the four quarterly sales to get the total sales for the year.

Step 1-5: Add code in lines 50 and 51 of the SelectedIndexChaned event from Figure 9-8.
Add the code in lines 55 and 56 of the SelectedIndexChanged event to compute the annual sales.

Step 1-6: Make sure the Sorted property of the ComboBox is set to its default value of False. Put a break on line 54. Run the form. Select product number 1005.
When the program breaks, display the Locals window (Debug, windows, Locals) and verify that the selectedIndex is 3 and the List displays the corresponding array of sales.
Continue running. The total sales should be 232.

Step 1-7: Set the Sorted property of ComboBox to True. Run the form and select the same product number, 1005. You will see the wrong total sales because the index of 1005 has changed to 1 in the sorted ComboBox, but in the List, the index of 1005 is still 3.

Finding an Item in the List Using foreach Loop

When either the ComboBox or the List is sorted, and the other is not sorted, the order of the two sets of product numbers is different. In this case, you need to use a loop to match the product number in each item (array) in the List to the selected product number, as shown in Figure 9-9.

Figure 9-9: Code to find an item using foreach loop

```
59    // If either the List or the ComboBox is sorted and the other not sorted, then
60    // loop through each item in sales collection to find the selected Product:
61    // This works also when both are sorted, and both are not sorted
62    string selectedProduct = cboProdNumber.SelectedItem.ToString();
63
64    // 1. Using foreach loop to search the List:
65    foreach (string[] prodSales in sales) // prodSale represents an item (array) in the List.
66    {
67        if (prodSales[0] == selectedProduct) // element 0 is the Product#
68        {
69            // compute total sales for quarters 1 - 4
70            annualSales = 0;
71            for (int quarter = 1; quarter <= 4; quarter++)
72                annualSales = annualSales + int.Parse(prodSales[quarter]);
73
74            lstDisplay.Items.Add(string.Format(fmtStr, cboProdNumber.SelectedItem, annualSales));
75            return;     // Exit the method, if a matching product is found
76        }
77    }
```

Line 65: foreach (string[] prodSales in sales)

Because each item in the List is an array of string types, each item is assigned to the variable prodSales, which also is an array of string types.

Line 67: if (prodSales[0] == selectedProduct)

prodSales[0] is the first element of the array, which is the product#. It is compared to the product number selected in the ComboBox.

Line 71 - 72: for (int quarter = 1; quarter <= 4; quarter++)
            annualSales = annualSales + int.Parse(prodSales[quarter]);

Elements 2 through 5 (index 1 - 4) of the prodSales array contain the quarterly sales, which are added together to compute the annual sales.

Step 1-8: Add code from Figure 9-9 to the SelectedIndexChanged event handler. Make sure the Sort property of the ComboBox is set to True. Comment out the code that uses the previous method in lines 50–58.

Step 1-9: Remove the break from line 54. Put a break on line 76 and run the form.
Observe the values of the prodSales array, including prodSales[0], which is the product number. Repeat clicking the Run button and observe the values. Note that this method gives the correct annual sales when the ComboBox is sorted, although the data in the product numbers in the sales array are not sorted.

It's time to practice! Do Exercise 9-1.

Next, you will create the same application using a Dictionary collection.

Review Questions

9.3 The file, Price.txt, contains the Product# and price of several products, as shown below:
Price.txt
"1033", 89.95
"1009", 21.50
"1014", 159.00
"1005", 2.50
"1001", 55.70
"1025", 33.45

The following code is intended to read the data from Price.txt and store them in a List of Arrays named prices:

```
List<string[]> prices = new List<string[]>();
private void GenericList_Load(object sender, EventArgs e)
{
        string currentLine;
        string[] fields;
        StreamReader PriceReader = new StreamReader("Price.txt");
        while (PriceReader.EndOfStream == false)
        {
                currentLine = PriceReader.ReadLine();
                fields = currentLine.Split(',');
                XXXXXX   // missing code to add product number and price to the List, prices
        }
        PricesReader.Close();
}
```

In the above code, "XXXXXX" represents missing code to add the product number and price to the List, prices. Write the missing code.

9.4   A List of Arrays named prices is created as follows:
      List List<string[]> prices = new List<string[]>();

Assume that each row from the text file, Price.txt (see Review Question 9.3), is read and stored in the List as a string array.
Write the code to find the price for product# "1001," and store the price in a string variable named price.

9.3 Dictionary<TKey, TValue> Collection

A Dictionary is a collection of keys and values. TKey and TValues specify the type of the key and of the value, respectively. For example, a dictionary that stores a set of product numbers and corresponding prices might look as follows:

Key (Prod#)	Value (Price)
1033	89.95
1009	21.50
1014	159.00
1005	2.50
1001	55.70
1025	33.45

Thus, a major difference between a List and a Dictionary is that a Dictionary requires a unique key that makes it easier to find an item using the key. Similar to a List, a Dictionary can store data of any type.

A Dictionary is implemented as a hash table that uses an algorithm to determine the location of an item based on the key. Overall, the Dictionary is efficient in finding items and in inserting/deleting items, though the performance may vary with the type of the key. However, it is less efficient than Lists for looping through with foreach. Figure 9-10 shows some commonly used properties and methods of the Dictionary.

Figure 9-10: Commonly used properties and methods of Dictionary

Property	Description
Count	The number of key/value pairs in the Dictionary
Item	The value associated with the specified key
Keys	A collection containing the keys in the Dictionary
Values	A collection containing the values in the Dictionary
Method	
Add(key, value)	Adds a key/value pair to the Dictionary
Clear	Removes all keys and values
Remove(key)	Removes the value for the specified key
ContainsKey(key)	Determines if the Dictionary contains the specified key
ContainsValue(value)	Determines if the Dictionary contains the specified value
TryGetValue(key, *out* variable)	Returns the value, if the key is found; if not, returns false

Creating a Dictionary

The syntax to create a Dictionary object, and assign it to a reference variable, is
> **Dictionary<*TKey, TValue*> *dictionayName* = new Dictionary <*TKey, TValue*>();**

where *TKey* and *TValue* specify the type of the key and of the value, respectively, and *dictionaryName* is the name of the variable that references the Dictionary.

A value can be of any data type. But using reference types, like array, for keys may pose a problem because this may not generate unique hash keys for storing the values using the standard method of generating keys. Similarly, using objects for keys becomes more complex.

Examples

> Dictionary<string, float> prices = new Dictionary<string, float>();
> —Creates a Dictionary named prices with key of string type and value of float type

> Dictionary<int, Array> sales = new Dictionary<int, Array>();
> —Creates a Dictionary named sales with key of int type and value of Array type

> Dictionary<short, Customer> customers = new Dictionary<int, Customer>();
> —Creates a Dictionary named customers with key of short type and value of Customer type

Names like prices and sales are the names of variables that reference the Dictionary object, although, for the sake of brevity, we describe them as the names of Dictionary.

In general, the term "Dictionary" is used in place of "variable that references the Dictionary object."

Adding Items to a Dictionary

To add an item to a Dictionary, you add a key and a value, as in the following examples:
> prices.Add("Litebook 15", 399.95);
> —Adds an item with "Litebook 15" as the key and 399.95 as the value

> float[] prodSales = {55.5, 29.0, 109.5, 85.0};
> sales.Add(4, prodSales);
> —Adds an item with 4 as the key and prodSales array as the value

Accessing an Item

An item in a Dictionary is accessed by specifying the key, as in the following examples:

> prices["Litebook 15"]
> —Returns the value 399.95

> sales[4]
> —Returns the array {55.5, 29.0, 109.5, 85.0};

Tutorial 2: Dictionary with Product# as Key and Price as Value

The application developed in this tutorial stores a string type product number as the key and a float type price as the value of the Dictionary. The product numbers and corresponding prices are read from the file Price.txt shown in Figure 9-11, and stored in the Dictionary, prices<string, float>.

Figure 9-11: Price.txt file

"1033", 89.95
"1009", 21.50
"1014", 159.00
"1005", 2.50
"1001", 55.70
"1025", 33.45

A ComboBox is used to let the user select a product number and display the corresponding price from the Dictionary, as shown in Figure 9-12.

Figure 9-12: The DictionaryValueType form to find product price

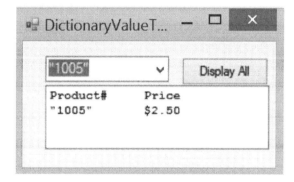

The coding has three parts:
1. Create a Dictionary named prices.
2. Read each row from Price.txt, split it and store it in a temporary array, and add the product number as the key and price as the value to the Dictionary. In addition, add the product number to the ComboBox. This is done in the Load event handler of the form.
3. When the user selects a product number from the ComboBox, find the corresponding item from the Dictionary. This is done in the SelectedIndexChanged event of the ComboBox.

Creating a Dictionary

Step 2-1: Open the form DictionaryValueType, shown in Figure 9-12.

Step 2-2: Add the code to create a Dictionary named prices with key of string type, representing the product number and value of float type, representing the price, as shown in line 21 in Figure 9-13. The Dictionary is declared at the class level so that the Dictionary can be used from the Load event and from the SelectedIndexChanged event.

Figure 9-13: Code to create the Dictionary

```
17      public DictionaryValueType()
18      {
19          InitializeComponent();
20      }
21      private Dictionary<string, float> prices = new Dictionary<string, float>();
```

Figure 9-14 shows part 2 of the code that reads each row from Prices.txt, splits it and stores it in a temporary array, and adds the product number as the key and price as the value. Except for line 36, the code is similar to what you used to store the data in a List. The code also adds the product number to the ComboBox.

Figure 9-14: Code to add items to the Dictionary and ComboBox

```
23      private void DictionaryValueType_Load(object sender, EventArgs e)
24      {
25          string currentLine;
26          string[] fields = new string[2];
27          //Create StreamReader
28          StreamReader PriceReader = new StreamReader("Price.txt");
29
30          //Read each line from the file, split and store in an array.
31          while (PriceReader.EndOfStream == false)
32          {
33              currentLine = PriceReader.ReadLine(); // Read product# & price
34              fields = currentLine.Split(',');
35              // Add the key (Prod#) and value (price) to the Dictionary:
36              prices.Add(fields[0], float.Parse(fields[1]));
37              cboProdNumber.Items.Add(fields[0]);   // Add product# to ComboBox
38          }
39          PriceReader.Close();
40      }
```

Line 36: prices.Add(fields[0], float.Parse(fields[1]));

Adds the first element (product#) of fields array as the key and the second element (price) as the value.

Step 2-3: Add the code in line 36 from Figure 9-14 to create the Dictionary.

Step 2-4: Make sure the file Price.txt is in the Debug folder of the project.
Put a break at line 38 at the end of the loop, and run the form.
When the program breaks at line 38, click *Continue* to add another item.
Display Watch window (Debug, Windows, Watch, Watch1) and type prices for Name.
Expand prices and the two key-value pairs (index 0 and 1) to view the first two keys and values, as shown in Figure 9-15.

Figure 9-15: prices Dictionary in Watch window after adding two items

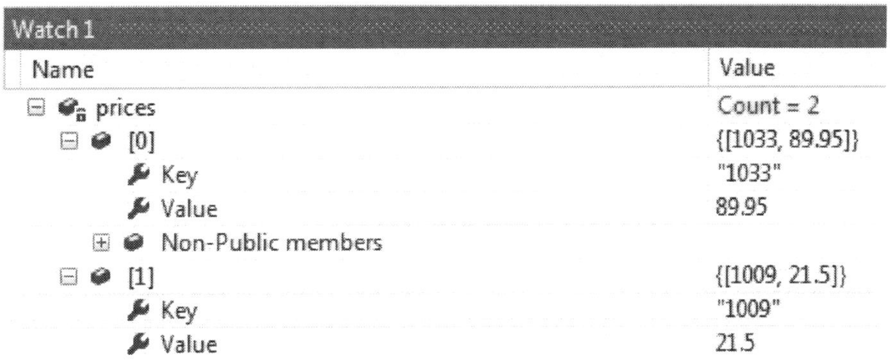

Name	Value
⊟ prices	Count = 2
⊟ [0]	{[1033, 89.95]}
Key	"1033"
Value	89.95
⊞ Non-Public members	
⊟ [1]	{[1009, 21.5]}
Key	"1009"
Value	21.5

Finding an Item from the Dictionary

Unlike a List, the value for an item in a Dictionary can be found by specifying the key. In this example, when the user selects a product number from the ComboBox, it is used as the key to find the corresponding price.

Figure 9-16 shows the code to find the price for the selected product number by using the product number as the key.

Figure 9-16: Code to find the price for a product

```
41
42    private void cboProdNumber_SelectedIndexChanged(object sender, EventArgs e)
43    {
44        string fmtStr = "{0,-10}{1,8:C}";
45        lstDisplay.Items.Clear();
46        lstDisplay.Items.Add(string.Format(fmtStr, "Product#", "Price"));
47
48        string selectedProdcut = cboProdNumber.SelectedItem.ToString();
49        // Use the key (Product#) to get the value (price).
50        float price = prices[selectedProdcut];
51        lstDisplay.Items.Add(string.Format(fmtStr, selectedProdcut, price));
52    }
```

Line 50: float price = prices[selectedProdcut];

The selected product number (selectedProduct) is used as the key in **prices[selectedProdcut]**, which returns the corresponding value (price).

Unlike in a List, this method of finding the value using the key works even when the ComboBox is sorted, because you are not using the SelectedIndex of the ComboBox to find an item.

Step 2-5: Add the code in line 50 from Figure 9-16 to the SelectedIndexChaned event of the ComboBox to find the price for the product number selected by the user. Uncomment line 51.

Step 2-6: Remove the break from line 38. Run the form, and test it by selecting product numbers from the ComboBox.

KeyValuePair Collection

Each item in the Dictionary is of KeyValuePair type and is a member of the KeyValuePair collection. A KeyValuePair consists of a key and a value. Figure 9-17 shows the code to display every key and value using a foreach loop to loop through the collection of KeyValuePairs.

Figure 9-17: Using a KeyValuePair to hold each item from the Dictionary

```csharp
        private void btnDisplayAll_Click(object sender, EventArgs e)
        {
            string fmtStr = "{0,-10}{1,8:C}";
            lstDisplay.Items.Add(string.Format(fmtStr, "Product#", "Price"));

            foreach (KeyValuePair<string, float> kvp in prices)
            {
                lstDisplay.Items.Add(string.Format(fmtStr, kvp.Key, kvp.Value));
            }
        }
```

Line 59: foreach (KeyValuePair<string, float> kvp in prices)

Each item in the Dictionary is of KeyValuePair type. Line 59 declares variable kvp to be of KeyValuePair type, with the key as string type and the value as float type. The foreach loop gets each KeyValuePair from prices and assigns it to the variable kvp, one at a time.

Line 61: lstDisplay.Items.Add(string.Format(fmtStr, kvp.Key, kvp.Value));

The Key and Value properties of the KeyValuePair type variable, kvp, are used to display the key and value in the ListBox, as follows:

```
Product#      Price
"1033"       $89.95
"1009"       $21.50
"1014"      $159.00
"1005"        $2.50
"1001"       $55.70
"1025"       $33.45
```

Step 2-7: Add the code in lines 59–62 from Figure 9-17 to the Click event handler of btnDisplayAll. Run the form and test the code.

Keys Collection and Values Collection

The Keys collection is a collection of keys from all items of the Dictionary. Similarly, there is a separate Values collection, which is a collection of values from all items. Figure 9-18 shows the first few items in the two collections of the Dictionary.

Figure 9-18: Keys collection and Values collection

Keys Collection	Values Collection
"1033"	89.95
"1009"	21.50
"1014"	159.00
"1005"	2.5

You may use the foreach loop to access the key or the value from each item directly without using a KeyValuePair to access the entire item. Figure 9-19 shows the code that uses a foreach loop to access each price from the Values collection, and compute the average price.

Figure 9-19: Using the Values collection to access each price and compute the average price

```
64
65      private void btnAvgPrice_Click(object sender, EventArgs e)
66      {
67          float totalPrice = 0;
68          foreach (float price in prices.Values)
69              totalPrice += price;
70
71          float avgPrice = totalPrice / prices.Count;
72          lstDisplay.Items.Add("Average Price: " + avgPrice.ToString("C2"));
73      }
```

Line 68: foreach (float price in prices.Values)

   totalPrice += price;

Gets individual prices from the Values collection, and computes the sum.

Step 2-8: Add the code from lines 68–71 of Figure 9-19 to the Click event handler of btnAvgPrice. Run the form and test the code.

You may use the same method to access each key from the Keys collection.

The following is the complete code that uses a Dictionary with product number as the key and price as the value.

```csharp
21      private Dictionary<string, float> prices = new Dictionary<string, float>();
22
23      private void DictionaryValueType_Load(object sender, EventArgs e)
24      {
25          string currentLine;
26          string[] fields = new string[2];
27          //Create StreamReader
28          StreamReader PriceReader = new StreamReader("Price.txt");
29
30          //Read each line from the file, split and store in an array.
31          while (PriceReader.EndOfStream == false)
32          {
33              currentLine = PriceReader.ReadLine(); // Read product# & price
34              fields = currentLine.Split(',');
35              // Add the key (Prod#) and value (price) to the Dictionary:
36              prices.Add(fields[0], float.Parse(fields[1]));
37              cboProdNumber.Items.Add(fields[0]);   // Add product# to ComboBox
38          }
39          PriceReader.Close();
40      }
41
42      private void cboProdNumber_SelectedIndexChanged(object sender, EventArgs e)
43      {
44          string fmtStr = "{0,-10}{1,8:C}";
45          lstDisplay.Items.Clear();
46          lstDisplay.Items.Add(string.Format(fmtStr, "Product#", "Price"));
47
48          string selectedProdcut = cboProdNumber.SelectedItem.ToString();
49          // Use the key (Product#) to get the value (price).
50          float price = prices[selectedProdcut];
51          lstDisplay.Items.Add(string.Format(fmtStr, selectedProdcut, price));
52      }
53
54      private void btnDisplayAll_Click(object sender, EventArgs e)
55      {
56          string fmtStr = "{0,-10}{1,8:C}";
57          lstDisplay.Items.Add(string.Format(fmtStr, "Product#", "Price"));
58
59          foreach (KeyValuePair<string, float> kvp in prices)
60          {
61              lstDisplay.Items.Add(string.Format(fmtStr, kvp.Key, kvp.Value));
62          }
63      }
64
65      private void btnAvgPrice_Click(object sender, EventArgs e)
66      {
67          float totalPrice = 0;
68          foreach (float price in prices.Values)
69              totalPrice += price;
70
71          float avgPrice = totalPrice / prices.Count;
72          lstDisplay.Items.Add("Average Price: " + avgPrice.ToString("C2"));
73      }
```

It's time to practice! Do Exercise 9.2.

Dictionary with an Array as Its Value

Next, we look at an example of a Dictionary with an array as its value.

Tutorial 3: Dictionary with Product# as Key and Sales Array as Value

In this application, you store a product number of string type as the key and an array of four quarterly sales as the value in a Dictionary. The product numbers and corresponding quarterly sales are read from the file Sales.txt, shown in Figure 9-20, and stored in the Dictionary, sales<string, Array>.

Figure 9-20: Sales.txt File

1033, 16, 32, 23, 40
1009, 144, 182, 169, 205
1014, 54, 49, 72, 63
1005, 38, 62, 55, 77
1001, 7, 9, 8, 15
1025, 98, 110, 105, 130

Figure 9-21 shows the Dictionary with the product number and sales.

Figure 9-21: sales Dictionary<string, array> showing sample data for key and value

Key	Value			
1033	16	32	23	40
1009	144	182	169	205
1014	54	49	72	63
1005	38	62	55	77
1001	7	9	8	15
1025	98	110	105	130

The product number is selected from the ComboBox, and the total sales displayed in a ListBox, as shown in Figure 9-22.

Figure 9-22: The DictionaryArrayType form

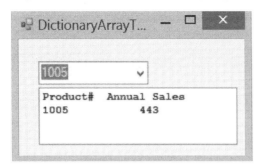

Similar to the last two applications that used Dictionary<string, float> and List<string[]>, the coding for this form has three parts:

1. Create a Dictionary named sales<string, array>.
2. Read each row from Sales.txt, split it and store it in a temporary array, and add the product number as the key and the array of four quarterly sales as the value to the Dictionary. Add the product number to the ComboBox. This is done in the Load event handler of the form.
3. When the user selects a product number from the ComboBox, find the corresponding value (array of quarterly sales) from the Dictionary, and compute the total sales. This is done in the SelectedIndexChanged event of the ComboBox.

Creating a Dictionary

Step 3-1: Open the form DictionaryArrayType, shown in Figure 9-22.

Step 3-2: Add the code in line 21 to create a Dictionary named sales with key of string type representing the product number and value of Array type representing the array of four sales, as shown in Figure 9-23.

The first half of the code that reads data from the file and stores each line in an array should be familiar code. We will focus on lines 36 and 40, which contain new statements.

Line 36: Array.Copy(fields, 1, salesArray, 0, 4);

This statement uses the Array.Copy method, discussed in Chapter 7, to copy the four quarterly sales from elements 1–4 of the array fields into a new array named salesArray.
To refresh your memory, the general syntax of Array.Copy is

    Array.Copy(source array, starting index in the source array, target array,
        starting index in target array, number of elements to copy)

Line 40: sales.Add(fields[0], salesArray);

This statement adds product number from fields[0] as the key and salesArray as the value.

Figure 9-23: Code to add items to the Dictionary and ComboBox

```
21      private Dictionary<string, string[]> sales = new Dictionary<string, string[]>();
22
23      private void DictionaryArrayType_Load(object sender, EventArgs e)
24      {
25          string currentLine;
26          string[] fields;
27          StreamReader SalesReader = new StreamReader("Sales.txt");
28          //Read each line from the file, split and store in an array.
29          while (SalesReader.EndOfStream == false)
30          {
31              currentLine = SalesReader.ReadLine(); // Read product# & four sales.
32              fields = currentLine.Split(',');
33
34              // Copy four sales from fields into a another array, salesArray.
35              string[] salesArray = new string[4];
36              Array.Copy(fields, 1, salesArray, 0, 4);
37              //Syntax: Array.Copy(fromArray, fromIndex, toArray, toIndex, length
38
39              // Add the key (Prod#) and value (salesArray) to the Dictionary:
40              sales.Add(fields[0], salesArray);
41              cboProdNumber.Items.Add(fields[0]);   // Add product# to the ComboBox
42          }
43          SalesReader.Close();
44      }
```

Step 3-3: Add the code in lines 36 and 40 from Figure 9-23 to the Load event of the form.

Step 3-4: Put a break at line 42. Run the form. When the code breaks at line 42, click *Continue* to add the second item. Display the Watch window (Debug, Windows, Watch) and type sales for Name. Expand sales, Key and Value to view their values, as shown in Figure 9-24.

Figure 9-24: sales Dictionary with the Key and value for first two items

Name	Value
⊟ sales	Count = 2
⊟ [0]	{[1033, System.String[]]}
Key	"1033"
⊟ Value	{string[4]}
[0]	" 16"
[1]	" 32"
[2]	" 23"
[3]	" 40"
⊞ Non-Public members	
⊟ [1]	{[1009, System.String[]]}
Key	"1009"
⊟ Value	{string[4]}
[0]	" 144"
[1]	" 182"
[2]	" 169"
[3]	" 205 "

Figure 9-25 shows the code to find the quarterly sales for a selected product number by using the product number as the key.

Figure 9-25: Code to find sales for a product

```
45
46          private void cboProdNumber_SelectedIndexChanged(object sender, EventArgs e)
47          {
48              string fmtStr = "{0,-10}{1,8}";
49              lstDisplay.Items.Clear();
50              lstDisplay.Items.Add(string.Format(fmtStr, "Product#", "Annual Sales"));
51
52              string selectedProduct = cboProdNumber.SelectedItem.ToString();
53              // Use the key (Product#) to get the value (quarterly sales array).
54              string[] salesArray = sales[selectedProduct];
55
56              int annualSales = 0;
57              foreach (string qtrSale in salesArray)
58                  annualSales = annualSales + int.Parse(qtrSale);
59              // Or, you may use a for loop:
60              //for (int quarter = 0; quarter <= 3; quarter++)
61              //    annualSales = annualSales + int.Parse(salesArray[quarter]);
62              lstDisplay.Items.Add(string.Format(fmtStr, selectedProduct, annualSales));
63          }
```

Line 54: sales[selectedProduct] uses selectedProduct as the key to get the value, which is an array. The array returned is stored in salesArray.

Lines 57–58 use a foreach loop to add all four elements of salesArray.

Step 3-5: Add the code in lines 54–58 from Figure 9-25 to the SelectedIndexChanged event of the ComboBox to find the sales for the selected product number. Run the form and test the code.

It's time to practice! Do Exercise 9.3.

Review Question

9.5   How does a Dictionary differ from a List?

Exercises

Figure 9-26 shows data from the Vendor.txt file that are stored in the Debug folder in Ch9_Collections project directory.

Figure 9-26: Vendor.txt file (vendor#, name, street, city, state, zip, phone)

1,FCC,445 12th Street SW, Washington,DC,20554,888-225-5322
2,U.S.Copyright Office,101 Independence Ave, Washington,DC,20559,202-707-3000
3,CDC,1600 Clifton Road,Atlanta,GA,30333,800-232-4636
4,White House, 1600 Pennsylvania Avenue NW, Washington,DC,20500,202-456-1111
5,Labor Statistics,Postal Sq Bldg,2 Mass Ave,Washington,DC,20212,800-877-8339
6,AMTRAK,60 Massachusetts Avenue NE, Washington,DC,20002,800-523-6590
7,U.S. Air Force,1690 Air Force Pentagon,Washington,DC,20330-1670,800-423-8723
8,National Guard,111 South George Mason Dr,Arlington,VA,22204,800-864-6264
9,Office of Govt Ethics,1201 New York Ave,Washington,DC,20005,202-482-9300
10,IRS,1111 Constitution Ave NW, Washington, DC,20224,800-829-1040

Exercise 9-1: List

Create a form named **VendorsList** that allows the user to find and display the zip code and phone number for any specified vendor, as follows:

Read vendor records from **Vendor.txt** (Figure 9-26), and load them into a List of arrays, named vendors, and add the vendor names into a sorted ComboBox.

When the user selects a vendor name from the ComboBox, find and display the name and corresponding zip code and phone number in a ListBox.

Exercise 9-2: Dictionary

Create a form named **VendorsDictionary1** that allows the user to find and display the phone number for any specified vendor, as follows:

Read each vendor record from the Vendor.txt file (Figure 9-26). Add the name as the **key**, and the phone number as the **value** into a Dictionary named vendorPhones. Add the name to a sorted ComboBox.

When the user selects a vendor name from the ComboBox, find and display the name and corresponding phone number in a ListBox.

Exercise 9-3: Dictionary

Create a form named **VendorsDictionary2** that allows the user to find and display the zip code and phone number for any specified vendor, as follows:

Read each vendor record from the Vendor.txt file (Figure 9-26), split it and store it in an array. Copy the zip code and phone into another array (you may use Array.Copy). Add the name as the **key,** and the array of zip code and phone number as the **value** into a Dictionary named vendors. Add the name to a sorted ComboBox.

When the user selects a Vendor name from the ComboBox, find and display the name and corresponding zip code and phone number in a ListBox.

Chapter 10

Graphical User Interface: Additional Controls

Chapter 5 introduced several controls that make it easy for users to enter data and help minimize data entry errors: RadioButton, CheckBox, ScrollBar, ComboBox and ListBox. In this chapter, you learn how to use two additional controls that help enhance the user interface:

1. The **ListView** control that allows you to display and work with multiple data items in a single row
2. The **Tab** control that lets you create a form with multiple pages, each page with its own set of controls

Three more user interface controls are introduced in Chapter 11: MenuStrip, ContextMenuStrip and ToolStrip controls.

Topics

10.1	ListView: Features	10.3	Accessing Data from ListView
10.2	Adding Items to ListView	10.4	Forms with Tab Pages

10.1 ListView: Features

A ListView, similar to the ComboBox and ListBox, lets you display multiple rows of data, and allows you to select the data from a row. A major difference is that ListView lets you add multiple data items to a row, as shown in Figure 10-1, which displays the id, name and unit price of each show in a theater using a ListView. By contrast, a row in a ComboBox and ListBox consists of only a single string. If you want to add multiple data items to a row, you must concatenate them together to form a single string, which makes it difficult to access individual data items from a row. A ListView lets you directly access an individual data item like the ticket price.

Similar to a ListBox, the ListView lets you select one or more items from a List. ListViews also can display items in different formats, including images and CheckBoxes.

To help understand the use of ListView, let's look at an application where the user selects a show from a ListView named **lvwShows** by clicking on the corresponding row, as shown in Figure 10-1. The total price (extended price) is computed by multiplying the unit price from the selected row by the number of tickets from the ScrollBar.

Figure 10-1: Example of a ListView

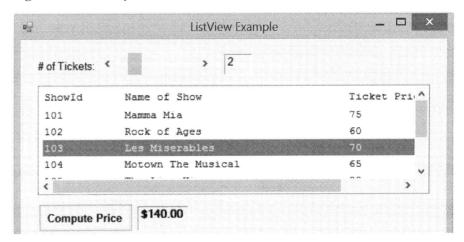

The data shown in Figure 10-1 are read from a Text file, Show.txt, shown in Figure 10-2, and added to the ListView, lvwShows.

Figure 10-2: Show.txt file

101,Mamma Mia,75
102,Rock of Ages,60
103,Les Miserables,70
104,Motown The Musical,65
105,The Lion King,90
106,The Phantom of the Opera,80
107,Pippin,75

Review Question

10.1   How does a ListViewItem differ from a ListBox?

10.2 Adding Items to ListView

Similar to the ComboBox and ListBox, the ListView has an **Items** collection that contains all items (rows) in the ListView.

ListViewItem

The items (rows) within the Items collection are of **ListViewItem** type (not string type), which may consist of **SubItems** that represent different data items (columns), like the id, name and ticket price for a show.

To add a row to a ListView, create a **ListViewItem** object and add it to Items collection.

To create a ListViewItem object named showsLVI, you may store the subitems in a string array and pass the array as an argument to the ListViewItem class, as follows:

```
String[] fieldsArray = {"101", "Mama Mia", "75"};
ListViewItem showsLVI = new ListViewItem(fieldsArray);
```

After you create the ListViewItem object, you add it to the Items collection of the ListView. The following statement adds the ListViewItem, showsLVI, to the ListView, lvwShows:

```
lvwShows.Items.Add(showsLVI);
```

You also may create a ListViewItem with just the first subitem, and then add the remaining subitems individually. But it is more convenient here to add all subitems together as a single array.

View Property

To display subitems of a row in tabular format with column headings, as shown in Figure 10-1, you need to set the View property to Details, as shown in Figure 10-3.

Figure 10-3: The View property of ListView

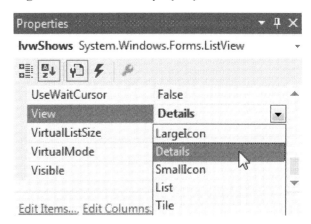

Column Collection

To display column headings, click on Columns (Collection) property of the ListView to open the ColumnHeader Collection Editor window. Click Add to add the first header and change its Text property to the desired heading. You also may change the Width property. You may repeat this for each heading, and click *OK* when done.

The code for reading each line from the Show.txt file using a loop, splitting it, storing it in a string array, and adding it to the Items collection is shown in Figure 10-4.

Figure 10-4: Code to add items to ListView

```csharp
private void ListView_Load(object sender, EventArgs e)
{
    string currentLine;
    string[] fields = new string[3];
    StreamReader PriceReader = new StreamReader("Show.txt");
    // Read each line, split it, and store in an array:
    while (PriceReader.EndOfStream == false)
    {
        currentLine = PriceReader.ReadLine();
        fields = currentLine.Split(',');

        // Use the array to create a ListViewItem (row).
        ListViewItem ShowsLVI = new ListViewItem(fields);

        // Add the entire ListViewItem to the ListView control
        lvwShows.Items.Add(ShowsLVI);
    }
    PriceReader.Close();
}
```

Line 33 creates a ListViewItem object using the array, fields, that contains the data from the current row. The ListViewItem object is referenced by the variable, showsLVI.

Line 36 adds the ListViewItem object to the ListView, lvwShows.

Table 10-1 presents a list of the properties and events that are important in working with ListViews.

Table 10-1: Properties and Events of ListView

Property	Description
CheckBoxes	If it is set to True, a CheckBox appears on each item (row), making it easier to select multiple items without using the control key. The default is False.
CheckedItems	If CheckBoxes property is set to True, CheckedItems collection would return the collection of checked items.
FullRowSelect	When set to True, you can select a row by clicking on any data item, and the entire row will be highlighted. The default is False, which requires you to click on the first data item of the row to be selected, and only that sub item is highlighted.
Items	The collection of ListViewItems.
SelectedItems	The collection of selected items.
SelectedIndices	The collection of indices of selected items.
Sorting	When set to Ascending (Descending), the items are sorted in ascending (descending) order using the data item in the first column. The default is None.
View	Determines the appearance of ListViewItems. When set to Details, data is displayed in a grid format. Other options include LargIcon, SmallIcon, List and Tile.
Events	
ColumnClick	Triggered when a column header is clicked.

Tutorial 1: Selecting a Show from a ListView

This tutorial creates the application that lets the user select a show, and computes the total price of tickets.

Step 1-1: Open the project named Ch10_GUIadditionalControls from Tutorial_Starts folder. Open the form, ListViewApp.

Add a ListView, as shown in Figure 10-1, and name it **lvwShows**.

Step 1-2: Set the View property of the ListView to Details, as shown in Figure 10-3. In addition, set the **FullRowSelect** property to True.

Step 1-3: Set column names using the Columns (Collection) property, as specified under "Column Collection."

Step 1-4: Open and view the Text file named Show.txt from the Debug folder of the project directory. Close the file.

Step 1-5: Add the code in lines 33 and 36 from Figure 10-4 to the Load event handler of the form. Run the project, and test the code.

Step 1-6: Make sure that the variable numOfTickets is declared at class level so that it can be accessed from multiple methods, as follows:

```
19
20          int numOfTickets;
```

Step 1-7: Open the ValueChanged event handler for the ScrollBar (Properties, events, ValueChanged). View the code that stores the value in the class level variable, numOfTickets, and displays the value in the Label.

Figure 10-5: The ValueChanged event handler of ScrollBar

```
40
41          private void hsbNumOfTickets_ValueChanged(object sender, EventArgs e)
42          {
43              numOfTickets = hsbNumOfTickets.Value;
44              lblNumOfTickets.Text = numOfTickets.ToString();
45          }
```

Review Question

10.2   What is the difference between a List View and a ListViewItem?

10.3 Accessing Data from a ListView

You may access a specific data item using the **Items** and **SubItems** collections by specifying the index of the item (row) and the index of the sub item (column). For example, to access the ticket price in the third column (index 2) for "Rock of Ages" in the second row (index 1), use the Items and SubItems collections as follows:

    lvwShows Items[1].SubItems[2].Text.

SelectedItems Collection

Similar to the ListBox, the ListView has the **SelectedItems** collection that includes all items (rows) selected by the user. Each selected item can be accessed using its index:

    SelectedItems[0] gives the first item (row) selected by the user
    SelectedItems[1] gives the second item selected by the user, and so forth.

For example, the ticket price from the single selected row (assuming only one show is selected) is given by

    lvwShows.SelectedItems[0].SubItems[2].Text;

If multiple data items from a row are to be accessed, you may copy the entire row to a variable of ListViewItem type, and access individual data items from the variable, as follows:

    ListViewItem showsLVI = lvwShows.SelectedItems[0];
    float unitPrice = float.Parse(showsLVI.SubItems[2].Text);

In the above code, showsLVI is a variable of ListViewItem type.

Computing the Extended Price for the Selected Show

The code to compute the extended price (number of tickets x ticket price) is given in Figure 10-6.

Figure 10-6: Code to get ticket price from selected row and compute extended price

```
46
47      private void btnComputePrice_Click(object sender, EventArgs e)
48      {
49          float unitPrice = float.Parse(lvwShows.SelectedItems[0].SubItems[2].Text);
50          // If the CheckBoxes property of the ListView is set to True,
51          // use CheckedItems collection instead of SelectedItems as follows:
52          // float unitPrice = float.Parse(lvwShows.CheckedItems[0].SubItems[2].Text);
53
54          // Alternate method:Store the selected row in a variable & get price from it:
55          ListViewItem showsLVI = lvwShows.SelectedItems[0];
56          unitPrice = float.Parse(showsLVI.SubItems[2].Text);
57                  // gets price from SubItems[2].
58          float totalPrice = numOfTickets * unitPrice;
59          lblAmtDue.Text = totalPrice.ToString("C2");
60      }
```

Step 1-8: Add the code from Figure 10-6 to the Click event handler of btnComputePrice. Run the form and test the code.

Use appropriate properties from Table 10-1 to add CheckBoxes to the items. Let the user select an item by checking the CheckBox for the item. Make necessary changes in the code to compute the total price. Test the code

It's time to practice! Do Exercise 10-1.

ColumnClick Event: Selecting a Column

In addition to selecting a row, you also may select a column by clicking the column header, which triggers the ColumnClick event of the ListView. You may use this event to take an action when the user selects a column. Figure 10-7 shows the code within the ColumnClick event handler of the ListBox, to display the data item at the selected column of the selected row, in a Label named lblCurrentCell.

Figure 10-7: Code to get data from selected column

```
61
62      private void lvwShows_ColumnClick(object sender, ColumnClickEventArgs e)
63      {
64          int index = e.Column;
65          if (lvwShows.SelectedItems.Count > 0)
66              lblCurrentCell.Text = lvwShows.SelectedItems[0].SubItems[index].Text;
67      }
```

In line 64, the Column property of the parameter **e** gives the index of the selected column. The content of the cell in the selected column of the selected row is displayed in the Label, lblcurrentCell, as shown in Figure 10-8.

Figure 10-8: The ListViewApp form with content of current cell displayed

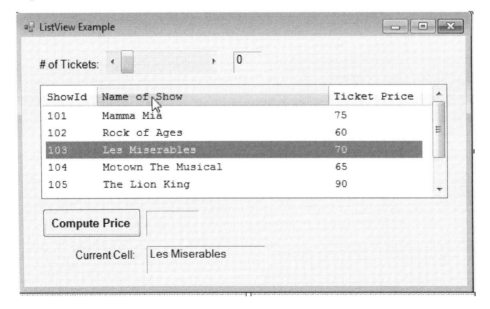

Step 1-9: Add the code from Figure 10-7 to the ColumnClick event handler of the ListView. Run the form. Select a row, and click on a column header to test the code.

Accessing Every Row of a ListView

A convenient way to access the data in every row of a column is to use a loop. For example, the code within the Click event handler of btnComputeAvg, shown in Figure 10-9, adds the price from each row using a foreach loop, computes the average price and displays the average in a Label named lblAvgPrice.

Figure 10-9: Using a foreach loop to access every row of ListView

```
68
69      private void btnComputeAvg_Click(object sender, EventArgs e)
70      {
71          float sumOfPrices = 0;
72          foreach (ListViewItem ShowsLVI in lvwShows.Items)
73              sumOfPrices = sumOfPrices + float.Parse(ShowsLVI.SubItems[2].Text);
74          float avgPrice = sumOfPrices / lvwShows.Items.Count;
75          lblAvgPrice.Text = avgPrice.ToString("C");
76      }
```

Figure 10-10 shows the form with the Button and Label to compute and display the average.

Figure 10-10: The ListViewApp form with a button to compute average

Step 1-10: Add the code shown in Figure 10-9 to the Click event handler of btnComputeAvg. Run the form, and test the code.

10.3 Accessing Data from a ListView

The following is the complete code for the ListViewApp form:

```csharp
    int numOfTickets;
    private void ListView_Load(object sender, EventArgs e)
    {
        string currentLine;
        string[] fields = new string[3];
        StreamReader PriceReader = new StreamReader("Show.txt");
        // Read each line, split it, and store in an array:
        while (PriceReader.EndOfStream == false)
        {
            currentLine = PriceReader.ReadLine();
            fields = currentLine.Split(',');

            // Use the array to create a ListViewItem (row).
            ListViewItem showsLVI = new ListViewItem(fields);

            // Add the entire ListViewItem to the ListView control
            lvwShows.Items.Add(showsLVI);
        }
        PriceReader.Close();
    }

    private void hsbNumOfTickets_ValueChanged(object sender, EventArgs e)
    {
        numOfTickets = hsbNumOfTickets.Value;
        lblNumOfTickets.Text = numOfTickets.ToString();
    }

    private void btnComputePrice_Click(object sender, EventArgs e)
    {
        float unitPrice = float.Parse(lvwShows.SelectedItems[0].SubItems[2].Text);
        // If the CheckBoxes property of the ListView is set to True,
        // use CheckedItems collection instead of SelectedItems as follows:
        // float unitPrice = float.Parse(lvwShows.CheckedItems[0].SubItems[2].Text);

        // Alternate method:Store the selected row in a variable & get price from it:
        ListViewItem showsLVI = lvwShows.SelectedItems[0];
        unitPrice = float.Parse(showsLVI.SubItems[2].Text);
                // gets price from SubItems[2].
        float totalPrice = numOfTickets * unitPrice;
        lblAmtDue.Text = totalPrice.ToString("C2");
    }

    private void lvwShows_ColumnClick(object sender, ColumnClickEventArgs e)
    {
        int index = e.Column;
        if (lvwShows.SelectedItems.Count > 0)
            lblCurrentCell.Text = lvwShows.SelectedItems[0].SubItems[index].Text;
    }
```

```
68
69      private void btnComputeAvg_Click(object sender, EventArgs e)
70      {
71          float sumOfPrices = 0;
72          foreach (ListViewItem ShowsLVI in lvwShows.Items)
73              sumOfPrices = sumOfPrices + float.Parse(ShowsLVI.SubItems[2].Text);
74          float avgPrice = sumOfPrices / lvwShows.Items.Count;
75          lblAvgPrice.Text = avgPrice.ToString("C");
76      }
```

It's time to practice! Do Exercise 10-2 and Exercise 10-3.

Review Questions

10.3   The following ListView named lvwShows highlights the item selected by the user at runtime.

ShowId	Name of Show	Ticket Price
101	Mamma Mia	75
102	Rock of Ages	60
103	Les Miserables	70
104	Motown The Musical	65
105	The Lion King	90
107	Pippin	75
108	The Phantom of the Opera	80

What would be the output displayed in the Label, lblDisplay, after executing the following statement?

lblDisplay.Text = lvwShows.SelectedItems[0].SubItems[1].Text;

10.4   The following ListView named lvwShows displays information on different shows:

ShowId	Name of Show	Ticket Price
101	Mamma Mia	75
102	Rock of Ages	60
103	Les Miserables	70
104	Motown The Musical	65
105	The Lion King	90
107	Pippin	75
108	The Phantom of the Opera	80

What would be the output displayed in the Label, lblDisplay, after executing the following code?

foreach (ListViewItem showsLVI in lvwShows.Items)
    lblDisplay.Text = showsLVI.SubItems[2].Text;

10.4 Forms with Tab Pages

The tab control lets you create a multiple pages on a form, each page with its own set of controls. For example, consider a form similar to the DictionaryValueType form from Chapter 9, which lets you find the price for a specified product, and lets you list every product that is priced above a specified limit. Rather than doing both tasks on a single page of the form, you can have the controls for the two different tasks on two different tab pages of a single form, as shown in Figure 10-11.

Figure 10-11: The ProductPrices form with two tab pages

An alternative to using two tab pages for two different tasks is to use two different forms. But there is a significant difference between using two different forms and using two pages on a single form. When you use two or more tab pages on a single form, you can access controls on any one tab page from another page by using the name of the control. That is, tab pages act like different sections of a form. By contrast, to access controls on one form from another form, you need to create an instance of the page, as discussed in Chapter 11.

Each tab page on a form is a member of the TabPages collection of the TabControl. So, you can refer to a tab page by its index, starting with 0 for the first page, 1 for the second page, and so on. The **SelectedIndex** property of the TabControl allows you to identify the tab page that is currently selected.

Selecting a page from a TabControl fires the **SelectedIndexChanged** event, just like when you select an item from a ComboBox. So, you can code this event handler to take actions, if any, when the user selects a tab page.

Trutorial 2: Multipage Form with Tab control

Let's create a form that has two pages, as shown in Figure 10-11.

Step 2-1: Create a new project named Ch10_MultiPageForms. Change the name of the form to ProductPrices.

Step 2-2: Add the TabControl to the form from Containers group on the Toolbar. The TabControl displays two TabPages, as shown in Figure 10-12.

Select the TabControl by clicking on the header area, which includes tabs, tabPage1 and tabPage2. When you click on the name tabPage1 (or tabPage2), it might look like that page is selected. But clicking anywhere on the header area selects only the TabControl.
Set the name property of the TabControl to tabProductPrices.

Figure 10-12: The TabControl

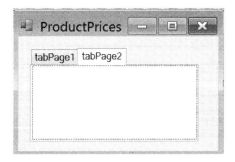

Step 2-3: To select tabPage1, click on tabPage1 in the header area, and then click inside the page (with dotted border).
Change the name of the page from tabPage1 to tbpLookUpPrices and set the Text property to Look Up Prices.
Similarly, change the name of tabPage2 to tbpDisplayProducts and set the Text property to Display Products.

Step 2-4: Select the page Look Up Prices. Add a ComboBox named cboProdNumber and a Label named lblPrice. The tab page should look as in Figure 10-11.

Step 2-5: Select the page Display Products. Add a button named DisplayProducts, a TextBox named txtPrice and a ListBox named lstDisplay. Add a Label to display the prompt "Enter Price Limit." The form should look as follows:

Because we already discussed the code in Chapter 9, we will not add the code for the two pages, but will focus on using tab control related features.

Next, we create the code to set focus to cboProdNumber, when the user selects Look Up Price page, and set focus to txtPrice, when the user selects Display Products page. This is done by coding the **SelectedIndexChanged** event of the TabControl, which is triggered when the user selects a tab page.

Step 2-6: Create the SelectedIndexChanged event handler of the TabControl (Select the TabControl by clicking the header area. Select events in Properties window and double click SelectedIndexChanged). Add the code from Figure 10-13 to the event handler.

Figure 10-13: SelectedIndexChanged event handler of tabProductPrices

```
private void tabProductPrices_SelectedIndexChanged(object sender, EventArgs e)
{
    if (tabProductPrices.SelectedIndex == 0)
        cboProdNumber.Focus();
    else if (tabProductPrices.SelectedIndex == 1)
        txtPriceLimit.Focus();
}
```

Step 2-7: Run the form. Click the page Display Products. The TextBox should have the focus. Click the tab page Look up Price. The ComboBox should have the focus.

Review Question

10.5   What is the difference between using multiple tab pages to do different tasks on a single form versus using multiple forms?

Exercises

Exercise 10-1

Modify ComputePrice_Click event procedure in the ListViewApp form so that the name of the selected show and the total price are displayed in the Label lblAmtDue, as in

   Les Miserables: $140.00

Exercise 10-2

Add a button on the ListViewApp form, and add the necessary code to the Click event handler to display all shows that start with the letter "M." Display the names of such shows in a ListBox.

Exercise 10-3

This exercise is a modified version of the IceCream form, shown below, from Exercise 6-1 in Chapter 6.

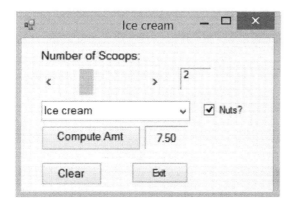

This exercise uses a ListView (instead of the ComboBox) to display sundae names and prices, as described below:

1. Create a text file named Sundaes.txt, with sundae names and the following prices:

    Ice Cream, 3.5
    Custard, 3.75
    Yogurt, 4.00
    ...

    Make up the rest of the sundae names and prices.

2. Replace the ComboBox with a ListView. Read the sundae names and unit prices from the Sundaes.txt file, and display them in the ListView in two columns (in place of the ComboBox used in Exercise 5-1). This should be done in the Load event handler of IceCream form. Let the user select a Sundae from the ListView.

3. Let the user select the number of scoops using a scroll bar, and select nuts using a CheckBox.

4. A clear button should clear the ScrollBar and selection from the ListView so that no item is selected.

5. Have the program use the price of the selected sundae from the ListView to compute and display the amount due in a Label.

    Amount due = Unit price x Number of scoops + Cost of nuts ($0.50), if any

Extra Credit

Let the user select more than one sundae. Compute the cost assuming the number of scoops selected by the user applies to each selected sundae.

Chapter 11

Multi-form Applications and Menus

So far, we have only looked at forms that work independently without interacting with other forms. However, it may be necessary for a form to interact with other forms in a variety of ways, including opening/closing another form, accessing a control on another form, or sharing data or methods between forms.

In this chapter, you will learn how to code one form to show another form, access controls on another form and share methods between forms. Sharing data between forms is not covered here because proper sharing of data involves knowledge of additional object oriented programming concepts.

Topics

11.1	Introduction		11.4	Closing a Parent Form
11.2	Accessing an Existing Form from Another		11.5	Menus
			11.6	Toolstrips
11.3	Sharing a Method between Multiple Forms			

11.1 Introduction

To help understand multi-form applications, consider an application that includes a form named **FinancialPlanner,** very similar to the FinancialPlanner form introduced in Chapter 4. Given yearly investments and growth rate, this form determines how long it takes for the investment value to reach the target amount. The investment value for every year until the value exceeds the target amount is displayed on a separate form named **DisplayInvestments**, as shown in Figure 11-1.

Figure 11-1: The DisplayInvestment form

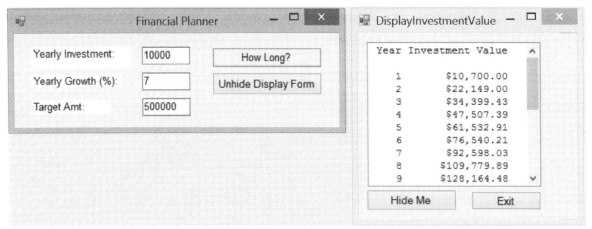

11.2 Accessing an Existing Form from Another

Accessing a form typically involves getting data from controls on the form, or displaying data in controls. Before you can access the controls, you need to open the form.

Opening an Existing Form from Another

Opening an existing form involves the following steps:

1. Declare a variable whose type is the form class you want to open. Note that the form you want to open, like DisplayInvestment, is a class.
   To declare a variable named displayInvestment of the type, DisplayInvestment form, use the following code:
   DisplayInvestment displayInvestment;

   Note that the variable starts with the lowercase letter "d," and the data type starts with the uppercase letter "D."

2. Create an instance of the form (an object) and store the reference to the object in the variable. For example,
   displayInvestment = new DisplayInvestment();

   The right side of the equal sign creates the object, whose reference is assigned to the variable displayInvestment.

   You may combine the statements from steps 1 and 2 as follows:
   DisplayInvestment displayInvestment = new DisplayInvestment();

3. Use the Show() method of the object to make it visible.
   displayInvestment.Show();

Accessing a Control on Another Form

To access a control on a form from other forms, you need to make the form public. To do this, open the Designer window for the form by double clicking the file name in Solution Explorer, as shown in Figure 11-2.

Figure 11-2: Opening the Designer window of a form

Next, change the control from "Private" to "Public," as shown in Figure 11-3.

11.2 Accessing an Existing Form from Another

Figure 11-3: Designer window of the form

```
23      Windows Form Designer generated code
83
84          public System.Windows.Forms.ListBox lstDisplay;
85          private System.Windows.Forms.Button btnHide;
86          private System.Windows.Forms.Button btnExit;
```

A public control on another form can be accessed using the variable that references the object. For example, the following code would add a blank line in the ListBox, lstDisplay, on the form, DisplayInvestment using the variable, displayInvestment:
    displayInvestment.lstDisplay.Items.Add("");

Figure 11-4 shows the complete code in the form FinancialPlanner to compute investment values and display them in the ListBox on the form DisplayInvestments.

Figure 11-4: Code to access a control on another form

```
19      //Declare a variable of type DisplayInvestment form.
20      DisplayInvestment displayInvestment;
21
22      private void btnHowlong_Click(object sender, EventArgs e)
23      {
24          // Create an instance of DisplayIvestment form and
25          // store its reference in the variable displayInvestment:
26          displayInvestment = new DisplayInvestment();
27          // Load and show the form:
28          displayInvestment.Show();
29
30          // Get input data from Text Boxes:
31          double yearlyInvestment = double.Parse(txtInvestment.Text);
32          float yearlyGrowthPercent = float.Parse(txtGrowthRate.Text);
33          double targetAmt = double.Parse(txtTargetAmt.Text);
34
35          // Print heading
36          string formatCode = "{0,5}{1,17}";
37          // Add heading to the ListBox lstDisplay in DisplayInvestment form.
38          displayInvestment.lstDisplay.Items.Add(string.Format(formatCode,
39                                          "Year", "Investment Value"));
40          displayInvestment.lstDisplay.Items.Add("");
41
42          // Set year and investment value to zero
43          int year = 0;
44          double investmentValue = 0;
45          // Use while loop to compute investment value for each year
46          while (investmentValue < targetAmt)
47          {
48              year = year + 1;
49              investmentValue = (investmentValue + yearlyInvestment) *
50                                          (1 + yearlyGrowthPercent / 100);
51              // Add year & investment value to the lstDisplay in DisplayInvestment form
52              displayInvestment.lstDisplay.Items.Add(string.Format(formatCode,
53                                          year, investmentValue.ToString("C")));
54          }
55      }
```

Line 20: The variable, displayInvestment, is declared at the class level so that it can be used in multiple methods.

Lines 26–28: Creates an instance of the DisplayInvestment form and opens it.

Lines 31–53: Computes investment values and displays them in the ListBox lstDisplay. This is essentially the same as the code we discussed in Chapter 4, except that the ListBox is on the form DisplayInvestment.

So, the statements in lines 38, 40 and 52 that display results in the ListBox use the variable displayInvestment to access the ListBox, as in
displayInvestment.lstDisplay.Items.Add(string.Format(formatCode,
    "Year", "Investment Value"));

Tutorial 1: Multi-form Financial Planning Application

Let's develop a multi-form fianancial planning application.

Step 1-1: Open the project, project Ch11_MultiFormApps1.

Step 1-2: Open the form, FinancialPlanner.

Step 1-3: Add the missing code in lines 20, 26 and 28 from Figure 11-4, to open the DisplayInvestment form.
Uncomment the statements in lines 38, 40 and 52, which write data into the DisplayInvestment form.

Step 1-4: Open Program.cs and change the startup form to FinancialPlanner. Run the form and test the code.

Step 1-5: Click the "How Long" button multiple times, each time with a different set of input data for yearly investment, growth rate and target amount. Note that every time you click the button, a new instance of the DisplayInvestment form is displayed.

Creating Multiple Instances of the Same Form

As observed in Step 1-5, every time you click the "How Long" button, the statements in line 26 and 28,
    displayInvestment = new DisplayInvestment();
    displayInvestment.Show();
will create and open a new instance of the DisplayInvestment form. This allows you to simultaneously display investment values for different sets of input data.

Hiding versus Closing a Form

You may hide a form without closing it. Hiding a form keeps the form and its properties and settings in memory. You don't lose the data entered/displayed in controls at runtime. It just is not visible.
You hide a form using the Hide method of the form. For example,
    this.Hide()
hides the current form.

The following code within the FinancialPlanner form hides an instance of the DisplayInvestment form:
    displayInvestment.Hide();

Closing a form removes the form from memory. So, you lose all current settings and data.

Unhiding a Form

The Show method makes a hidden form visible. The statement
   displayInvestment.Show();
shows the instance of the DisplayInvestment form represented by the variable displayInvestment.

Step 1-6: Add the following code to the Click event handler of btnHide button on the DisplayInvestment form.
   this.Hide();

Step 1-7: Add the following code to the Click event handler of btnUnHideDisplayForm button on the FinancialPlanner form.
   displayInvestment.Show();

Step 1-8: Run the FinancialPlanner form. Click "How Long?" button to open DisplayInvestment, and hide it by clicking the "Hide Me" button. Unhide it by clicking "Unhide Display Form." Note that the contents of the ListBox are not lost when you hide the form. Keep the form open.

Step 1-9: Change the target amount to 10000 and open another instance of the DisplayInvestment form. Now, hide both instances of DisplayInvestment. Next, Click "Unhide Display Form" button. Note that only the latest (second) instance of the DisplayInvestment form is visible. This is because the variable displayInvestment in the statement displayInvestment.Show() holds reference only to the latest instance.

Show versus ShowDialog

You may open a form using the Show method or the ShowDialog method.

When C# processes a statement that calls the Show method, it opens the form and continues processing the statements that follow the call. However, with the ShowDialog method, C# opens the form and stops processing the statements that follow the call until the dialog window is closed. So, ShowDialog is not appropriate if a form has to open a form and then access a control on that form, as in the current example.

You may use the ShowDialog method if you want to display information for the user or get input from the user before processing other statements in the code. For example, you may use a dialog window to ask the user for authentication information, like a password, before proceeding further.

In addition, if a form is opened using ShowDialog, the form is opened in **modal** mode, which means that the user cannot work with other open forms until the dialog window is closed, thus forcing the user to take an action. The Show method, on the other hand, opens a form in **nonmodal** mode, which makes other forms accessible to the user.

Step 1-10: Change the Show method to ShowDialog in Line 28, and put a break in line 31, as shown in Figure 11-5.

Figure 11-5: The ShowDialog method

```
28          displayInvestment.ShowDialog();
29
30          // Get input data from Text Boxes:
31          double yearlyInvestment = double.Parse(txtInvestment.Text);
```

Step 1-11: Run the form and click the "How Long?" button. The DisplayInvestment form opens, but no data are displayed because processing stops at line 28.
Close the DisplayInvestment form. Note that the program breaks at line 31, indicating that processing continues when you close the dialog window.
Change line 28 back to displayInvestment.Show();.

Review Questions

11.1 A form named DisplayOrder has a Label named lblTotalCost that displays the total cost. Another form within the same project needs to get the data from the Label, lblTotalCost, at runtime. Write the statement(s) to open DisplayOrders, get the total cost and save it in a variable named totalCost. Assume that lblTotalCost is Public.

11.2 How does hiding a form differ from closing a form?

11.3 What is the difference between the Show and ShowDialog methods of a form?

11.3 Sharing a Method between Multiple Forms

When a method needs to be used by more than one form within a project, it would be easier to maintain it if there is only one copy of the method within the project, rather than one in each form. Keeping such methods within any one form makes maintenance difficult, if the form is deleted due to future changes in the application. It would be better to keep shared methods in a separate class within the project.

The application shown in Figure 11-6 is used to illustrate how a method is shared between multiple forms. It uses three different forms:

1. A form named LookUpPrice that finds the price for a specified product number. To make repeated searches efficient, the data are stored in a Dictionary when the form is loaded, as we did in the form DictionaryValueType that was discussed in Chapter 9.
2. A second form named DisplayProducts that displays the product number and prices for a group of products with prices below a specified limit. Again, the data are read into a Dictionary when the form is loaded.
3. A form named ProductsMain that serves as a switchboard to let users open the other two forms.

Figure 11-6: A main form that opens two other forms

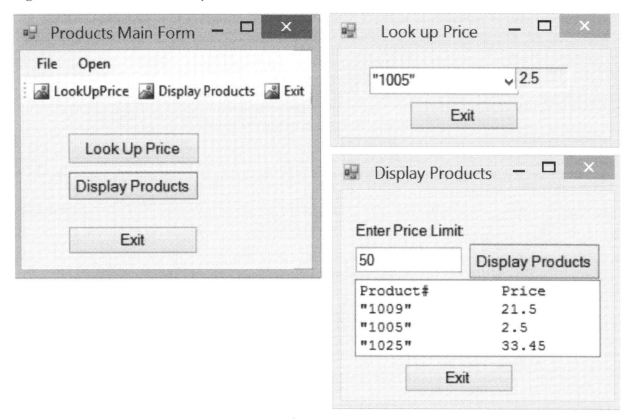

Because both LookUpPrice and DisplayProducts need to read and store the data in a Dictionary, we use a common method called LoadDictionary within a separate class and called from both forms, as shown in Figure 11-7.

Figure 11-7: Sharing a method between two forms

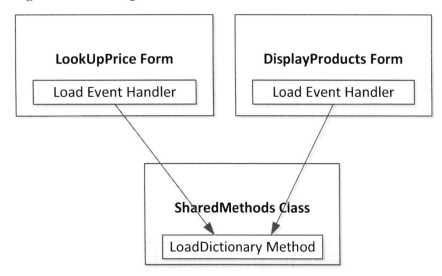

Tutorial 2: Sharing the LoadDictionary Method between Forms

This tutorial develops the application shown in Figure 11-7, which shares a method between two forms.

Step 2-1: Open the project, Ch11_MultiFormApps2, from the Tutorial_Starts folder.
The Price.txt file shown in Figure 11-8 is stored in the Debug folder of the current project.

Figure 11-8: Price.txt file

```
"1033", 89.95
"1009", 21.50
"1014", 159.00
"1005", 2.50
"1001", 55.70
"1025", 33.45
"1040", 130.50
"1035", 109.00
```

Step 2-2: Add a class named SharedMethods (select *Project, Add Class*).
Add a new method named LoadDictionary that reads the product numbers and prices from the text file Price.txt, and load them into a Dictionary<string, float>, as shown in Figure 11-9. You may copy the code from Load event handler of DictionaryValueType form in Ch9_Collections project.
Make sure you add the using directive **using System.IO;**, as shown in line 6, which specifies the namespace that contains the StreamReader class.

Figure 11-9: Code to add items to the Dictionary

```csharp
using System.IO;
namespace Ch11_MultiFormApps2
{
    class SharedMethods
    {
        public static void LoadDictionary(Dictionary<string, float> prices)
        {
            string currentLine;
            string[] fields = new string[2];
            //Create StreamReader
            StreamReader PriceReader = new StreamReader("Prices.txt");

            //Read each line from the file, split and store in an array.
            while (PriceReader.EndOfStream == false)
            {
                currentLine = PriceReader.ReadLine(); // Read a product# and price
                fields = currentLine.Split(','); // Store in an array
                // Add the key (Prod#) and value (price) to the Dictionary:
                prices.Add(fields[0], float.Parse(fields[1]));
            }
            PriceReader.Close();
        }
    }
}
```

Static Methods

Note that the method LoadDictionary is declared as a static method. A static method is not associated with a specific instance (object) of the class. It is associated only with the class. So, it is invoked by the class name, followed by the member access operator (.) and the method name, as in

> SharedMethods.LoadDictionary

To call a **nonstatic method**, you have to create an object of the class and use the object name in place of the class name.

Public Methods

LoadDictionary is also a public method. Public methods can be called from other forms. The LoadDictionary method needs to be called from the forms LookUpPrice and DisplayProducts.

Passing Dictionary as a Parameter to a Method

The header of the LoadDictionary method in line 11 declares a parameter (a reference variable) named prices that is of Dictionary type, so that prices can be shared with the calling program. The rest of the code to read data and load into the Dictionary is essentially same as the code in the form DictionaryValueType that we discussed in Chapter 9.

Next, create the two forms that invoke this method.

The coding for the **LookUpPrice** form has the following three steps:

1. Declare a variable named prices that references a Dictionary at the class level. The variable needs to be declared at the class level so that it can be used in the Load event to add product numbers to the ComboBox from the Dictionary, and used in the SelectedIndexChanged event of the ComboBox to find the price for a selected product.
2. Call the LoadDictionary method (from SharedMethods class) to load data into the Dictionary. This is done within the Load event handler of the form. In addition, add the product number to the ComboBox.
3. When the user selects a product number from the ComboBox, find the corresponding item from the Dictionary. To do this, you use the SelectedIndexChanged event of the ComboBox.

Step 2-3: Add the code from Figure 11-10 to the LookUpPrices form.

> Specify LookUpPrice as the startup form in the Program.cs file. Run the form and test the code.

Figure 11-10: Code to load data into the Dictionary and find price

```
19      private Dictionary<string, float> prices = new Dictionary<string, float>();
20
21      private void LookUpPrice_Load(object sender, EventArgs e)
22      {
23          // Call LoadDictionary method from sharedMethods class.
24          SharedMethods.LoadDictionary(prices);
25          // Add product numbers (key) to the ComboBox:
26          foreach (string prodNumber in prices.Keys)
27              cboProdNumber.Items.Add(prodNumber);
28      }
29
30      private void cboProdNumber_SelectedIndexChanged(object sender, EventArgs e)
31      {
32          string selectedProduct = cboProdNumber.SelectedItem.ToString();
33          // Use the key (Product#) to get the value (price).
34          float price = prices[selectedProduct];
35          lblPrice.Text = price.ToString();
36      }
```

Line 19 creates a Dictionary, and assigns its reference to a variable named prices.

Line 24 invokes the LoadDictionary method, passing the variable, prices, as an argument. When control returns to the Load event handler, the Dictionary referenced by prices will have the data stored in it.

Lines 26 and 27 add the product numbers to the ComboBox by looping through each item in the Keys collection of the Dictionary.

Line 34 finds the price from the Dictionary using the selected product as the key.

It's time to practice! Do Step 2-4.

Step 2-4: Do Exercise 11-1 to add the code to the DisplayProducts form to do the following:
  Declare a variable to reference the Dictionary, at the class level.
  Call the LoadDictionary method to load the data into the Dictionary.
  Display products that have prices below the limit.
  Run the form and test the code.

Step 2-5: Open ProductsMain form, which opens the other two forms.

Step 2-6: Add the code from Figure 11-11 to the Click event handlers of the buttons.

Figure 11-11: Code to open forms

```csharp
        private void btnLookUp_Click(object sender, EventArgs e)
        {
            LookUpPrice lookUpPrice = new LookUpPrice();
            lookUpPrice.Show();
        }
        private void btnDisplayPrices_Click(object sender, EventArgs e)
        {
            DisplayProducts displayProducts = new DisplayProducts();
            displayProducts.Show();
        }
        private void mnuExit_Click(object sender, EventArgs e)
        {
            this.Close();
        }
```

Lines 21 and 22 create an instance of the form LookUpPrice, and use the Show method of the object to show the form. Similarly, lines 26 and 27 open the DisplayProducts form.

Step 2-7: Specify PrductsMain as the start up form in the the Program.cs file.
Run the main form and test the code.

The complete code for the three forms (ProductsMain, DisplayProducts and LookUpPrice) and the SharedMethods class that contains the LoadDictionary method is shown below:

Code for ProductsMain form

```csharp
    public partial class ProductsMain : Form
    {
        public ProductsMain()
        {
            InitializeComponent();
        }
        private void btnLookUp_Click(object sender, EventArgs e)
        {
            LookUpPrice lookUpPrice = new LookUpPrice();
            lookUpPrice.Show();
        }
        private void btnDisplayPrices_Click(object sender, EventArgs e)
        {
            DisplayProducts displayProducts = new DisplayProducts();
            displayProducts.Show();
        }
```

Code for the DisplayProducts form

```csharp
13  public partial class DisplayProducts : Form
14  {
15      public DisplayProducts()
16      {
17          InitializeComponent();
18      }
19      private Dictionary<string, float> prices = new Dictionary<string, float>();
20
21      private void DisplayPrices_Load(object sender, EventArgs e)
22      {
23          // Call LoadDictionary method from sharedMethods class.
24          SharedMethods.LoadDictionary(prices);
25      }
26
27      private void btnDisplayProducts_Click(object sender, EventArgs e)
28      {
29          lstDisplay.Items.Add("Product#        Price");
30          foreach (KeyValuePair<string, float> kvp in prices)
31          {
32              float priceLimit = float.Parse(txtPriceLimit.Text);
33              if (kvp.Value < priceLimit)
34                  lstDisplay.Items.Add(kvp.Key + "           " + kvp.Value);
35          }
36      }
```

Code for the LookUpPrice form

```csharp
13  public partial class LookUpPrice : Form
14  {
15      public LookUpPrice()
16      {
17          InitializeComponent();
18      }
19      private Dictionary<string, float> prices = new Dictionary<string, float>();
20
21      private void LookUpPrice_Load(object sender, EventArgs e)
22      {
23          // Call LoadDictionary method from sharedMethods class.
24          SharedMethods.LoadDictionary(prices);
25          // Add product numbers (key) to the ComboBox:
26          foreach (string prodNumber in prices.Keys)
27              cboProdNumber.Items.Add(prodNumber);
28      }
29
30      private void cboProdNumber_SelectedIndexChanged(object sender, EventArgs e)
31      {
32          string selectedProdcut = cboProdNumber.SelectedItem.ToString();
33          // Use the key (Product#) to get the value (price).
34          float price = prices[selectedProdcut];
35          lblPrice.Text = price.ToString();
36      }
```

Code for SharedMethods class with the LoadDictionary method

```csharp
 9    class SharedMethods
10    {
11        public static void LoadDictionary(Dictionary<string, float> prices)
12        {
13            string currentLine;
14            string[] fields = new string[2];
15            //Create StreamReader
16            StreamReader PriceReader = new StreamReader("Prices.txt");
17
18            //Read each line from the file, split and store in an array.
19            while (PriceReader.EndOfStream == false)
20            {
21                currentLine = PriceReader.ReadLine(); // Read a product# and price
22                fields = currentLine.Split(','); // Store in an array
23                // Add the key (Prod#) and value (price) to the Dictionary:
24                prices.Add(fields[0], float.Parse(fields[1]));
25            }
26            PriceReader.Close();
27        }
28    }
```

Review Questions

11.4 How would you share a method between different forms in a project? That is, what change(s) do you need to make in the method, and where will you create the method?

11.5 What is the difference between a static method and a nonstatic method?

11.4 Closing a Parent Form

When you close a parent form (a form that creates an instance of another form), like ProductsMain, the instances of all other forms that were created in the main form also will be closed. This could be a problem in some situations. Let's look at an example.

Login Form

As in many applications, assume that the user needs to enter a valid password before accessing the form ProductsMain. We use a form named Login, as shown in Figure 11-12, to let the user enter a password.

Figure 11-12: Login form

The user enters a password and clicks the OK button. If the password is valid, we want the form ProductsMain to be displayed and the Login form to be closed; if not, a message is to be displayed. Cancel should close the Login form.

If you create and show an instance of the ProductsMain form from the Login form, you cannot close the Login form without also closing the ProductsMain form. Closing the Login form will automatically close the ProductsMain form.

So, how do you close the Login form but keep ProductsMain open? You open both the Login form and the ProductsMain form from the Main method in Program class, as shown in Figure 11-13.

Figure 11-13: Main method in Program class

```
 9      static class Program
10      {
11          static void Main()
12          {
13              Application.EnableVisualStyles();
14              Application.SetCompatibleTextRenderingDefault(false);
15
16              Login login = new Login();
17              login.ShowDialog(); // Open Login form as a dialog box.
18              // Processing of Main stops till Login form is closed
19
20              if (login.DialogResult == DialogResult.OK)
21                  Application.Run(new ProductsMain());
22          }
23      }
```

The statements in Lines 16 and 17 create an instance of the Login form and open it as a dialog box. So, after the Login form is opened, processing of Main stops until the Login is closed. Let's look at what happens on the Login form. Figure 11-14 shows the code in the Login form.

11.4 Closing a Parent Form

Figure 11-14: Code to verify password on the Login form

```csharp
20      private void btnOK_Click(object sender, EventArgs e)
21      {
22          if (txtPassword.Text == "Csharp101")
23          {
24              this.DialogResult = DialogResult.OK;
25              this.Close();
26          }
27          else
28          {
29              MessageBox.Show("Sorry, invalid passwd");
30              txtPassword.Focus();
31              txtPassword.SelectAll();
32          }
33      }
34      private void btnCancel_Click(object sender, EventArgs e)
35      {
36          this.Close();
37      }
```

If the user enters the correct password and clicks the OK button on the Login form, the **DialogResult** property of the Login form is assigned the value **DialogResult.OK**, as shown in Line 24 in Figure 11-14, and the form is closed. Processing of Main method would continue at line 20, which checks the value of the DialogResult property of the Login form.

If the password is invalid, the user is given the option to re-enter it. If the user closes the Login form without entering the correct password, the DialogResult property would have the default value **DialogResult.Cancel**.

When the Login form is closed, processing of Main continues at line 20 of the Main method:
    if (login.DialogResult == DialogResult.OK)
This statement checks whether the value of DialogResult property of the Login form is DialogResult.OK. If it is, the statement in line 21,
    Application.Run(new ProductsMain());
opens the ProductsMain form.

Note that the statement Application.Run(new ProductsMain()) may be replaced by the following pair of statements:
    ProductsMain ProductsMain = new ProductsMain();
    ProductsMain.ShowDialog();

In summary, if the user enters a valid password, the ProductsMain form is opened by the method Main() rather than by the Login form. So, closing the Login form will not have any effect on the ProductsMain form.

Step 2-8: Open the Login form, shown in Figure 11-15.

Figure 11-15: The Login form

Step 2-9: Add the code in lines 24 and 25 from Figure 11-14 to the Click event handlers of the OK button.

Step 2-10: Add the code in lines 16–21 from Figure 11-13 to the Main method in the class, Program.

Step 2-11: Run the project and test the code with both valid and invalid passwords, and by clicking the Cancel button.

In summary, to open the ProductsMain form based on authentication in the Login form, do not open the parent form, ProductsMain, from the Login form. Open both from the Main method in the Program class. That allows closing the Login form without closing ProductsMain.

11.5 Menus

Buttons on a form provide a convenient way to access a limited number of major functions that are performed frequently. Menus provide access to additional functions. We will look at two types of menus:

1. The top-level menu (main menu) that is displayed at the top of the form to let you select form-wide functions
2. Context menus (shortcut menus) that are displayed when you right click an object, to let you select functions that apply to that object

A top-level menu is added to a form by dragging and dropping the MenuStrip object from the Toolbox to the form. Let's create the menus shown in Figure 11-16 on the ProductsMain form to understand the process.

Figure 11-16: Menus on ProductsMain form

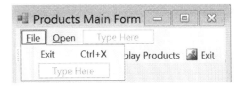

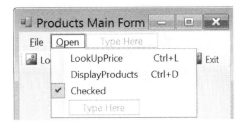

Step 2-12: Add the MenuStrip object from the *Menus and Toolbars* group on the Toolbox. You will see the object in the component tray. Change the name of the MenuStrip from menuStrip1 to MainMenu.

Step 2-13: Click on MainMenu in the component tray. The Menu Designer is displayed at the top of the form with the *Type Here* field, as shown in Figure 11-17.

Figure 11-17: The Menu Designer

Step 2-14: Type **File** in the *Type Here* field. This creates a new menu item of the type ToolStripMenuItem. Change the Text property to &File to make "F" the **access key** for this menu. That means you can press Alt+F to access this menu.

Step 2-15: Type **Exit** in the *Type Here* field in the dropdown menu below File. Because we add code to the Click event handler of this menu item, change its name to **mnuExit** to make it more meaningful. For the **ShortCutKeys** property, select Ctrl and x. That means pressing Ctrl+x will have the same effect as selecting this menu item.

Adding Code to the Click Event Handler of Menu Items

Each menu item acts like a button. You can add code to the Click event handlers of the menu items.

Step 2-16: Double click Exit menu item to open the Click event handler. For code, type
this.Close();

Next, you create the second menu, **Open**, and its dropdown menu with two options: **Look Up Price** and **Display Products**.

Step 2-17: Type **Open** in the *Type Here* field. In the Text property, change the Text to &Open to make "O" the **access key** for this menu. That means you can press Alt+O to access this menu.

Step 2-18: Type **Look Up Price** in the *Type Here* field in the dropdown menu below Open. Because we add code to the Click event handler of this menu item, change its name to **mnuLookUpPrice**. For the **ShortCutKeys** property, select Ctrl and L. That means pressing Ctrl+L will have the same effect as selecting this menu item.

Step 2-19: Type **Display Products** in the *Type Here* field below **Look Up Price.** Change the name of the menu item to **mnuDisplayProducts**. Set Ctrl+D as the ShortCut for this menu item.

When you have different groups of menu items, you may add a **separator** between groups by selecting it from the dropdown menu of the *Type Here* field, as shown in Figure 11-18:

Figure 11-18: Menu with a separator

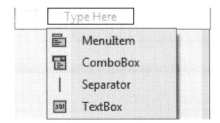

Adding Code to the Click Event Handlers of the Menu Items Look Up Price and Display Products

Note that the menu item mnuLookUpPrice and the button btnLookUpPrice perform the same function of opening the form, LookUpPrice. So, rather than duplicating the code in the Click event handlers of both the button and the menu, we will create a method named LookUpPrice() and call it from both event handlers.

Calling Event Handlers

You may wonder, why not call the the Click event handler btnLookUp_Click directly from the Click event handler mnuLookUpPrice_Click, instead of creating a new method and calling it from both event handlers? This is not desireable because if future changes in the user interface require deleting the btnLookUp button, the menu won't work. In general, calling event handlers from other programs makes maintenance more difficult.

Step 2-20: Create a method named OpenLookUpPrice(), and call it from the Click event handler of the menu item, mnuLookUpPrice, as shown in Figure 11-19:

Figure 11-19: The OpenLookUpPrice method

```
33      private void OpenLookUpPrice()
34      {
35          LookUpPrice lookUpPrice = new LookUpPrice();
36          lookUpPrice.Show();
37      }
38      private void mnuLookUpPrice_Click(object sender, EventArgs e)
39      {
40          OpenLookUpPrice();
41      }
```

Step 2-21: Delete the code from the Click event handler of btnLookUpPrice and add the code to call the OpenLookUpPrice method, as shown in Figure 11-20:

Figure 11-20: The modified event handler

```
19      private void btnLookUp_Click(object sender, EventArgs e)
20      {
21          // Modified code:
22          OpenLookUpPrice();
23      }
```

Submenus

You may create a submenu for a menu item like Display Products by typing the submenu item in the *Type Here* field displayed to the right.

Checked Menu Items

A checked menu item has a CheckBox next to it. When the user clicks the menu item, the CheckBox alternates between the states checked and unchecked. To create a checked menu item, set the Checked property of a standard menu item to True.

A checked menu item may be used for on/off type of actions, like hiding and unhiding a control, and to indicate which item is selected from a group of menu items.

You may add code to the Click event handler of a checked menu item just like any other event handler.

It's time to practice! Do Exercise 11-2.

Context Menus

Typically, a context menu is associated with a control, and it is displayed by right clicking the control. For example, right clicking a TextBox at runtime displays its standard context menu items, like undo, cut, copy, and paste.

The process of creating a ContextMenu is very similar to the process of creating a top-level menu (main menu). A major difference is that you need to associate the context menu to a specific control.

The context menu is created using the ContextMenuStrip control. Let's create a context menu to let the user change the BackColor and ForeColor of the ListBox on the form DisplayProducts.

Step 2-22: Drag and drop the ContextMenuStrip object from the *Menus and Toolbars* group on the DisplayProducts form. You will see the object in the component tray. Change the name of the ContextMenuStrip from ContextMenuStrip1 to DisplayContextMenu.

Step 2-23: To associate the ContextMenu to the ListBox, set the ContextMenu property of the ListBox, lstDisplay, to DisplayContextMenu.

Step 2-24: Click on DisplayContextMenu in the component tray. The Menu Designer is displayed at the top of the form with the *Type Here* field.
Type BackColor in the *Type Here* field. Change the name of the field to cnuBackColor (the prefix "cnu" represents ContextMenu).
Set ShortCutKeys to Ctrl+B.

The ContextMenu is used to change the BackColor and ForeColor of the ListBox, lstDisplay, using the ColorDialog control, which should be added to form.

Step 2-25: Add the ColorDialog control to the form, and change its name to DisplayColorDialog. Double click BackColor menu item from the ContextMenu and enter the code shown in Figure 11-21 to the Click event handler of cnuBackColor.

Figure 11-21: Click event handler of ContextMenu

```
37
38              private void cnuBackColor_Click(object sender, EventArgs e)
39              {
40                  DisplayColorDialog.ShowDialog();
41                  lstDisplay.BackColor = DisplayColorDialog.Color;
42              }
```

Line 40 displays the ColorDialog window, and line 41 assigns the selected color to the BackColor property of the ListBox.

Step 2-26: Run the project and test the code.

It's time to Practice! Do Exercises 11-3.

Review Question

11.6   Describe a context menu.

11.6 ToolStrips

The ToolStrip control lets you create a dockable Toolbar that acts as a container for buttons and a variety of other controls, including Labels, TextBoxes and ComboBoxes. Unlike menu items in a top-level menu (main menu), the buttons on a Toolbar are readily visible.

Both an image and Text may be displayed on a button, as shown in Figure 11-22. By default, the ToolStrip is docked at the top of the form. You may use the Dock property of ToolStrip to choose other options, including *Left* and *Right*. Another feature of ToolStrip is that when there is not enough room to display all controls on a ToolStrip, controls are moved to a dropdown menu in the overflow area. In Figure 11-22, the Exit button in the overflow area is available by clicking the dropdown arrow at the end.

Figure 11-22: The ToolStrip control

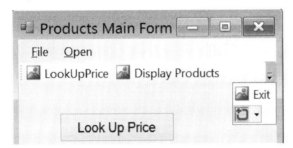

To add a Toolbar, drag and drop the ToolStrip control on the form.

Let's create a Toolbar that essentially duplicates the functionality of the three buttons on the ProductsMain form, as shown in Figure 11-22.

Step 2-27: Drag and drop the ToolStrip control on the form. Change the name of the ToolStrip object in the component tray to ProductToolStrip. By default, the ToolStrip is docked at the top of the form. You may use the Dock.

Step 2-28: Click on the ToolStrip to display the designer at the top of the form. Click on dropdown handler on the control to view the list of controls that can be added, as shown in Figure 11-23.

Figure 11-23: Controls on ToolStrip

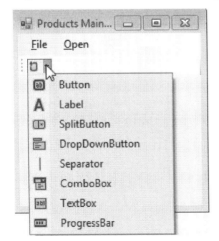

Select the Button control. Change the name of the button to tsbLookUpPrice and the Text property to Look Up Price.

To display both image and Text, change **DisplayStyle** to ImageAndText.

To display an image on the button, click on the ellipses on the right of Image property to open the *Select Resources* window, and select an image that represents the task of opening.

Adding Code to the Click Event Handler of ToolStrip Buttons

The Click event handler is created by double clicking the button.

Step 2-29: Add the code from Figure 11-24 to call the OpenLookUpPrice method from the Click event handler of the button.

Figure 11-24: Click event handler of tsbLookUpPrice

```
42
43      private void tsbLookUpPrice_Click(object sender, EventArgs e)
44      {
45          OpenLookUpPrice();
46      }
```

The modified code for ProductsMain form with menus and ToolStrip is shown below:

```
19      private void btnLookUp_Click(object sender, EventArgs e)
20      {
21          // Modified code:
22          OpenLookUpPrice();
23      }
24      private void btnDisplayPrices_Click(object sender, EventArgs e)
25      {
26          DisplayProducts displayProducts = new DisplayProducts();
27          displayProducts.Show();
28      }
29
30      private void btnExit_Click(object sender, EventArgs e)
31      {
32          this.Close();
33      }
34      private void mnuExit_Click(object sender, EventArgs e)
35      {
36          this.Close();
37      }
38      private void OpenLookUpPrice()
39      {
40          LookUpPrice lookUpPrice = new LookUpPrice();
41          lookUpPrice.Show();
42      }
43      private void mnuLookUpPrice_Click(object sender, EventArgs e)
44      {
45          OpenLookUpPrice();
46      }
47      private void tsbLookUpPrice_Click(object sender, EventArgs e)
48      {
49          OpenLookUpPrice();
50      }
```

Review Question

11.7 Different methods that can be used to let the user perform a task in a C# form include (1) clicking a standard button, (2) selecting an item from a standard top-level menu (main menu), (3) selecting an item from a Context menu, and (4) clicking a button on a custom Toolbar.

Consider a form that allows the user to perform different tasks associated with processing payroll (compute pay, deductions, etc.) and save the data. These tasks are performed by clicking standard buttons on the form. In addition, it is desired to let the user change the appearance (color and font size) of a ListBox on the form.

Which one of the above methods is appropriate to let the user change the back color and/or the font size of the ListBox?

Exercises

Exercise 11-1

Create the form DisplayProducts, shown in Figure 11-25. Create a Dictionary in this form, and call LoadDictionary method to load the Dictionary with data. Use the Dictionary to display all products with price below the specified limit.

Figure 11-25: The Display Products form

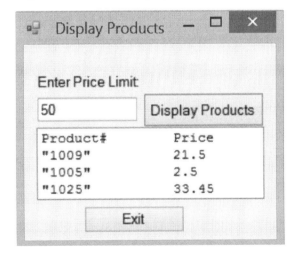

Exercise 11-2

Create a new method named OpenDisplayProducts to open the DisplayProducts form in Ch11_MultiformApps2. Call this method from the Click event handlers of btnDisplayProducts button and mnuDisplayProducts menu item.

Exercise 11-3

Add another menu item, ForeColor, to the Context menu on the DisplayProducts form to let the user change the ForeColor of the ListBox

Chapter 12

Databases

Database systems, in general, make data access faster and easier than file-based systems. They incorporate features that make it easier to share data, reduce redundancy and enhance security. In this chapter, you will learn how to access databases from C# programs.

Popular database management systems (software that manage databases) include Oracle, Microsoft SQL Server, DB2 and Microsoft Access. We use SQL Server database in this book; however, the techniques and steps presented in this chapter generally are applicable to all databases.

Topics

12.1	Introduction	12.7	Selecting Records Using Binding Source
12.2	Creating Database Objects	12.8	Selecting Records Using Table Adapter
12.3	Displaying Data in a DataGridView	12.9	Untyped Datasets: Displaying Records
12.4	Accessing Data from a DataGridView	12.10	Untyped Datasets: Select Records
12.5	Displaying Data in Details View	12.11	Untyped Datasets: Add/Edit/Delete Record
12.6	Finding the Record for a Selected Key	12.12	Command Object and DataReader

12.1 Introduction

A relatively easy way to work with data from database tables is to display data in controls like TextBoxes, ComboBoxes and DataGridViews using a two-step process:

1. Transfer the data from **database tables,** which typically are stored on the disk, to **data tables**, which are temporary tables maintained in memory.
2. Bind (link) controls on a form to the data table to display the data.

A data table may contain data from one or more database tables. Data tables generally are placed inside a container called **dataset**. Thus, a dataset is a collection of one or more data tables.

The **table adapter** and **binding source** are two important .Net Framework classes that help you to work with databases.

The **table adapter** connects the data table to the database, and retrieves data from and saves data to the database.

The **binding source** links the user interface controls on the form to a data table. In addition, it provides methods and properties to perform additional functions like selecting records from the data table, and moving between records.

Figure 12-1 represents the relationship between database tables, data table and controls on a form.

Figure 12-1: Relationship between database tables, data table and controls on a form

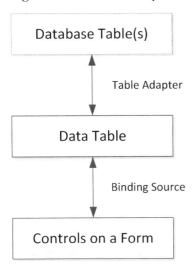

We will use the wizard to create the necessary objects and to help you work with the data.

12.2 Creating Database Objects

To create the database object, first, you specify the **data source**—that is, the source of data, its type and location. The data source could be a database on the local computer or on a server, or a web service or other object.

Next, you drag the data table to the form to create the controls that display the data and objects that bind the controls to the data table.

Specifying the Data Source

You specify the data source using the **Data Source Configuration Wizard**, which in turn generates the data tables and other objects, like the binding source.

Tutorial 1: Display Employee Records Using the Wizard

Let's create an application that lets you work with a human resource database that includes an Employee table.

Step 1-1: Create a new project named Ch12_Databases. Rename the form to EmployeeTabularView.

Step 1-2: Copy the SQL Server database file HR.mdf from Tutorial_Starts/DataFiles folder to a local directory outside the project directory, Ch12_Database. **Appendix B** provides instructions on how to create a new database in Visual Studio.

Step 1-3: To start the Data Source Configuration Wizard, click *Project* from the menu, and click *Add New Data Source*. You will see the Configuration Wizard window, as shown in Figure 12-2.

Figure 12-2: Selecting data source type

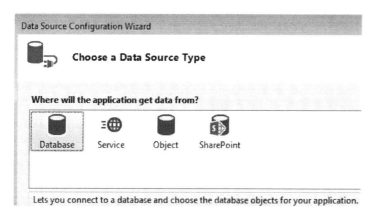

Step 1-4: Select *Database* and click *Next*. You will be asked to select a database model.

Step 1-5: For database model, select *Dataset*. You will be asked to choose the data connection, as shown in Figure 12-3.

Figure 12-3: Choosing data connection

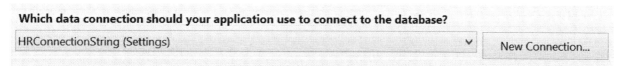

Step 1-6: If the dropdown list shows the database file name, HR.mdf, you may go to step 1-8. If not, click the *New Connection* button. If the *Add Connection* window is displayed, click the *Change* button. The *Choose Data Source* window, shown in Figure 12-4, opens.

Figure 12-4: Adding connection

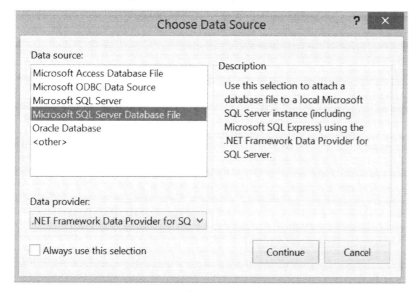

Step 1-7: Select ***Microsoft SQL Server Database File*** for *Data Source*. The dropdown list for *Data Provider* should show *.Net Framework Data Provider for Microsoft SQL Server*.

Click *Continue*. The following *Add Connection* window appears:

Click *Browse*, and select HR.mdf from the local directory. Click *Open*.

Click *OK*. If the database is not compatible with the current instance of SQL Server, you will be asked whether you want to upgrade the database. Select *Yes*.

Step 1-8: Click *Next*. You may be asked whether you want to copy the file to your project. **Make sure you select *No*.**

If you answer *Yes*, Visual Studio will copy the database file to the project directory, and by default, changes made to the database at runtime will be lost the next time the project is run.

Click *Next*. You will be asked whether you want to save the connection string.

Accept the default (Yes), and Click *Next*. You will be asked to select the database objects, as shown in Figure 12-5.

Figure 12-5: Selecting database objects

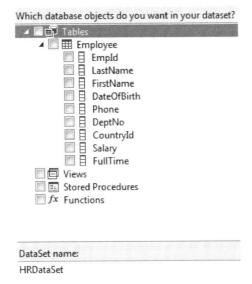

Chapter 12: Databases

Step 1-9: Select Employee table by checking the CheckBox next to it. This will select all fields of the Employee table.

If you want to display only certain fields, uncheck the fields to be excluded. You also may select a view that combines data from multiple tables.

Leave the dataset name as HRDataSet. Though the current data are only from the Employee table, we will be adding data from other tables later. So, the dataset will have data tables from the entire HR database.

Click *Finish*.

If the Server Exaplorer window is not displayed, select *View, Server Explorer*. The Server Explorer window will show HR.mdf under Data Connections, as shown in Figure 12-6.

Figure 12-6: Server Explorer

▲ 🗄 Data Connections
　▷ 🗄 HR.mdf

Step 1-10: Display the DataSource window: Click *View, Other Windows, Data Sources*. The DataSource window will be displayed on the left, as shown in Figure 12-7.

Figure 12-7: DataSource window

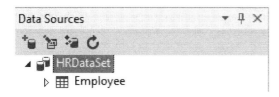

To view the fields and their properties, right click Employee table in Server Explorer window, and select *Open Table Definition*. The T-SQL window displays the SQL that creates the table. Close dbo.Employee[Design] window.

12.3 Displaying Data in a DataGridView

The DataGridView control provides a simple way to display and edit an entire table in a scrollable window. An alternative is to view one record at a time in controls like TextBoxes and Labels that display a single data item. To display a data table and work with it on a form, you follow two simple steps:

1. Choose between DataGridView (to view the entire table) and Details (to view one record at a time) from the dropdown list for the data table, as shown in Figure 12-8.
2. Drag and drop the data table on the form.

Figure 12-8: Display options

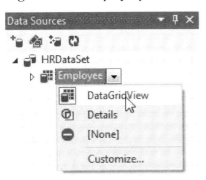

First, we will display employee data in a DataGridView.

Step 1-11: Choose DataGridView from the dropdown list for Employee data table.

Step 1-12: Drag and drop Employee table on the form.

You will see a grid with the field names, a navigation bar at the top, and several objects, including the HRDataSet, employeeTableAdapter and employeeBindingSource, in the section called component tray, as shown in Figure 12-9. All these objects were created automatically by the wizard when you dropped the Employee table on the form.

Figure 12-9: DataGridView bound to the Employee table

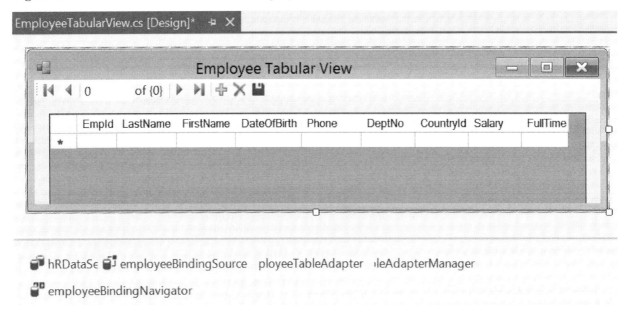

You may change the width, header, name, format or any other properties of columns by clicking on *edit column* (at the bottom of Properties window) to open the edit column window.

To understand the functions of the various objects displayed in the component tray, let's take another look at the relationship between them. The relationship, introduced earlier in Figure 1-10, is presented in Figure 12-10.

Figure 12-10: Relationship between components

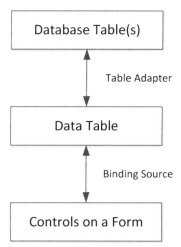

The objects on the form include

HRDataSet: A dataset contains one or more data tables. Currently, HRDataSet contains only the Employee data table. When you run the project, the employeeTableAdapter will load the data from Employee database table into Employee data table.

employeeTableAdapter: Moves data between the database table (Employee), and the data table that also is named Employee, using **Fill** and **Update** (or **UpdateAll**) methods:

> The **Fill** method of the table adapter copies data from the database table into the data table.

> The **Update/UpdateAll** method saves the data from the data table into the database table.

employeeBindingSource: Moves data between Employee data table and the DataGridView on the form.

employeeBindingNavigator: Provides the navigation bar at the top of the form.

TableAdapterManager: If you are editing multiple related tables on a form, you can use the UpdateAll method of TableAdapterManager to update all tables in the correct order.

Step 1-13: First, specify Currency format for Salary column as follows:

Select the DataGridView, and select *Columns* from the Properites window. Click the ellipses (…) on the right side of the property to display the *Edit Columns* window.

Select Salary, *DefaultCellStyle*, and click the ellipses (…) on the right side of the property, as shown in the following figure.

12.3 Displaying Data in a DataGridView

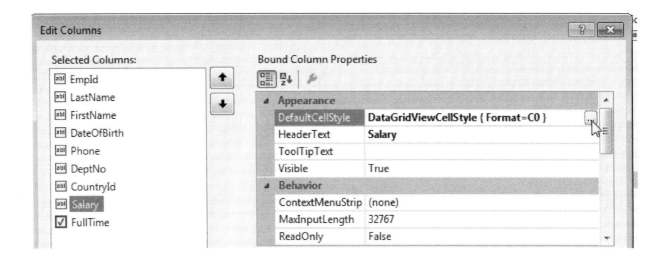

Select *Format* from the *CellStyle Builder* window, and click the ellipses on the right to open the *Format String Dialog* window.

Select *Currency*, and specify 0 for *Decimal Places*.

Next, run the project. You will see the employee records displayed in the grid, as in Figure 12-11.

To test the form, make changes to a record, click Save and close the form. Now reopen the form to make sure the changes were saved. Test to make sure adding a new record and deleting a record work correctly.

Figure 12-11: Employee table in the DataGridView

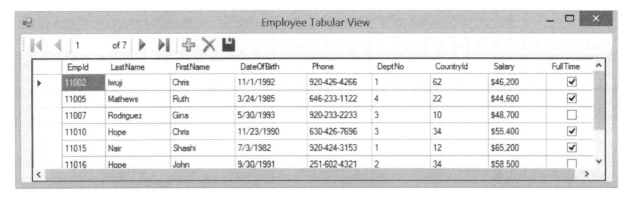

Data-Bound Controls

A data-bound control is a user interface control that is linked to a data source. When you dragged and dropped the Employee data table to the form, the wizard added the DataGridView to the form and bound it to the Employee data table through the binding source. As you can see from Figure 12-12, the Data Source property of the DataGrid is set to employeeBindingSource.

Figure 12-12: DataSource property of DataGridView

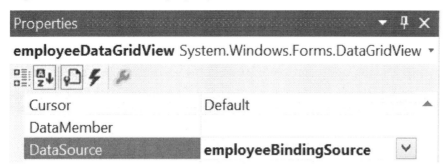

The employeeBindingSource, in turn, is linked to the Employee table in the HRDataSet, as indicated by the DataMember and DataSource properties of the binding source, shown in Figure 12-13.

Figure 12-13: DataMember and DataSource properties of binding source

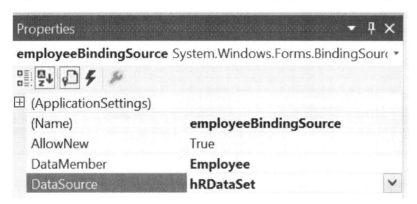

Auto-generated Code

Figure 12-14 shows the code generated by the wizard.

Figure 12-14: Code generated by the wizard

```
20      private void EmployeeTabularView_Load(object sender, EventArgs e)
21      {
22          // TODO: This line of code loads data into the 'hRDataSet.Employee' table.
23          this.employeeTableAdapter.Fill(this.hRDataSet.Employee);
24      }
25
26      private void employeeBindingNavigatorSaveItem_Click(object sender, EventArgs e)
27      {
28          this.Validate();
29          this.employeeBindingSource.EndEdit();
30          this.tableAdapterManager.UpdateAll(this.hRDataSet);
31      }
```

The statement in line 23,
   this.employeeTableAdapter.Fill(this.hRDataSet.Employee);
uses the Fill method of the table adapter to load data from the database table into the data table, Employee.

Saving changes from the DataGridView to the database table is a two-step process:

1. Save changes from the DataViewGrid to the data table, Employee. This is done using the EndEdit method of the binding source in line 29:
   this.employeeBindingSource.EndEdit(),
2. Save changes from the data table to the database table. This is done using the UpdateAll method of tableAdapterManager in line 30:
   this.tableAdapterManager.UpdateAll(this.hRDataSet);

View or Edit Connection String

Step 1-14: To view or edit connection string information, double click Settings under properties in the Solution Explorer, as shown in Figure 12-15.

Figure 12-15: Viewing ConnectionString information

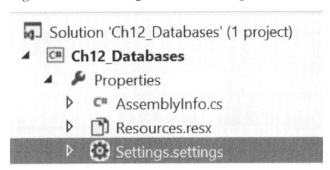

You will see the connection string information, as shown in Figure 12-16.

Figure 12-16: ConnectionString information

Name	Type	Scope	Value
HRConnectionString	(Connectio...	Application	Data Source=(LocalDB)\v11.0;AttachDbFilename=|DataDirectory|

Step 1-15: Click in the right-most column under *Value*, and click on the ellipses to open the Connection Properties window that allows you to edit the information.
Click *Cancel*, and close the Settings window.

The Structure of the Dataset

Adding a data source also adds the file HRDataset.xsd (xsd for XML Schema Document) to the Solution Explorer window.

Step 1-16: To view the structure of the tables in the dataset, double click HRDataset.xsd in Solution Explorer. You will see HRDataSet.xsd window that shows the Employee data table, and the table adapter associated with it, as shown in Figure 12-17.

Figure 12-17: Dataset Schema window

Step 1-17: Select EmployeeTableAdapter to view its properties.
The properties window shows the connection string and the Select, Update, Delete and Insert commands used by the table adapter.
Click the "+" sign on the left of Select Command to view the SQL used by the table adapter to select records from the database. You also may view the SQL for other commands.

Review Questions

12.1   What is the difference between a dataset table and a database table?

12.2   What is the function of the table adapter?

12.3   What is the function of the binding source?

12.4   What does the Fill method of the table adapter do?

12.5   What does the Update method of the table adapter do?

12.6   What is a data bound control?

12.7   How is the data in a data bound control linked to a field in a data table?

12.4 Accessing Data Items from a DataGridView

In addition to viewing and editing database tables, you can access individual data items using the properties and events of the DataGridView. Figure 12-18 shows some of the major properties and events of the DataGridView.

Figure 12-18: Properties and events of DataGridView

Property	Description
ColumnCount	The number of columns.
Columns	Collection of all columns.
CurrentRow	Row containing the current cell. Each row consists of a collection of cells represented by the **Cells** collection.
CurrentCell	Current cell. The **Value** property gives the value of a cell.
Enable Adding (Editing, Deleting) - This is an item in the **task panel** that is displayed when you click the arrow (smart tag) at the upper right corner.	When checked, user can add (edit, delete) columns. By default, it is checked.
RowCount	The number of rows, including the row for new records.
Rows	Collection of all rows. Access a cell using the column index or the column name (e.g., Rows[3].Cells[4]; or Rows[3].Cells["Phone"]). Get the value (of object type) associated with a cell using Value property (e.g., Rows[3].Cells["Phone"].Value).
SelectedColumns	Collection of columns selected by the user.
SelectedRows	Collection of rows selected by the user.
Events	
CellClick	When a cell is clicked.
CurrentCellChanged	When the current cell changes.

Let's look at some examples of how you would use these properties and events to work with the table.

Rows and Cells Collections

You may access any cell using the **Rows** and **Cells** collections by specifying its column and row indices. For example, the data item in row 2 and column 8 is given by

employeeDataGridView.Rows[1].Cells[7].Value.ToString()

Note that the index of row 1 is 0, index of row 2 is 1, and so on.

Value Property

The **Value** property gives the value associated with a cell. The ToString method converts the Value, which is of object type, to string.

You also may use the column name in place of the column index, as in

employeeDataGridView.Rows[1].Cells["Salary"].Value.ToString()

Column Names

The above statement assumes that the column name for Salary field has been changed to Salary from its original name **salaryDataGridViewTextBoxColumn**, which is shown in Figure 12-19. Though the column headers (HeaderText property) in the DataGridView are the same as the field names, the column names are not, as shown in Figure 12-19. So, if you use column names in code, you may want to change the names of those columns so that the column names are the same as the field names and column headers. As described earlier, you display the *Edit Columns* window by clicking the ellipses to the right of Columns property of the DataGridView.

Select the DataGridView, and select *Columns* from the Properites window. Click the ellipses (…) on the right side of the property to display the *Edit Columns* window.

Figure 12-19: Column properties of DataGridView

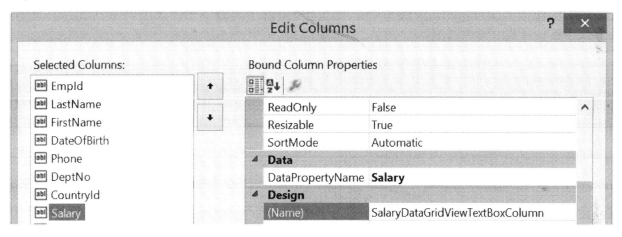

Displaying the Value of a Selected Cell

You may automatically display the value of a cell when the user selects it by clicking it, using the **CurrentCellChanged** event handler and the **CurrentCell** property. Figure 12-20 shows the code to display the value of the current cell in a Label.

Figure 12-20: CurrentCellChanged event handler of DataGridView

```
33      private void employeeDataGridView_CurrentCellChanged(object sender, EventArgs e)
34      {
35          lblCurrentCell.Text = employeeDataGridView.CurrentCell.Value.ToString();
36      }
```

Step 1-18: Add a Label named lblCurrentCell and another Label that displays the prompt, "Current Cell," as shown in Figure 12-21.

Create the CurrentCellChanged event handler of the DataGridView and add the code from Figure 12-20. Click a cell to test the code. The Label should show the data item from the selected cell.

12.4 Accessing Data Items from a DataGridView

Figure 12-21: Employee form showing the average salary

You may display the value from any column in the selected row using the **CurrentRow** property of the DataGridView, and the **Columns** collection of the CurrentRow. For example, the following statement displays the phone number from the current row, in a Label named lblPhone:

   lblPhone.Text = employeeDataGridView.CurrentRow.Cells[4].Value.ToString();

Step 1-19: Add a Label named lblPhone, and a button named btnDisplayPhone, as shown in Figure 12-21. Add the above statement to the Click event handler of the button. See Figure 12-22. Test the code.

Accessing Data from Every Row in a Column

Now let's look at accessing every row in a column, using a loop. You will compute the average salary and display it in a Label when the user clicks a button, as shown in Figure 12-21. The code to display the phone number and compute the average is shown in Figure 12-22.

Figure 12-22: Code to access every row

```
38          private void btnDisplayPhone_Click(object sender, EventArgs e)
39          {
40              lblPhone.Text = employeeDataGridView.CurrentRow.Cells[4].Value.ToString();
41          }
42          private void ComputeAvg_Click(object sender, EventArgs e)
43          {
44              float sumOfSalary = 0, avgSalary;
45              int numOfRows = employeeDataGridView.RowCount-1;
46                          // RowCount includes the row for new record.
47              // Use for loop to vary row index and get salary from Salary column.
48              for (int rowIndex = 0; rowIndex < numOfRows; rowIndex ++)
49                  sumOfSalary = sumOfSalary + float.Parse(employeeDataGridView.
50                              Rows[rowIndex].Cells["Salary"].Value.ToString());
51              // To compute total only for selected rows, use SelectedRows[]
52              avgSalary = sumOfSalary/numOfRows;
53              lblAvg.Text = avgSalary.ToString("C");
54          }
```

The statement in line 50 assumes that the name of the column for salary has been changed to Salary.

Further, note that the RowCount property in line 45 includes the row for new record.

Step 1-20: Change the name of the Salary column to salary: Click edit columns under properties to open the edit column window. Select Salary, and change the *Name* property (under *Design*) to salary.

Step 1-21: Add a Label named lblAvg and a button named btnComputeAvg. Add the code from Figure 12-22 to the Click event handler of btnComputeAvg.

Run the project and test the code.

12.5 Displaying Data in Details View

The DataGridView displays an entire table in a scrollable window. This is convenient when the number of records and fields in a table is relatively small.

The details view displays one record at a time, with each data item displayed in a separate control, like Label, TextBox or CheckBox. This makes editing/adding/deleting individual records easier, and is particularly suitable when there is a large number of records and fields.

To display employee data in details view, select Details from the dropdown list for Employee data table, as shown in Figure 12-23.

Figure 12-23: Display options

For each field of Employee table, you may select the type of control (TextBox, NumericUpDown, etc.) from the dropdown list for the field, shown in Figure 12-24. The default control is TextBox, except for DateOfBirth, which has DateTimePicker, and FullTime, which has CheckBox. We will use the default controls for this form.

Figure 12-24: Selecting the type of control to display a field

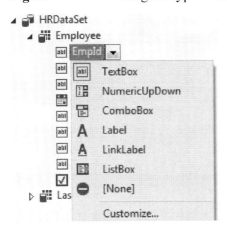

Next, drag and drop the Employee table to the form to add the controls and the components needed to bind the controls to the database table.

Step 1-22: Add a new form named EmployeeDetailsView.

Step 1-23: Select Details from the dropdown list for Employee data table.

Step 1-24: Drag and drop the Employee table on the form. Change the Text property of the CheckBox for FullTime field to "Full Time?"

Rather than dragging and dropping an entire table, you may drag and drop one field at a time if you do not want all fields on the form.

Controls are added to the form, as shown in Figure 12-25. In addition, the five components, including the DataSet and table adapter, are added to the component tray.

Figure 12-25: Controls and components added to the form by the wizard

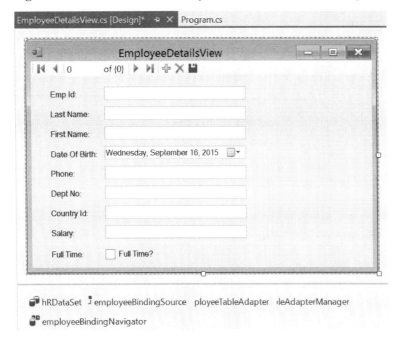

Note that each form needs its own set of five components.

Auto-generated Code

The generated code to Fill the data table and to Update the database table are identical to that for the DataGridView.

Step 1-25: Specify EmployeeDetailsView as the start-up form in the Program.cs file. Run the form. Use the navigation bar to view records, add, delete, and save changes.

12.6 Finding the Record for a Selected Key

When a table contains a large number of records, searching and finding specific records is particularly important in Details view, which displays one record at a time.

We will create a separate form that focuses on selecting records rather than adding/editing/deleting records.

Searching a Key (Unique) Field

A simple way to let the user select the value for a key field (or for any unique field) to find the corresponding record is by displaying the values of the key in a ComboBox or ListBox.

Tutorial 2: Find Employee Records Using the Wizard

In this tutorial, you create a form that allows the user to select an employee's id from a ComboBox and display the employee record, as shown below:

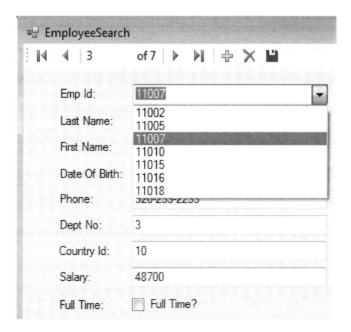

To create the form, drag and drop the Employee table in Details mode as before, except that we use a ComboBox (instead of the standard TextBox) for EmpId.

Step 2-1: Add a new form named **EmployeeSearch**.

Step 2-2: Select Details from the dropdown list for Employee data table.

For EmpId field, select ComboBox for type of control, as shown in Figure 12-26.

Figure 12-26: Selecting ComboBox to display employee ids

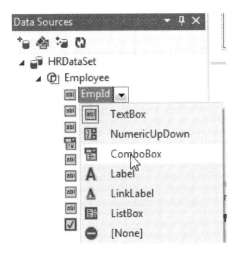

Step 2-3: Drag and drop the Employee table on the form.

The form looks the same as the EmployeeDetails form you created previously, except that the EmpId field is a ComboBox.

Step 2-4: Run the new form. The ComboBox for EmpId shows the id for the current record, but the ComboBox does not have the list of EmpId's from all records, as we want.

DataBindings Property of Controls

The reason each control displays the data from a field of the current record is that each control is bound to a field using its DataBindings property. The DataBindings property of the ComboBox that displays EmpId is shown in Figure 12-27.

Figure 12-27: DataBindings property of ComboBox

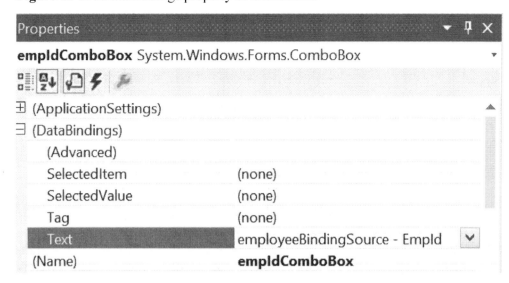

Figure 12-27 shows that the Text property of the ComboBox is bound to the EmpId field in the Employee data table (through employeeBindingSource). That means the data from EmpId field in the current record would be assigned to the Text property of the ComboBox, and any changes to the Text property are saved in the EmpId field of the table.

Similarly, the Text property of the TextBox for LastName is bound to the LastName field in the Employee data table. The Text property of each TextBox generated from the table is bound to the corresponding field.

Step 2-5: Select the ComboBox for EmpId and view the DataBindings property. Repeat for LastName TextBox.

View the DataBindings property of the DateTimePicker for DateOfBirth. Note that the Value of the control is bound to DateOfBirth field.

View the DataBindings property of the CheckBox. Note that the CheckState property is bound to the FullTime field.

Displaying a List of Values in a ComboBox

Binding the Text property of a ComboBox to a field shows only the data from the current record. **The purpose of the EmpId ComboBox is not to show the EmpId from the current record, but to display all EmpIds so the user can select one to find the corresponding record**. How do you do that?

First, you need to set two important properties of the ComboBox (or ListBox):
Set **DataSource** to the binding source.
Set **DisplayMember** to the field.
Setting these two properties would display a list of all EmpIds in the ComboBox.

Binding tables that do not have a binding source: Note that data tables that are not dragged and dropped to a form won't have a corresponding binding source. If there is no binding source for a table, you select
  Other Data Source, Project Data Sources, HRDataSet, Employee.
Selecting the table for DataSource, in turn, will create the binding source for the table.

Next, you need to remove the DataBindings for the ComboBox by setting the DataBindings to *None*. If not, when you select an EmpId from the list, the new value will replace the existing value for current record, resulting in an error because there is already another record that has the selected value of EmpId.

Step 2-6: Select the ComboBox that displays EmpId, and set the DataSource property to employeeBindingSource (that links to Employee data table), as shown in Figure 12-28.

Figure 12-28: DataSource property of ComboBox

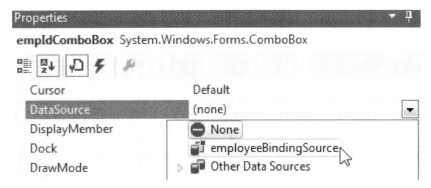

Next, set the Display Member to EmpId, as shown in Figure 12-29.

Figure 12-29: DisplayMember property of ComboBox

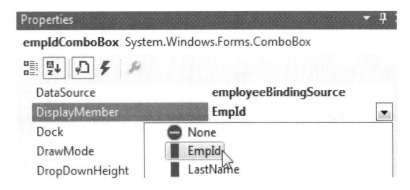

Step 2-7: Set the DataBindings property of EmpId ComboBox to **None,** as shown in Figure 12-30.

Figure 12-30: Setting the DataBindings property of EmpId ComboBox to **None**

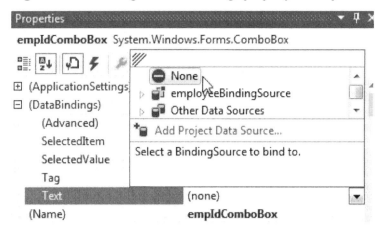

Step 2-8: Run the form. Select an id from the EmpId ComboBox and verify that the record for the selected id is displayed.

How does it find the record when there is no code to do it? The secret is that the ComboBox is getting the EmpIds from the Employee data table that also is bound to other controls on the form. So, when you select an EmpId from the ComboBox, you are changing the current record in the Employee data table, which in turn changes the data in all bound controls. This won't happen if you bind the ComboBox to one data table and the other controls to a different data table.

It's time to practice! Do Exercise 12-1.

Review Question

12.8   The data from a table named Customer is displayed on a form in Details View. The controls on the form were created by dragging and dropping the data table on the form after changing CustomerId field to a ComboBox. The purpose of the ComboBox is to let the user select a CustomerId from a list of all CustomerIds and display the record for the selected id. What properties of the ComboBox need to be changed from the default setting? What should be the setting for those properties?

12.7 Selecting a Group of Records Using the Binding Source

When a table contains a large number of records, it is important to be able to look at a group of records, like employees from a department. To understand ways to limit the records displayed on a form, let's take another look at the objects (components) that are used in moving data from the database to the form. As shown in Figure 12-31, the table adapter controls what is displayed in the data table in memory. Similarly, the binding source controls what is displayed on the form from the data table.

Figure 12-31: Relationship between database components

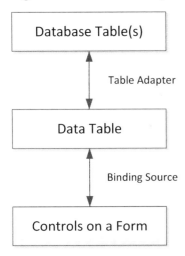

Figure 12-31 suggests that there are two ways you can select the records to be displayed on a form:

1. Use the binding source to control what is displayed in the controls on the form. For example, you may load the entire Employee database table into the Employee data table, but filter what is displayed in the controls on a form. You use the Filter property of the binding source to do it.

2. Use the table adapter to select a group of records from the database and load it into the data table. To do this, you specify the criterion to select records using the SQL language.

In this section, we look at selecting records using the binding source.

To select records using the binding source, assign the criteria (for selecting records) to the Filter property of the binding source. The general syntax is
      BindingSource.Filter = "criterion";
The criterion could be any Boolean expression like
      "LastName = 'Hope' AND DeptNo = 2"

Here are examples of complete statements:
      employeeBindingSource.Filter = "LastName = 'Hope'";
      employeeBindingSource.Filter = "Salary > 40000";
      employeeBindingSource.Filter = "DateOfBirth > #01-01-1990#";
      employeeBindingSource.Filter = "FullTime = True";

Note that the string data are enclosed in a pair of apostrophes, and Date type is enclosed in # signs. If the field is of numeric or Binary type, then do not enclose the data in apostrophes.

The whole criterion must be enclosed in quotes.

Filtering Records Using the Last Name from a TextBox

Rather than hard-coding the value of the field (like 'Hope' and 40000), you can make the criterion more flexible by getting the value from a TextBox or ComboBox.

Consider the first example:
    employeeBindingSource.Filter = "LastName = 'Hope'";

How does the criterion look if the last name is taken from the TextBox, txtLastName?
You need to concatenate three different strings:
    Everything before the last name Hope: LastName = '
    The last name from the TextBox: txtLastName.Text
    The last apostrophe: '

The concatenated string looks as follows: "LastName = '" + txtLastName.Text + "'"
The last part of the criterion, "'", is an apostrophe enclosed in quotes.

The complete statement would be
    employeeBindingSource.Filter = "LastName = '" + txtLastName.Text + "'";

Or, you may split the criterion into four parts, as in

    employeeBindingSource.Filter = "LastName = " + "'"+ txtLastName.Text + "'"

Similarly, if the Salary comes from a TextBox, txtSalary, the statement would look as follows:
    employeeBindingSource.Filter = "Salary > " + txtSalary.Text;

Step 2-9: Add a TextBox named txtLastName and Button named btnFilterByLastName to the form, as shown in Figure 12-32.

Figure 12-32: The EmployeeSearch form that filters records using a last name

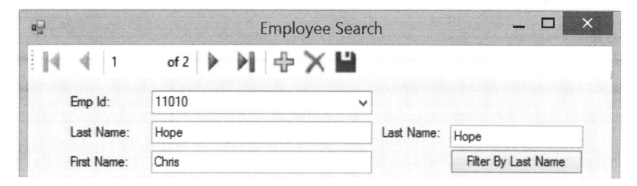

Add the code from Figure 12-33 to the Click event handler of the button.

Figure 12-33: Code to filter employee records using a last name

```
37      private void btnFilterByLastName_Click(object sender, EventArgs e)
38      {
39          employeeBindingSource.Filter = "LastName = '" + txtLastName.Text + "'";
40          // Or, use the wild card character "*" for to specify incomplete strings
41          employeeBindingSource.Filter = "LastName LIKE '" + txtLastName.Text + "*'";
42      }
```

Step 2-10: Run the form. Enter the name Hope into the TextBox and click the button. As the navigation bar shows (1 of 2), only the two records with the last name Hope are available on the form.

Next, enter "M" to display all employees whose last name starts with "M." Verify the result.

Filtering Records Using the Last Name Selected from a ComboBox

Rather than typing the last name into a TextBox, it would be easier to select a last name from a list of all valid unique last names. You could display all last names in the ComboBox by setting the Data Source to employeeBindingSource and Display Member to LastName. However, this would display the same last name multiple times if there are multiple records with that last name. So, we will create a new data table called LastNames that contains only unique last names, no duplicate names.

Adding a New Data Table and Corresponding Table Adapter

To add a new data table in the DataSet window, you go through many of the same steps you followed when you added the data source. Creating a data table also creates the table adapter for the table.

Step 2-11: Open the DataSet window by double clicking on the file HRDataSet.xsd in Solution Explorer. You will see the Employee data table and its table adapter.

Step 2-12: Right click anywhere in the window and select *Add, TableAdaper* (not Data Table). You will be asked to choose the connection. Because you are using the same connection, click *Next*. You will be asked to choose the Command Type.

Step 2-13: Keep the default, *Use SQL Statements*. Click *Next*. You will be asked to enter a SQL statement.

Step 2-14: Click *Query Builder* instead of typing in the query. Select Employee table from the *Add Table* window and click *Add*. Close *Add Table* window.

Step 2-15: Check LastName field, and click OK.
You will see the following SQL to select last names:
SELECT   LastName
FROM   Employee

Step 2-16: Change the SQL by inserting the key word DISTINCT before LastName, as follows:
SELECT   DISTINCT LastName
FROM   Employee

Step 2-17: Click *Advanced Options*. Uncheck the CheckBox for Insert, Update and Delete statements because the Last Names data table is used only for displaying data. Click OK.

Step 2-18: Make sure that *Fill a Data Table* is checked. Click *Next*. Click *Finish*. You will see the new data table with the name DataTable1.

Step 2-19: Right click the table name Employee1 and rename it to LastNames, as shown in Figure 12-34. Rename Employee1TableAdapter to LastNamesTableAdapter.

Figure 12-34: Renaming the data table, Employee1

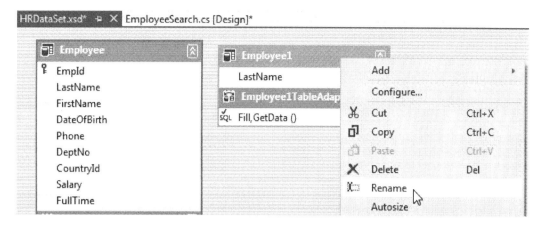

So, now you have a data table named LastNames that contains only unique last names from Employee database table. You also have the table adapter that gets the data from the database.

Now you can display the last names in a ComboBox by setting its Data Source and Display Member properties. Because no control on the form is yet bound to LastNames table, no binding source was created for LastNames. So, the DataSource is to be set to the LastNames table, which in turn will create the binding source.

Step 2-20: Add a ComboBox by the name cboLastNames. Note that the purpose of this ComboBox is to provide a list of last names, not to display the last name from the current record. So, you do not set the DataBindings property to LastName. Do not delete the txtLastName TextBox.

Step 2-21: Set the Data Source property of the ComboBox to LastNames table by selecting *Other Data Source, Project Data Source, HRDataSet, LastNames*. The DataSource property now shows the newly created binding source that has the name LastNamesBindingSource, as shown in Figure 12-35.

Figure 12-35: Setting the Data Source property of the ComboBox to LastNames

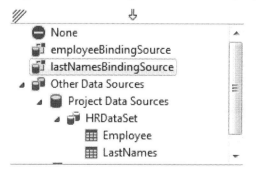

Set the Display Member property to LastName, as shown in Figure 12-36.

Figure 12-36: Display Member property of ComboBox

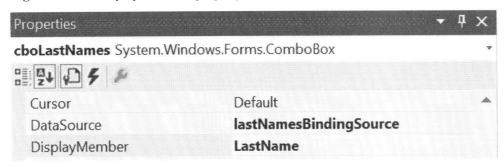

The ValueMember Property of ComboBox

You can display data from one field in a ComboBox, but use the corresponding data from another field to find a record. For example, to filter Employee records using department number, you may display department names (from Department table) in the ComboBox to make selection easier, but use the corresponding DeptNo to filter records. In this case, you would specify department name for DisplayMember and DeptNo for ValueMember property. The criterion for filtering would use the **SelectedValue** property that has the value (DeptNo), not the Text property (department name) that has the text displayed in the ComboBox.

Because using the ValueMember property generally gives more flexibility, we will use it in this example for demonstration, though it is not necessary.

Step 2-22: Set the ValueMember property of the ComboBox to LastName field.

Step 2-23: Run the form, and make sure the ComboBox displays all unique last names.

Next, write the code to filter records using the selected name, and run this code automatically whenever the user selects a name from the ComboBox. What event of the ComboBox should you use?

SelectedIndexChanged versus SelectionChangeCommitted Events

Both events are fired when the item selected in a ComboBox changes. However, there are some differences that are important for a data bound ComboBox.

SelectedIndexChanged event also is fired when SelectedIndex is set programmatically, like when you set the SelectedIndex to -1 to reset the ComboBox, and when the DataSource property of the ComboBox is set at runtime. Executing the code to filter records when these two events take place can have some unintended consequences.

But the **SelectionChangeCommitted** is not fired under the above two conditions. It is raised only when a user selects an item from the ComboBox.

Neither event is fired when items are added through the **Items** property of the ComboBox. So, in general, it is better to use the SelectionChangeCommitted event for data bound ComboBoxes. The code to filter Employee records using the selected name is shown in Figure 12-37.

Figure 12-37: Code to filter employee records

```
44    private void cboLastNames_SelectionChangeCommitted(object sender, EventArgs e)
45    {
46        employeeBindingSource.Filter = "LastName = '" + cboLastNames.SelectedValue + "'";
47    }
```

Note that you should use the **SelectedValue** (not the Text property) of the ComboBox, which gives the value of the field specified by the ValueMember property. The Text property is not changed at the time this event is raised, because this event is raised before the SelectedIndexChanged event; so, you cannot use the Text property in this event handler.

Step 2-24: Create the SelectionChangeCommitted event handler for cboLastNames, and add the code from Figure 12-37. Run the form and test the code.

Cancelling a Filter

To cancel a filter specified using the BindingSource.Filter property, set the Filter to a null string.

Step 2-25: Add a button named btnCancelFilter, and add the code from Figure 12-38 to the Click event handler. Run the form and test the code.

Figure 12-38: Code to cancel filter

```
48
49        private void btnCancelFilter_Click(object sender, EventArgs e)
50        {
51            employeeBindingSource.Filter = "";
52        }
```

It's time to practice! Do Exercise 12-2

Selecting a List from One Table to Filter Another Table

Suppose you want to select a department number from a ComboBox and filter employee records to view only employees from that department. The HR database includes a second table named department, as shown in Figure 12-39.

Figure 12-39: Department table

DeptNo	DeptName	DeptPhone
1	Accounting	920-424-1414
2	Finance	920-424-3377
3	Marketing	920-424-2442
4	Purchasing	920-424-0405
5	Human Resources	920-424-1166
6	Information Technology	920-424-3020

A list of unique department numbers for display in the ComboBox can be obtained from the Employee table by altering the SQL, or it can be obtained from the Department table. Because the Employee table is significantly larger than the Department table, it would be more efficient to get such a list from the Department table. Getting department numbers from the Department table also gives you the option to display department names in the ComboBox but use department numbers to filter records.

To get the department numbers (and names, if needed), you need to first create a data table and table adapter for Department table using the same steps that you followed to create the LastNames data table.

It's time to practice! Do Exercise 12-3.

Review Question

12.9   The data from a table named Employee is displayed on a form in Details View in TextBoxes. The controls on the form were created by dragging and dropping the data table on the form.

What would be the effect of executing the following statement on the set of records that the user can see on the form?
employeeBindingSource.Filter = "CountryId = 22";

What would be the effect of executing the above statement on the records stored in the Employee data table in memory?

12.8 Selecting Records Using the Table Adapter

In the previous section, you selected records using the Filter property of the binding source that controls what is transferred between a data table and the form. Another way to select records is to use the table adapter to change the SQL language that it uses to transfer data from the database to the dataset. This is fundamentally different from using the binding source.

Binding Source versus Table Adapter to Select Records

When you use binding source to filter records, typically the entire data from a table(s) is loaded from the database into a data table in memory. The binding source is used to help the user select different groups of data to work with. After the data are loaded into the data table, any change made to the database is not reflected in the data that the user sees. So, the user may not be seeing the latest data. Further, large tables consume a large amount of memory. A benefit of loading the entire table into memory and then filtering records using the binding source is that the database needs to be accessed only once in the beginning, thus reducing the network traffic.

When you use the table adapter, every time the user wants a group of data, it is loaded from the database into a data table and displayed on a form. So, each time a group of data is loaded, it will include the latest changes. The memory usage is lower because only selected records are held in data tables. This also means that the database is accessed every time the user requests a group of records.

Next, we look at the use of the table adapter to select records.

Fill Method of Table Adapter

So far, the auto-generated code on all forms used the Fill method of the table adapter to load data from the database to the Employee data table. The fill method uses the following SQL query:

> SELECT EmpId, LastName, FirstName, DateOfBirth, Phone, DeptNo, CountryId, Salary, FullTime
> FROM dbo.Employee

Note: To see this SQL, double click HRDataSet.xsd and click on *Fill, GetData()*, under *EmployeeTable Adapter*. You will see the above SQL in the CommandText property of the Fill method.

Creating a Parameter Query

To load only a subset of the database table, you need to create a new Fill method with a modified form of the standard SQL query, to include a criterion with a parameter for the last name to be specified, as follows:

SELECT EmpId, LastName, FirstName, DateOfBirth, Phone, DeptNo, CountryId, Salary, FullTime
FROM dbo.Employee
WHERE LastName = @LastName

The @ symbol indicates that **LastName** is a parameter whose value should be provided when you run the SQL.

For Access databases, use the question mark (?) in place of @ symbol, as in ?LastName.
For Oracle databases, use a colon (:), as in :LastName.

To **add a new query** (and a Fill method that uses the query), you right click the table adapter in the dataset schema window.

Tutorial 3: Parameter Query to Select Records

The parameter query, discussed above, is created in a new form, EmployeeSearch_ParmQry.

Step 3-1: Add a new form named **EmployeeSearch_ParmQry**.

Step 3-2: Choose Details mode for Employee data table, and ComboBox style for EmpId field. Drag and drop Employee table on the form.

Step 3-3: Open dataset Schema window by double clicking on HRDataSet.xsd in Solution Explorer.

Right click on employeetable adapter and select *AddQuery*.

Make sure that "Use SQL Statement" is selected. Click *Next*.

Make sure that "Select which returns rows" is selected. Click *Next*.

Note: If the option to "Select which returns rows" is grayed out, then you did not right click the Employee Table Adapter to create the query. Press *Cancel* and right click the Employee Table Adapter to add a query.

Change the SQL by adding the criterion as follows:

SELECT EmpId, LastName, FirstName, DateOfBirth, Phone, DeptNo, CountryId, Salary, FullTime
FROM   dbo.Employee
WHERE LastName = @LastName

Click *Next*.

Change the name of the method to **FillByLastName**. Click *Next*. Click *Finish*.

You will see the new method FillByLastName added to employeeTableAdapter, as shown in Figure 12-40.

Figure 12-40: The Parameter query for EmployeeTableAdapter

Using a Parameter Query

When you use a parameter query, you need to provide a value for the parameter(s). You will use data from a TextBox, txtLastName, for FillByLastName query as follows:

> employeeTableAdapter.FillByLastName(hRDataSet.Employee, txtLastName.Text);

Step 3-4: Add a TextBox named txtLastName and a Button named btnFillByLastName with the caption "Select Records," as shown in Figure 12-41.

Figure 12-41: The EmployeeSearch_ParmQry form

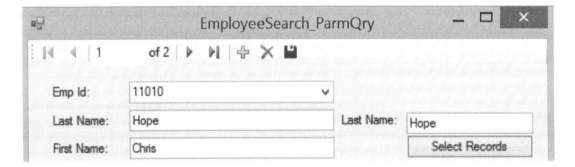

Step 3-5: Comment out the statement in the Load event handler, which uses the standard Fill method, as shown in Figure 12-42. Add the code from Figure 12-42 to the Click event handler of the button.

Figure 12-42: Load event handler for the EmployeeSearch_ParmQry form

```
28          private void EmployeeSearch_ParmQry_Load(object sender, EventArgs e)
29          {
30              // TODO: This line of code loads data into the 'hRDataSet.Employee' table.
31              //this.employeeTableAdapter.Fill(this.hRDataSet.Employee);
32          }
33          private void btnSelectRecords_Click(object sender, EventArgs e)
34          {
35              employeeTableAdapter.FillByLastName(hRDataSet.Employee, txtLastName.Text);
36          }
```

Step 3-6: Run the form and test the code.

Step 3-7: To give the user an option to display all records, add a button, btnAllRecords, with the caption "Display All." Add the code to use the standard query, shown in Figure 12-43. Run the form and test the code.

Figure 12-43: Standard Fill method to display all records

```
37
38          private void btnAllRecords_Click(object sender, EventArgs e)
39          {
40              this.employeeTableAdapter.Fill(this.hRDataSet.Employee);
41          }
```

It's time to practice! Do Exercise 12-4.

Review Question

12.10 The data from a table named Employee are displayed on a form in Details View in TextBoxes. The controls on the form were created by dragging and dropping the data table on the form. A parameter query named FillByLastName was created using the following SQL:

SELECT EmpId, LastName, FirstName, DateOfBirth, Phone, DeptNo, CountryId, Salary, FullTime
FROM   dbo.Employee
WHERE DeptNo = @DeptNo

What would be the effect of executing the following statement on the Employee data table?
    employeeTableAdapter.FillByLastName(hRDataSet.Employee, 11)

12.9 Untyped Datasets: Displaying Records

As discussed earlier in this chapter, the relationship between database tables and controls on a form can be represented by the illustration in Figure 12-43.

Figure 12-43: Relationship between database tables and controls on a form

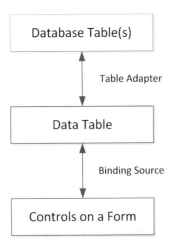

In the examples presented in previous sections, the data tables and other data access objects (data adapter and binding source), which are required to link database tables to controls on a form, are generated by the wizard at design time. The data types of the fields in the data table are determined at design time based on the data type of the corresponding fields in the database table. So, data tables created at design time contain fields that are of the same type as the fields in the database table; therefore, they are called **typed data tables**. The datasets that contain such tables are called **typed datasets**.

Typed datasets make it easy to check mismatch of data types between fields in the database table and the data that the application tries to store in those fields. However, generating typed datasets and other objects, like the data adapter, at design time makes it difficult to reuse the code in other forms or projects because the code is tied to objects within a form.

Bound versus Unbound Controls

Previous sections also used **data bound controls** that are easily generated by the wizard when you drag and drop the data table on the form. Data bound controls are bound to specific fields in the data table through the binding source. So, data on the form is linked tightly to the fields in the data table. The current row is displayed automatically in the controls. However, bound controls provide less flexibility in working with the data and validating it, compared to **unbound controls**, which are not directly bound to any field.

Creating objects at runtime using code, rather than at design time using the wizard, and displaying data in unbound controls require more code. However, this approach makes the code more flexible, efficient and reusable.

In this section, you learn how to develop applications that use code to create the dataset, table adapter and navigation bar at runtime, and display data in unbound controls. The datasets that are created at runtime using code are called **untyped datasets** because the data tables do not contain fields whose types are defined.

Tutorial 4: Display Records Using Untyped Datasets

This tutorial develops a form to display employee records using **untyped datasets** and **unbound controls**. You work with Employee table from HR2.mdf database file found in the DataFiles folder. You will develop a simple form to display employee records, as shown in Figure 12-44.

Figure 12-44: The Employee_Display form

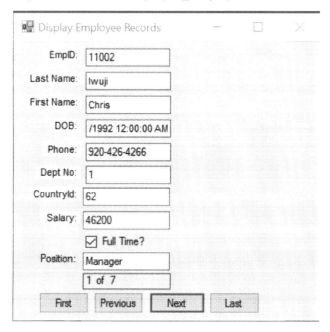

Step 4-1: Open the project, Ch12_Databases_UntypedDatasets. Open the form, Employee_Display. This form contains all necessary controls, but you need to add the code.

Note that the controls were not generated by dragging and dropping data tables; they were manually added to the form. Therefore, they are **unbound** controls that are not bound to fields in the data table, as indicated by the DataBindings property of the controls.

Displaying Records from a Table

Displaying records from a table involves the following steps:

1. Create the data table, data adapter and binding source objects.
2. Fill data from the database table into the data table using the Fill method of the data adapter.
3. Link the binding source to the data table. The binding source has methods/properties to move between rows, get the position/count of rows and select rows using the Filter property.
4. Copy data from the current row of the binding source to the controls on the form.

These are the same steps that were taken to create typed datasets using the wizard, which generated the necessary code. In this section, you need to develop the code because you are not using the wizard.

Creating the Data Table(s)

A data table object is created from the DataTable class using the standard syntax for creating objects:
   DataTable *variableName* = new DataTable()
where *variableName* is any valid variable name.

For example, the following statement creates a data table object for the Employee table, and references it using the variable, dtEmployee.
   DataTable dtEmployee = new DataTable();

Creating the Data Adapter

A data adapter is created from the DataAdapter class:
   DataAdapter *variableName* = new DataAdapter(*connectionString, Sql*)

The first parameter, *connectionString*, provides the necessary information to connect to the database. To connect to a local database, the syntax of the connection string is
   @"Data Source = (LocalDB)\MSSQLLocalDB; AttachDbFilename = *path\fileName*";

Here is an example:
   @"Data Source = (LocalDB)\MSSQLLocalDB; AttachDbFilename = C:\VC\HR2.mdf";

If the database is to be accessed from a server, the Data Source would provide the server address. The structure of the connection string may vary with the version of SQL Server you are using. The connection string may include additional information on security, user id and password.

The second parameter, **Sql**, represents the SQL statement to select data. The syntax to select data from a single table is
   SELECT *fieldNames* FROM *tableName* WHERE *criterion*

In the above statement, *fieldNames* represents the list of field names separated by commas, *tableName* represents the table name, and *criterion* represents the criterion, if any, to select records. If all fields are to be selected, you may replace field names by an asterisk. For example, to select all fields from all records of the Employee table, use the following SQL statement:

   SELECT * FROM Employee

The WHERE clause is discussed briefly later in this section.

Thus, the following code creates a data adapter to access all Employee records, and references it using the variable, daEmployee.
   string connString = @"Data Source = (LocalDB)\MSSQLLocalDB; AttachDbFilename =
          C\VC\HR2.mdf";
   string empSQLStr = "SELECT * FROM Employee";
   SqlDataAdapter daEmployee = new SqlDataAdapter(empSQLStr, connString);

Creating the Binding Source

The binding source is created from the BindingSource class:
   BindingSource *variableName* = new BindingSource;

Here is an example that creates a binding source, which is referenced by the variable, bsEmployee:
   BindingSource bsEmployee = new BindingSource();

Filling the Data Table

Filling the data table consists of loading data from the database table (or from a view that might have data from multiple tables) to the corresponding data table. The following statement fills the data table, dtEmployee, using the table adapter daEmployee:

daEmployee.Fill(dtEmployee);

The data adapter knows what data should be loaded into the data table, because it contains the necessary SQL statement that was specified when the data adapter was created.

Linking the Binding Source to the Data Table

To link a binding source to a data table, the **DataSource** property of the binding source is set to the data table, as in

bsEmployee.DataSource = dtEmployee;

The complete code to create the create the data access objects, fill the data table and link the data table to the binding source is shown in Figure 12-45.

Figure 12-45: Code to create the data access objects

```
10      using System.Data.SqlClient;    // add this using directive
11
12      namespace Ch12_Databases_UntypedDatasets
13      {
14          public partial class Employee_Display : Form
15          {
16              public Employee_Display()
17              {
18                  InitializeComponent();
19              }
20              DataTable dtEmployee = new DataTable(); // creates the data table
21              SqlDataAdapter daEmployee;  // declares a reference variable for the data adapter;
22              BindingSource bsEmployee = new BindingSource(); // creates the binding source
23
24              private void Employee_Display_Load(object sender, EventArgs e)
25              {
26                  // Create the data adapter
27                  string connString = @"Data Source = (LocalDB)\MSSQLLocalDB; AttachDbFilename =
28                                        C:\VC\HR2.mdf";
29                  string empSQLStr = "SELECT * FROM Employee";
30                  daEmployee = new SqlDataAdapter(empSQLStr, connString);
31                  // Fill the data table, dtEmployere
32                  daEmployee.Fill(dtEmployee);
33                  // Link binding source to the data table
34                  bsEmployee.DataSource = dtEmployee;
```

Line 10: The using directive specifies the namespace, SqlClient, which contains the data access objects, like data table and data adapter.

Lines 20–22: The reference variables for the three objects are declared at the class level so that they can be used in multiple methods. The data table and binding source are created here, but the data adapter is created within the Load event handler of the form (lines 26–30) because it requires creating the connection and SQL strings.

Lines 27–30: The data adapter is created using the connection string and SQL statement, as discussed earlier.

Lines 31–34: The Fill method of the data adapter is used to fill the data table, and the DataSource property of the binding source is set to the data table.

In the next three steps of the tutorial, you add the above code to your form, and copy the database.

Step 4-2: Add the using directive shown in line 10, Figure 12-45, to the code window.

Step 4-3: Copy the database file, HR2.mdf, to a local folder outside the project folder.
Add the code from lines 20–22 at the class level.
Add the code from lines 26–30 to the Load event handler of the form to create the data adapter.
Make sure you specify the correct path for your database file.

Step 4-4: Add the code from lines 31–34 to fill the data table and link the binding source.

Copying Data from the Data Table into Controls on the Form

Because there is no concept of current row in a data table, the binding source is used to navigate through the rows and display records on the form.

The data in a binding source is just a view of the data from the data table. The view could consist of the entire table, or a subset of the data selected using the Filter property of the binding source. That is, the binding source helps you to look at the data table through different windows. A view also could present the data sorted in a certain order.

The data types of the binding source rows and the data table rows are not the same. The data viewed through a binding source is of **DataView** type, and each row of the binding source is of **DataRowView** type. But each row of a DataTable is of **DataRow** type.

The data from different columns of the current row of the data source is displayed in controls on the form using two steps:

1. Store the entire row in a DataRowView object.
2. Copy data from each column of the DataRowView into a control on the form.

Accessing a Binding Source Row

You access a row using its index. For example, bsEmployee[2] gets the third row in the binding source.

The index of the current row in the binding source is given by the **Position** property of the binding source, as in
    int currentRow = bsEmployee.Position;

The index is used in the following statement to get the entire current row. The row is cast to DataRowView type and its reference is assigned to the variable, drvEmployee, which is of DataRowView type.
    DataRowView drvEmployee = (DataRowView) bsEmployee[currentRow];

Alternate method: If the binding source doesn't filter rows, and therefore includes all rows from the data table, you may use the index of the current row to get the row from the **Rows** collection of the data table, as follows:

    DataRow drEmployee = dtEmployee.Rows[currentRow];

This doesn't work when the data displayed on the form are filtered by the binding source because the first row of the binding source, for example, may not be the first row in the data table.

Copying Data from DataRowView to Controls on a Form

You can access individual fields of the DataRowView using the field name or the index of the field. For example, you may get the employee id from the DataRowView, drvEmployee, using one of the following two options:

    drvEmployee["EmpId"].ToString();

or

    drvEmployee[0].ToString();

Because records are to be displayed repeatedly when you move from one row to another, you will create a separate method named DisplayCurrentRow to copy data from DataRowView to controls.

Figure 12-46 shows the code within the method to store the current row in a DataRowView, drvEmployee, and to copy data from the DataRowView to controls on the form.

Figure 12-46: Displaying data from a data table in controls on a form

```
45    private void DisplayCurrentRow()
46    {
47        // store the current row of the data table in a DataRowView
48        int currentRow = bsEmployee.Position;
49        DataRowView drvEmployee = (DataRowView) bsEmployee[currentRow];
50
51        // copy data from the DataRowView to controls on the form:
52        txtEmpId.Text = drvEmployee["EmpId"].ToString();
53            // Or, use the index as in, drvEmployee[0].ToString();
54        txtFirstName.Text = drvEmployee["FirstName"].ToString();
55        txtLastName.Text = drvEmployee["LastName"].ToString();
56        txtDOB.Text = drvEmployee["DateofBirth"].ToString();
57        txtPhone.Text = drvEmployee["Phone"].ToString();
58        txtCountryId.Text = drvEmployee["CountryId"].ToString();
59        txtDepNo.Text = drvEmployee["DeptNo"].ToString();
60        txtSalary.Text = drvEmployee["Salary"].ToString();
61        chkFullTime.Checked = bool.Parse(drvEmployee["FullTime"].ToString());
62        txtPosition.Text = drvEmployee["Position"].ToString();
63    }
```

Line 52: The data from the EmpId column of the DataRowView, drvEmployee, is displayed in the TextBox, txtEmpId.

Lines 54–62: Data from the remaining columns are displayed in the controls on the form.

Step 4-5: Add the code from Figure 12-46 to display the current row.
    Add the code from Figure 12-47 to call DisplayCurrentRow method.
    Run the project and test the code. The form should display the first record.

Figure 12-47: Statement to call DisplayCurrentRow method

```
35
36                  // call DisplayCurrentRow method
37                  DisplayCurrentRow();
```

Navigating through the Rows

You use the MoveFirst, MovePrevious, MoveNext and MoveLast methods of the binding source to move between rows.

Step 4-6: Add code to the Click event handlers of the navigation buttons, as shown in Figure 12-48.

Figure 12-48: Code to move between rows

```
64          private void btnFirst_Click(object sender, EventArgs e)
65          {
66              bsEmployee.MoveFirst();
67          }
68          private void btnLast_Click(object sender, EventArgs e)
69          {
70              bsEmployee.MoveLast();
71          }
72          private void btnPrevious_Click(object sender, EventArgs e)
73          {
74              bsEmployee.MovePrevious();
75          }
76          private void btnNext_Click(object sender, EventArgs e)
77          {
78              bsEmployee.MoveNext();
79          }
```

PositionChanged Event of the Binding Source

To display the current row, the DisplayCurrentRow method needs to be called every time you move to a new row. Rather than calling this method from each Click event handler shown in Figure 12-48, we will call this method whenever the position of the current row changes. To do this, you will create a method named bsEmployee_PositionChanged and link (subscribe) this method to the PositionChanged event of the binding source.

The PositionChanged event is fired whenever the position of the current row changes—that is, whenever you move to a new row. The PositionChanged method that called DisplayCurrentRow, and displays the current row number, is shown in Figure 12-49.

Figure 12-49: PositionChanged method

```csharp
81      private void bsEmployee_PositionChanged(object sender, System.EventArgs e)
82      {
83          // display data & row number whenever a the current record position changes
84          if (bsEmployee.Count > 0)
85          {
86              DisplayCurrentRow();
87              lblRecordNo.Text = (bsEmployee.Position + 1) + " of " + bsEmployee.Count;
88          }
89      }
```

Lines 84 uses the Count property to check whether the binding source is empty.

Line 87 uses the Position and Count properties to display the row number of the current row and the number of rows in the set.

Next, you need to link (subscribe) the above method to the PositionChanged event of the binding source, as shown in line 43 in Figure 12-50.

Figure 12-50: Linking PositionChanged event to bsEmployee_PositionChanged method

```csharp
38
39          // Display row number and count of rows
40          lblRecordNo.Text = (bsEmployee.Position + 1) + " of " + bsEmployee.Count;
41
42          // Link (subscribe) PositionChanged event to bsEmployee_PositionChanged method
43          bsEmployee.PositionChanged += bsEmployee_PositionChanged;
```

Step 4-7: Add the code from Figure 12-49 to create bsEmployee_PositionChanged method.
Add the code from line 43 in Figure 12-50 to link the method to the PositionChanged event of the binding source.
Add the statement from line 40 to display the row number and row count, the first time data are displayed on the form.

Step 4-8: Run the project and test the code.

The following is the complete code for the Employee_Display form:

```csharp
using System.Data.SqlClient;    // add this using directive

namespace Ch12_Databases_UntypedDatasets
{
    public partial class Employee_Display : Form
    {
        public Employee_Display()
        {
            InitializeComponent();
        }
        DataTable dtEmployee = new DataTable(); // creates data table
        SqlDataAdapter daEmployee; //declares a ref. variable for data adapter;
        BindingSource bsEmployee = new BindingSource(); //creates binding source

        private void Employee_Display_Load(object sender, EventArgs e)
        {
            // Create the data adapter
            string connString = @"Data Source = (LocalDB)\MSSQLLocalDB; AttachDbFilename = C:\VC\HR2.mdf";
            string empSQLStr = "SELECT * FROM Employee";
            daEmployee = new SqlDataAdapter(empSQLStr, connString);
            // Fill the data table, dtEmployere
            daEmployee.Fill(dtEmployee);
            // Link binding source to the data table
            bsEmployee.DataSource = dtEmployee;

            // call DisplayCurrentRow method to display the current row
            DisplayCurrentRow();

            // Display row number and count of rows
            lblRecordNo.Text = (bsEmployee.Position + 1) + " of " + bsEmployee.Count;

            // Link (subscribe) PositionChanged event to bsEmployee_PositionChanged method
            bsEmployee.PositionChanged += bsEmployee_PositionChanged;
        }
```

```csharp
45      private void DisplayCurrentRow()
46      {
47          // store the current row of the data table in a DataRowView
48          int currentRow = bsEmployee.Position;
49          DataRowView drvEmployee = (DataRowView) bsEmployee[currentRow];
50
51          // copy data from the DataRowView to controls on the form:
52          txtEmpId.Text = drvEmployee["EmpId"].ToString();
53              // Or, use the index as in, drvEmployee[0].ToString();
54          txtFirstName.Text = drvEmployee["FirstName"].ToString();
55          txtLastName.Text = drvEmployee["LastName"].ToString();
56          txtDOB.Text = drvEmployee["DateofBirth"].ToString();
57          txtPhone.Text = drvEmployee["Phone"].ToString();
58          txtCountryId.Text = drvEmployee["CountryId"].ToString();
59          txtDepNo.Text = drvEmployee["DeptNo"].ToString();
60          txtSalary.Text = drvEmployee["Salary"].ToString();
61          chkFullTime.Checked = bool.Parse(drvEmployee["FullTime"].ToString());
62          txtPosition.Text = drvEmployee["Position"].ToString();
63      }
64      private void btnFirst_Click(object sender, EventArgs e)
65      {
66          bsEmployee.MoveFirst();
67      }
68      private void btnLast_Click(object sender, EventArgs e)
69      {
70          bsEmployee.MoveLast();
71      }
72      private void btnPrevious_Click(object sender, EventArgs e)
73      {
74          bsEmployee.MovePrevious();
75      }
76      private void btnNext_Click(object sender, EventArgs e)
77      {
78          bsEmployee.MoveNext();
79      }
80
81      private void bsEmployee_PositionChanged(object sender, System.EventArgs e)
82      {
83          // display data & row number whenever a the current record position changes
84          if (bsEmployee.Count > 0)
85          {
86              DisplayCurrentRow();
87              lblRecordNo.Text = (bsEmployee.Position + 1) + " of " + bsEmployee.Count;
88          }
89      }
```

Review Questions

12.11 To create a table adapter object using code, you need to specify the _____ and _____.

12.12 What are the benefits of creating data access objects (data adapter, data table and binding source) using code at runtime, rather than at design time using the wizard?

12.13 How does an untyped data table differ from a typed data table?

12.14 What is the difference between a data bound control and an unbound control?

12.15 An application creates the data access objects (data adapter, data table and binding source) using code at runtime. What object and method do you use to load the data from the database table to the untyped data table?

12.16 How do you link a binding source to a data table?

12.17 List three major functions of a binding source.

12.18 An application creates the data access objects (data adapter, data table and binding source) using code at runtime. After loading data from a database table to an untyped data table, what are the major steps involved in displaying the current row in controls on the form?

12.10 Untyped Datasets: Selecting Records

Sections 12.7 and 12.8 discussed selecting records using the typed dataset, binding source and data adapter objects that were created at design time by the wizard. In this section, you will learn how to select records from **untyped** datasets using the binding source and data adapter objects that are created by code at runtime.

Consider the form, Employee_Select, shown in Figure 12-51. The form uses the Employee and Department tables from the HR2 database. The purpose of the form is to let you select employee records using two different methods: (1) by filtering records from the data table, and (2) by changing the SQL to select records from the database table.

Figure 12-51: The Employee_Select form

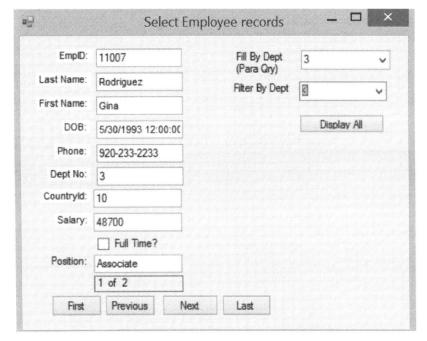

The form has two ComboBoxes on the right side; both let the user select a department number from a list of department numbers from the Department table.

The department number selected from the first ComboBox (labelled "Fill By Dept") provides the value for a parameter query that selects employee records from the database and loads them into the data table. The department number selected from the second ComboBox (labeled "Filter By Dept") provides the value to filter rows from the data table using the binding source.

The "Display All" button selects and displays all records.

The code to display Employee records is the same as that in the Employee_Display form discussed previously. Additional code is needed to create the data table and the data adapter for the Department table so that a list of DeptNo's can be added to the two ComboBoxes.

Figure 12-52 shows the class-level code that includes the code to create data access objects for the Department table. Note that the Department table doesn't need a binding source, because the only use of the table is to add DeptNo's to the ComboBox, which can be done directly from the data table.

Figure 12-52: Creating data access objects for Employee and Department tables

```
21              // data table, data adapter and binding source for Employee table
22              DataTable dtEmployee = new DataTable();
23              SqlDataAdapter daEmployee;
24              BindingSource bsEmployee = new BindingSource();
25
26              // data table and data adapter for Department table
27              DataTable dtDept = new DataTable();
28              SqlDataAdapter daDept;
```

Figure 12-53 shows the code to display Employee records on the form. This is the same as in the Employee_Display form.

Figure 12-53: Code to display Employee records

```
29          private void Employee_Select_Load(object sender, EventArgs e)
30          {
31              // Fill employee data table, dtEmployee
32              string connString = @"Data Source = (LocalDB)\MSSQLLocalDB; AttachDbFilename =
33                                    C:\VC\HR2.mdf";
34              string empSQLStr = "SELECT * FROM Employee";
35              daEmployee = new SqlDataAdapter(empSQLStr, connString);
36              daEmployee.Fill(dtEmployee);
37
38              // Display employee data
39              bsEmployee.DataSource = dtEmployee; // links binding source to the data table
40              DisplayCurrentRow();
41              lblRecordNo.Text = (bsEmployee.Position + 1) + " of " + bsEmployee.Count;
42              bsEmployee.PositionChanged += bsEmployee_PositionChanged;
```

Adding DeptNo's to ComboBoxes

In Section 12.6, which used typed datasets to display values in a ComboBox from a data table column, you set the following two properties of the ComboBox at design time:

> Set the DataSource property of the ComboBox to the binding source.
> Set the DisplayMember property to the column name.

In this form, you will set an additional property of the binding source:

> Set the ValueMember property to the column name.

If an application uses the SelectedValue of the ComboBox, the ValueMember must be set to a column name. The ValueMember could be set to a column different from the DisplayMember. For example, the Display Member could be the DeptName, and the ValueMember could be the DeptNo. That means the DeptNames are displayed in the ComboBox, but when the user selects a DeptName, the corresponding DeptNo is available through the SelectedValue property of the ComboBox.

Figure 12-54 shows the code to fill the data table, dtDept, and to add the list of DeptNo's from dtDept to the two ComboBoxes. Note that, as stated earlier, the Department table doesn't need a binding source; so, the DataSouce for the ComboBox is set to the data table.

Figure 12-54: Code to display DeptNo's in the ComboBoxes

```
45              // Fill department data table, dtDept
46              string DeptSQLStr = "SELECT DeptNo, DeptName FROM Department";
47              daDept = new SqlDataAdapter(DeptSQLStr, connString);
48              daDept.Fill(dtDept);
49
50              //Display DeptNo's from dtDept, in the ComboBox, cboFillByDept:
51              cboFillByDept.DataSource = dtDept;
52              cboFillByDept.DisplayMember = "DEPTNO";
53              cboFillByDept.ValueMember = "DEPTNO";
54
55              //Display DeptNo's from dtDept, in the ComboBox, cboFilterByDept
56              cboFilterByDept.DataSource = dtDept;
57              cboFilterByDept.DisplayMember = "DEPTNO";
58              cboFilterByDept.ValueMember = "DEPTNO";
```

Tutorial 5: Select Records Using Untyped Datasets

This tutorial develops the Employee_Select form shown in Figure 12-51.

Step 5-1: Open the project Ch12_Databases_UntypedDatasets. Open the form, Employee_Select.
Currently, the form has all the controls, and the code to display Employee records and to navigate between them.
Make sure that the connection string in line 33 points to the right folder.
Run the form to make sure that the employee records are displayed.

Next, you will add the code to display DeptNo's in the ComboBoxes, and to select Employee records.

Step 5-2: Open the code window and add the missing class-level code from Figure 12-52 to create the data access objects for Department.

Step 5-3: Add the missing code from Figure 12-54 to the Load event handler of the form to display DeptNo's in the ComboBoxes.

Step 5-4: Run the form and test the code. The ComboBoxes should display the DeptNo's.

Filtering Rows Using the Binding Source

Filtering involves using the Filter property of the binding source to select rows from the data table for display in controls on the form. Section 12.7 discussed filtering using data objects created at design time using the wizard. The process is the same for untyped data tables. To refresh your memory, the syntax of the statement to filter rows from the data table is

bindingSource.Filter = "*criterion*";

where *bindingSource* is the name of the binding source object, and *criterion* is any Boolean expression. Examples of the statement for string and numeric fields can be found in Section 12.7.

In the Employee_Search form, the criterion is that the employee's DeptNo should be equal to the value selected in the cboFilterByDept ComboBox. Figure 12-55 shows the code that specifies the criterion.

Because we want the selected rows to be displayed whenever the user selects a DeptNo from the ComboBox, the code is placed in the SelectionChangeCommitted event handler of the ComboBox. The SelectionChangeCommitted event is fired when the user selects an item from the ComboBox.

As discussed in Section 12.7, the SelectedIndexChanged event is less suitable because it is fired under other conditions, too.

Figure 12-55: Filtering using the binding source

```
61      private void cboFilterByDept_SelectionChangeCommitted(object sender, EventArgs e)
62      {
63          bsEmployee.Filter = "DeptNo = " + cboFilterByDept.SelectedValue.ToString();
64          if (bsEmployee.Count > 0)
65          {
66              lblRecordNo.Text = (bsEmployee.Position + 1) + " of " + bsEmployee.Count;
67              DisplayCurrentRow();
68          }
```

Step 5-5: Generate the SelectionChangeCommitted event handler for cboFilterByDept ComboBox. (Select cboFilterByDept. In the properties window, click on the **events** button [lightning bolt]. Double click the SelectionChangeCommitted event).

Add the code from Figure 12-55 to the event handler to filter records.
Run the form and test the code by selecting the value 3 from the cboFilterByDept. You should be able to see only two employee records.

Selecting Records from the Database Table using the Table Adapter

Selecting records from a database table using the table adapter involves changing the SQL statements to specify the criterion. The selected records are loaded into a data table. As discussed in Section 12.8, this is different from filtering, which selects rows from the data table for display on the form.

In Section 12.8, where the data access objects were created at design time, you used the wizard to create a separate Fill method that includes a criterion in its SQL. Thus, the table adapter had two different Fill methods: one to select all records, and another to select specified records.

The method you use in this section for untyped dataset is different. You change the original SQL by assigning the new SQL to the data adapter's **SelectCommand** property (which is of SqlCommand type, and is generated when you create the data adapter). This is done by assigning the new SQL to the CommandText property of the SelectCommand, as follows:

dataAdapterName.SelectCommand.CommandText = "*sqlString*";

where *dataAdapterName* represents the name of the data adapter, and *sqlString* represents the SQL statement. Figure 12-56 shows the code to select records for the DeptNo value selected in the ComboBox.

Figure 12-56: Using the table adapter to select records

```
72   private void cboFillByDept_SelectionChangeCommitted(object sender, EventArgs e)
73   {
74       //Specify the SQL
75       daEmployee.SelectCommand.CommandText = "SELECT * FROM Employee where DeptNo = " +
76                                               cboFillByDept.SelectedValue.ToString();
77       //Clear and fill the data table
78       dtEmployee.Clear();
79       daEmployee.Fill(dtEmployee);
80   }
```

Alternatively, you may use a **parameter** (a variable name preceded by the "@" sign) in place of the value of DeptNo in the SQL, and provide the value in a separate statement, as follows:

```
daEmployee.SelectCommand.CommandText = "SELECT * FROM Employee WHERE DeptNo = @deptNo";
daEmployee.SelectCommand.Parameters.AddWithValue("@deptNo",
                                                  cboFillByDept.SelectedValue.ToString());
```

The parameter, @deptNo, within the SQL, is replaced by the value, cboFillByDept.SelectedValue.ToString(), specified in the AddWithValue method in the second statement.

Step 5-6: Add the code from Figure 12-56 to select records using the table adapter.
Run the form and test the code by selecting the value 3 from the cboFillByDept ComboBox. You should be able to see only two employee records.

Selecting All Records

To select all records, you change the SQL so that no criterion is included, as follows:
　　　　daEmployee.SelectCommand.CommandText = "SELECT * FROM Employee";

In addition, you cancel the filter by setting the Filter property of the binding source to null. The complete code to display all records is shown in Figure 12-57.

Figure 12-57: Displaying all records

```
82      private void btnDisplayAll_Click(object sender, EventArgs e)
83      {
84          dtEmployee.Clear();
85          daEmployee.SelectCommand.CommandText = "SELECT * FROM Employee";
86          daEmployee.Fill(dtEmployee);
87
88          bsEmployee.Filter = ""; // remove the filter
89
90          lblRecordNo.Text = (bsEmployee.Position + 1) + " of " + bsEmployee.Count;
91          DisplayCurrentRow();
92      }
```

Step 5-7: Add the statements from Figure 12-57 to the Click event handler of btnDisplayAll.
　　Run the form and test the code by clicking the DisplayAll button after filtering records, and after selecting records using the data adapter.

The following is the complete code for Employee_Select form:

```csharp
21      // data table, data adapter and binding source for Employee table
22      DataTable dtEmployee = new DataTable();
23      SqlDataAdapter daEmployee;
24      BindingSource bsEmployee = new BindingSource();
25
26      // data table and data adapter for Department table
27      DataTable dtDept = new DataTable();
28      SqlDataAdapter daDept;
29
30      private void Employee_Select_Load(object sender, EventArgs e)
31      {
32          // Fill employee data table, dtEmployee
33          string connString = @"Data Source = (LocalDB)\MSSQLLocalDB; AttachDbFilename =
34                                                C:\VC\HR2.mdf";
35          string empSQLStr = "SELECT * FROM Employee";
36          daEmployee = new SqlDataAdapter(empSQLStr, connString);
37          daEmployee.Fill(dtEmployee);
38
39          // Display employee data
40          bsEmployee.DataSource = dtEmployee; // links binding source to the data table
41          DisplayCurrentRow();
42          lblRecordNo.Text = (bsEmployee.Position + 1) + " of " + bsEmployee.Count;
43          bsEmployee.PositionChanged += bsEmployee_PositionChanged;
44
45          // Fill department data table, dtDept
46          string DeptSQLStr = "SELECT DeptNo, DeptName FROM Department";
47          daDept = new SqlDataAdapter(DeptSQLStr, connString);
48          daDept.Fill(dtDept);
49
50          //Display DeptNo's from dtDept in the ComboBox, cboFillByDept:
51          cboFillByDept.DataSource = dtDept;
52          cboFillByDept.DisplayMember = "DEPTNO";
53          cboFillByDept.ValueMember = "DEPTNO";
54
55          //Display DeptNo's from dtDept in the ComboBox, cboFilterByDept:
56          cboFilterByDept.DataSource = dtDept;
57          cboFilterByDept.DisplayMember = "DEPTNO";
58          cboFilterByDept.ValueMember = "DEPTNO";
59      }
60
61      private void cboFilterByDept_SelectionChangeCommitted(object sender, EventArgs e)
62      {
63          bsEmployee.Filter = "DeptNo = " + cboFilterByDept.SelectedValue.ToString();
64          if (bsEmployee.Count > 0)
65          {
66              lblRecordNo.Text = (bsEmployee.Position + 1) + " of " + bsEmployee.Count;
67              DisplayCurrentRow();
68          }
69
70      }
```

```csharp
        private void cboFillByDept_SelectionChangeCommitted(object sender, EventArgs e)
        {
            //Specify the SQL
            daEmployee.SelectCommand.CommandText = "SELECT * FROM Employee where DeptNo = " +
                                        cboFillByDept.SelectedValue.ToString();
            //Clear and fill the data table
            dtEmployee.Clear();
            daEmployee.Fill(dtEmployee);
        }

        private void btnDisplayAll_Click(object sender, EventArgs e)
        {
            dtEmployee.Clear();
            daEmployee.SelectCommand.CommandText = "SELECT * FROM Employee";
            daEmployee.Fill(dtEmployee);

            bsEmployee.Filter = ""; // remove the filter

            lblRecordNo.Text = (bsEmployee.Position + 1) + "  of  " + bsEmployee.Count;
            DisplayCurrentRow();
        }

        private void DisplayCurrentRow()
        {
            // store the current row of the data table in a DataRow
            int currentRow = bsEmployee.Position;
            DataRowView drvEmployee = (DataRowView)bsEmployee[currentRow];

            // copy data from the datarow to controls on the form:
            txtEmpId.Text = drvEmployee["EmpId"].ToString();
            // Or, use the index as in, drvEmployee[0].ToString();
            txtFirstName.Text = drvEmployee["FirstName"].ToString();
            txtLastName.Text = drvEmployee["LastName"].ToString();
            txtDOB.Text = drvEmployee["DateofBirth"].ToString();
            txtPhone.Text = drvEmployee["Phone"].ToString();
            txtCountryId.Text = drvEmployee["CountryId"].ToString();
            txtDepNo.Text = drvEmployee["DeptNo"].ToString();
            txtSalary.Text = drvEmployee["Salary"].ToString();
            chkFullTime.Checked = bool.Parse(drvEmployee["FullTime"].ToString());
            txtPosition.Text = drvEmployee["Position"].ToString();
        }
```

```csharp
113
114    private void btnFirst_Click(object sender, EventArgs e)
115    {
116        bsEmployee.MoveFirst();
117    }
118    private void btnPrevious_Click(object sender, EventArgs e)
119    {
120        bsEmployee.MovePrevious();
121    }
122    private void btnLast_Click(object sender, EventArgs e)
123    {
124        bsEmployee.MoveLast();
125    }
126    private void btnNext_Click(object sender, EventArgs e)
127    {
128        bsEmployee.MoveNext();
129    }
130
131    private void bsEmployee_PositionChanged(object sender, System.EventArgs e)
132    {
133        // display row number and data, whenever a the current record position changes
134        if (bsEmployee.Count > 0)
135        {
136            lblRecordNo.Text = (bsEmployee.Position + 1) + "  of  " + bsEmployee.Count;
137            DisplayCurrentRow();
138        }
139    }
```

Review Question

12.19 A set of **unbound** textboxes is used to display records from the database table, Student (Id, Name, Major, Advisor, Phone), using the following objects:

        data adapter   - daStudent
        data table   - dtStudent
        binding source - bsStudent

The binding source is linked to the data table using appropriate statements in the code.
Two different code segments are shown below:

```
// code segment #1
daStudent.SelectCommand.CommandText = "SELECT * FROM Student WHERE Major = 'IS' "
daStudent.Fill(dtStudent);
DisplayCurrentRow();   // a method that displays current row, in the TextBoxes
```

```
// code segment #2
daStudent.Fill(dtStudent);
bsStudent.Filter = "Major = 'IS' "
DisplayCurrentRow();   // a method that displays current row, in the TextBoxes
```

How is the effect of executing code segment #1 different from the effect of executing code segment #2?

12.11 Untyped Datasets: Add/Edit/Delete Records

In this section, you will learn how to add, edit and delete Employee records using untyped dataset, data adapter and binding source that are created at runtime using code. Compared to using the Add, Save and Delete buttons on the navigation bar generated by the wizard, this method gives you more control on validating the data and controlling user actions.

As discussed in the previous section, the SelectCommand property of the data adapter is generated when you create the data adapter, and the SQL to select records was assigned to the SelectCommand. To add, edit and delete records, you need to generate the **InsertCommand**, **UpdateCommand** and **DeleteCommand** properties for the data adapter using the SqlCommandBuilder.
The following statement builds these commands for the data adapter, daEmployee:

SqlCommandBuilder *commandBuilder* = new SqlCommandBuilder(daEmployee)

where *commandBuilder* is any valid variable name.

The generated commands will have the necessary SQL. For example, the UpdateCommand will have the SQL to update the database table.

You will use the form shown in Figure 12-58 to learn how to develop the code to add, edit and delete records using untyped data tables.

Figure 12-58: The form to add, edit and delete Employee records

A difference in the user interface between the current version of the form and previous versions is the use of RadioButtons, rather than a TextBox, to make it easy to enter or change an employee's job position. In addition, a ComboBox is used to let the user select a DeptNo from a ComboBox to make it easy to enter a DeptNo while adding a new record or changing an existing record.

The **Add** button just clears the controls on the form so that you can enter the data for a new record and save the data.

The **Cancel** button displays the current row from the binding source so that any changes you made in the controls are removed.

The **Save** button saves new records as well as changes to the current record.

Tutorial 6: Add/Edit/Delete Using Untyped Datasets

This tutorial develops the code for the form Employee_AddEdDel.

Step 6-1: Open the project Ch12_Databases_UntypedDatasets. Open the form, Employee_AddEdDel.
This form includes all controls and the code to display Employee records without the Position field.
Open the code window and view the code.
Make sure that the connection string in line 38 specifies the right directory.
You will add the remaining code, including the code to add, save and delete records, and cancel changes made to the data.

Currently, the Load event handler includes the code to display employee records. In addition, the event handler includes the code to fill the data table, dtDept. You need to add the code to display DeptNo's in the ComboBox from the data table. The Department table doesn't need a binding source, because the only use of the table is to add DeptNo's to the ComboBox, which can be done directly from the data table.

Step 6-2: Add the following code to the Load event handler to display DeptNo's in the ComboBox.
Run the form, and make sure the ComboBox displays the DeptNo's.

```
52              //Display DeptNo's from data table, in cboDept:
53              cboDeptNo.DataSource = dtDept;
54              cboDeptNo.DisplayMember = "DEPTNO";
55              cboDeptNo.ValueMember = "DEPTNO";
```

Displaying Employee's Position in RadioButtons

As in previous forms, data from the current row is first stored in a DataRowView, and then copied from the DataRowView to controls on the form. This is done in the Display_Records method. The code to be added to display the Position field in RadioButtons is shown in Figure 12-59.

Figure 12-59: Code to display employee's position in RadioButtons

```
81                  switch (drvEmployee["Position"].ToString())
82                  {
83                      case "VP":
84                          rdbVP.Checked = true;
85                          break;
86                      case "Manager":
87                          rdbMgr.Checked = true;
88                          break;
89                      case "Staff":
90                          rdbStaff.Checked = true;
91                          break;
92                  }
```

Step 6-3: Add the code from Figure 12-59 to the Display_Records method.
Run the form and make sure the Position data are displayed in RadioButtons.

Adding New Records

Adding new records involves clearing the controls, entering the data, and saving the data. Figure 12-60 shows the Click event handler of the Add button, and the ClearControls method that it calls.

The event handler also sets the Boolean variable adding. This flag is used to determine whether the data to be saved is a new record or an existing record.

The ClearControls clears all the TextBoxes using a foreach loop to access each control from the Controls collection.

Figure 12-60: Clearing controls

```
95              private void btnAdd_Click(object sender, EventArgs e)
96              {
97                  ClearControls();    // clear all controls
98                  adding = true;
99              }

214             private void ClearControls()
215             {
216                 foreach (Control control in Controls)
217                 {
218                     if (control is TextBox)
219                         control.Text = null;    //or, control.Text = "";
220                 }
221                 cboDeptNo.Text = null;
222                 rdbVP.Checked = false;
223                 rdbMgr.Checked = false;
224                 rdbStaff.Checked = false;
225                 chkFullTime.Checked = false;
226             }
```

Step 6-4: Add the code from Figure 12-60 to the Click event handler of the Add button. The Boolean variable, adding, is already declared at the class level.

Saving Records

Saving a record is a two-step process: First, save the data from the controls into the **data table**, and then save the data from the data table to the **database table** using Update method of the table adapter.

The same **Save** button is used to save changes to existing records and to add new records. Figure 12-61 shows the event handler. Two methods, CopyToDataRowView and CopyToDataRow, that are called from the event handler (lines 109 and 123) are shown in Figure 12-62 and Figure 12-63, respectively.

The Boolean variable adding is used to check whether the data in the controls is a new record. Saving new records into the **data table** involves the following steps:

- Create a blank DataRow (line 108).
- Copy data from the controls to the DataRow by calling the method, CopyToDataRow (line 109). Note that the data table is a collection of rows of the type, DataRow.
- Add the DataRow to the data table (line 111).

Saving changes to an existing record into the **data table** also involves three steps:

- Create a variable of **DataRowView** type that references the current row of the binding source (line 121). The binding source rows are of DataRowView type.
- Copy data from the controls to the DataRowView by calling CopyToDataRowView (line 123).
- Apply the changes in the DataRowView to the data table using the **EndEdit** method of the binding source (line 124).

Figure 12-61: Saving data

```
101       private void btnSave_Click(object sender, EventArgs e)
102       {
103           int noOfRecords = 0;
104           // Save data to the data table:
105           if (adding == true)    // if it is a new record
106           {
107               // Create a blank DataRow by calling the NewRow() method of data table.
108               drEmployee = dtEmployee.NewRow();
109               CopyToDataRow();    // copy all data, into DataRow
110               // Add the new DataRow to the Rows collection of the data table
111               dtEmployee.Rows.Add(drEmployee);
112
113               lblRecordNo.Text = (bsEmployee.Position + 1) + " of " + bsEmployee.Count;
114               adding = false;
115           }
116           else // it's an existing record
117           {
118               // Create a DataRowView that references the current row of
119               // the binding source:
120               int currentRow = bsEmployee.Position;
121               drvEmployee = (DataRowView)bsEmployee[currentRow];
122
123               CopyToDataRowView();    // Copy data from controls the DataRowView:
124               bsEmployee.EndEdit();    //Save the DataRowView into data table
125           }
126           // Update the database table
127           noOfRecords = daEmployee.Update(dtEmployee); // Save to database table
128           MessageBox.Show("Changes saved to database");
129       }
```

The EndEdit Method of the Binding Source

The EndEdit method (line 124) applies the changes in the binding source rows to the data table it is linked to. When a binding source row is updated using unbound controls, as in the current example, it is important to copy the data from the controls into the binding source row, and call the **EndEdit** method to save the changes in the binding source row to the data table.

The Update Method of the Data Adapter

The Update method of the data adapter saves the changes in the data table to the database table, as coded in line 127:

noOfRecords = daEmployee.Update(dtEmployee);

The method returns the number of records updated.

Step 6-5: Add the code from Figure 12-61 to the Click event handler of btnSave. (The methods, CopyToDataRowView and CopyToDataRow, discussed below, are already created.)

The CopyToDataRowView Method

Saving an edited record involves copying data from all controls, except the primary key, into a DataRowView by calling the method, CopyToDataRowView, shown in Figure 12-62. The process of copying data from the controls into a DataRowView essentially is the opposite of displaying data from a DataRowView. The value of Position column is determined based on which RadioButton is checked (lines 141–147).

Figure 12-62: Copying data from controls to DataRowView

```
131     private void CopyToDataRowView()
132     {
133         // Copy data from controls into columns in the DataRowView
134         drvEmployee["LastName"] = txtLastName.Text;
135         drvEmployee["FirstName"] = txtFirstName.Text;
136         drvEmployee["DateOfBirth"] = txtDOB.Text;
137         drvEmployee["Phone"] = txtPhone.Text;
138         drvEmployee["CountryId"] = txtCountryId.Text;
139         drvEmployee["DeptNo"] = cboDeptNo.Text;
140         drvEmployee["Salary"] = txtSalary.Text;
141         drvEmployee["FullTime"] = chkFullTime.Checked;
142         string empPosition = null; //empPosition represents the job title
143         if (rdbVP.Checked == true)
144             empPosition = "VP";
145         else if (rdbMgr.Checked)
146             empPosition = "Manager";
147         else if (rdbStaff.Checked)
148             empPosition = "Staff";
149         drvEmployee["Position"] = empPosition;
150     }
```

Saving a new record involves copying data from all controls, including the primary key, to a DataRow by calling the method, CopyToDataRow, shown in Figure 12-63.

Figure 12-63: Copying data from controls to DataRow

```csharp
152     private void CopyToDataRow()
153     {
154         // Copy data from controls into columns of the DataRow
155         drEmployee["EmpId"] = txtEmpId.Text;
156         drEmployee["LastName"] = txtLastName.Text;
157         drEmployee["FirstName"] = txtFirstName.Text;
158         drEmployee["DateofBirth"] = txtDOB.Text;
159         drEmployee["CountryId"] = txtCountryId.Text;
160         drEmployee["DeptNo"] = cboDeptNo.Text;
161         drvEmployee["Salary"] = txtSalary.Text;
162         drEmployee["FullTime"] = chkFullTime.Checked;
163         string empPosition = null;
164         if (rdbVP.Checked == true)
165             empPosition = "VP";
166         else if (rdbMgr.Checked)
167             empPosition = "Manager";
168         else if (rdbStaff.Checked)
169             empPosition = "Staff";
170         drEmployee["Position"] = empPosition;
171     }
```

Deleting Records

Deleting a record from the database involves two steps:

1. Remove it from the data table using the RemoveCurrent method of the binding source.
2. Update the database table using the Update method of the data adapter.

Figure 12-64 shows the code. The code uses a MessageBox to confirm deletion of the record. The MessageBox returns a value of DialogResult type, which is an enumeration. The value, which depends on the button selected by the user, is cast to short type, and stored in the variable, response. The record is deleted if the value of response is the number corresponding to the *Yes* button.

Step 6-6: Add the code from Figure 12-64 to the Click event handler of btnDelete. Run the form.

Figure 12-64: Deleting a record

```csharp
173         private void btnDelete_Click(object sender, EventArgs e)
174         {
175             short response = 0;
176             response = (short)MessageBox.Show("Delete Record? ", "Confirm ",
177                         MessageBoxButtons.YesNo, MessageBoxIcon.Question);
178             if (response == (short)System.Windows.Forms.DialogResult.Yes)
179             {
180                 //Remove current record from dataset table
181                 bsEmployee.RemoveCurrent();
182
183                 //Update the database table with the dataset table
184                 this.daEmployee.Update(dtEmployee);
185                 MessageBox.Show("Record deleted from the database");
186             }
187         }
```

The following is the complete code for the Employee_AddEdDel form:

```csharp
22      DataTable dtEmployee = new DataTable();
23      SqlDataAdapter daEmployee;
24      BindingSource bsEmployee = new BindingSource();
25
26      // data table, data adapter and binding source Department table:
27      DataTable dtDept = new DataTable();
28      SqlDataAdapter daDept;
29      BindingSource bsDept = new BindingSource();
30
31      bool adding = false;
32      DataRowView drvEmployee;
33      DataRow drEmployee;
34
35      private void Employee_AddEdDel_Load(object sender, EventArgs e)
36      {
37          // Fill employee data table, dtEmployee
38          string connString = @"Data Source =
39                  (LocalDB)\MSSQLLocalDB; AttachDbFilename = C:\VC\HR2.mdf";
40          string empSQLStr = "SELECT * FROM Employee";
41          daEmployee = new SqlDataAdapter(empSQLStr, connString);
42          //Build UpdateCommand, DeleteCommand, and InserCommand
43          SqlCommandBuilder commandBuilder = new SqlCommandBuilder(daEmployee);
44
45          daEmployee.Fill(dtEmployee);
46
47          // Fill department data table, dtDept
48          string DeptSQLStr = "SELECT DeptNo, DeptName FROM Department";
49          daDept = new SqlDataAdapter(DeptSQLStr, connString); //Create data adapter
50          daDept.Fill(dtDept);
51
52          //Display DeptNo's from Department table, in cboDept:
53          cboDeptNo.DataSource = dtDept;  // DeptNo's are obtained directly from data table
54          cboDeptNo.DisplayMember = "DEPTNO";
55          cboDeptNo.ValueMember = "DEPTNO";
56
57          // Display employee data
58          bsEmployee.DataSource = dtEmployee; // links the binding source to the data table
59          DisplayCurrentRow();
60          lblRecordNo.Text = (bsEmployee.Position + 1) + " of " + bsEmployee.Count;
61          bsEmployee.PositionChanged += bsEmployee_PositionChanged;
62      }
```

```csharp
private void DisplayCurrentRow()
{
    // Copy the current row into a DataRowView:
    int currentRow = bsEmployee.Position;
    DataRowView drvEmployee = (DataRowView)bsEmployee[currentRow];

    // Copy data from the datarowview to controls on the form:
    txtEmpId.Text = drvEmployee["EmpId"].ToString();
    txtLastName.Text = drvEmployee["LastName"].ToString();
    txtFirstName.Text = drvEmployee["FirstName"].ToString();
    txtDOB.Text = drvEmployee["DateofBirth"].ToString();
    txtPhone.Text = drvEmployee["Phone"].ToString();
    txtCountryId.Text = drvEmployee["CountryId"].ToString();
    txtPhone.Text = drvEmployee["Phone"].ToString();
    cboDeptNo.Text = drvEmployee["DeptNo"].ToString();
    txtSalary.Text = drvEmployee["Salary"].ToString();
    chkFullTime.Checked = bool.Parse(drvEmployee["FullTime"].ToString());
    switch (drvEmployee["Position"].ToString())
    {
        case "VP":
            rdbVP.Checked = true;
            break;
        case "Manager":
            rdbMgr.Checked = true;
            break;
        case "Staff":
            rdbStaff.Checked = true;
            break;
    }
}

private void btnAdd_Click(object sender, EventArgs e)
{
    ClearControls();    // clear all controls
    adding = true;
}
```

```csharp
private void btnSave_Click(object sender, EventArgs e)
{
    int noOfRecords = 0;
        // Save data to the data table:
        if (adding == true)     // if it is a new record
        {
            // Create a blank DataRow by calling the NewRow() method of data table.
            drEmployee = dtEmployee.NewRow();
            CopyToDataRow();    // copy all data, into DataRow
            // Add the new DataRow to the Rows collection of the data table
            dtEmployee.Rows.Add(drEmployee);

            lblRecordNo.Text = (bsEmployee.Position + 1) + "  of  " + bsEmployee.Count;
            adding = false;
        }
        else // it's an existing record
        {
            // Create a DataRowView that references the current row of
            // the binding source:
            int currentRow = bsEmployee.Position;
            drvEmployee = (DataRowView)bsEmployee[currentRow];

            CopyToDataRowView(); //copy data from controls into columns of DataRowView:
            bsEmployee.EndEdit();   //Save the DataRowView into data table
        }
        // Update the database table
        noOfRecords = daEmployee.Update(dtEmployee); // Save to database table
        MessageBox.Show("Changes saved to database");
}

private void CopyToDataRowView()
{
    // Copy data from controls into columns in the DataRowView
    drvEmployee["LastName"] = txtLastName.Text;
    drvEmployee["FirstName"] = txtFirstName.Text;
    drvEmployee["DateOfBirth"] = txtDOB.Text;
    drvEmployee["Phone"] = txtPhone.Text;
    drvEmployee["CountryId"] = txtCountryId.Text;
    drvEmployee["DeptNo"] = cboDeptNo.Text;
    drvEmployee["Salary"] = txtSalary.Text;
    drvEmployee["FullTime"] = chkFullTime.Checked;
    string empPosition = null; //empPosition represents the job title
    if (rdbVP.Checked == true)
        empPosition = "VP";
    else if (rdbMgr.Checked)
        empPosition = "Manager";
    else if (rdbStaff.Checked)
        empPosition = "Staff";
    drvEmployee["Position"] = empPosition;
}
```

```csharp
        private void CopyToDataRow()
        {
            // Copy data from controls into columns of the DataRow
            drEmployee["EmpId"] = txtEmpId.Text;
            drEmployee["LastName"] = txtLastName.Text;
            drEmployee["FirstName"] = txtFirstName.Text;
            drEmployee["DateofBirth"] = txtDOB.Text;
            drEmployee["CountryId"] = txtCountryId.Text;
            drEmployee["DeptNo"] = cboDeptNo.Text;
            drvEmployee["Salary"] = txtSalary.Text;
            drEmployee["FullTime"] = chkFullTime.Checked;
            string empPosition = null;
            if (rdbVP.Checked == true)
                empPosition = "VP";
            else if (rdbMgr.Checked)
                empPosition = "Manager";
            else if (rdbStaff.Checked)
                empPosition = "Staff";
            drEmployee["Position"] = empPosition;
        }

        private void btnDelete_Click(object sender, EventArgs e)
        {
            short response = 0;
            response = (short)MessageBox.Show("Delete Record? ", "Confirm ",
                        MessageBoxButtons.YesNo, MessageBoxIcon.Question);
            if (response == (short)System.Windows.Forms.DialogResult.Yes)
            {
                //Remove current record from dataset table
                bsEmployee.RemoveCurrent();

                //Update the database table with the dataset table
                this.daEmployee.Update(dtEmployee);
                MessageBox.Show("Record deleted from the database");
            }
        }

        private void btnFirst_Click(object sender, EventArgs e)
        {
            bsEmployee.MoveFirst();
        }
        private void btnNext_Click(object sender, EventArgs e)
        {
            bsEmployee.MoveNext();
        }
        private void btnPrevious_Click(object sender, EventArgs e)
        {
            bsEmployee.MovePrevious();
        }
        private void btnLast_Click(object sender, EventArgs e)
        {
            bsEmployee.MoveLast();
        }
        private void bsEmployee_PositionChanged(object sender, System.EventArgs e)
        {
            if (bsEmployee.Count > 0)
            {
                lblRecordNo.Text = (bsEmployee.Position + 1) + " of " + bsEmployee.Count;
                DisplayCurrentRow();
            }
        }
```

```
213
214     private void ClearControls()
215     {
216         foreach (Control control in Controls)
217         {
218             if (control is TextBox)
219                 control.Text = null;      //or, control.Text = "";
220         }
221         cboDeptNo.Text = null;
222         rdbVP.Checked = false;
223         rdbMgr.Checked = false;
224         rdbStaff.Checked = false;
225         chkFullTime.Checked = false;
226     }
227
228     private void btnCancel_Click(object sender, EventArgs e)
229     {
230         DisplayCurrentRow();
231         adding = false;
232     }
```

Review Questions

12.20   What does the EndEdit method of binding source do?

12.21   True or false: The Rows of a data table are of DataRowView type.

12.22   What does the Update method of the data adapter do?

12.12 Command Object and DataReader

In the previous sections, you used the data adapter in combination with the data table to retrieve, update and delete records from a database. The data adapter has various SqlCommand objects like SelectCommand and UpdateCommand to do these tasks. In this section, you will create and use SqlCommand objects independent of data adapter and data table, and use the objects to execute SQL statements.

The data table was used in previous applications to keep the data in the memory while you work with the data. The data table, in combination with the data adapter and binding source, provides a convenient way to navigate through the records, and update, delete and add records. However, data tables take up memory. Further, updates to the database after filling a data table are not reflected in the data table because the connection to the database server is closed as soon as the data are loaded into the data table.

A **DataReader** (SqlDataReader, OdbcDataReader, or OracleDataReader) provides an efficient way to sequentially access data retrieved by an SQL statement. As the SQL executes, each row retrieved by the SQL is stored temporarily in the client's buffer, accessed sequentially by the Read method of the DataReader, and then removed from the buffer. So, DataReader can help reduce memory usage and improve performance. But, you cannot move back to previous records; it's a forward-only access, and the DataReader does not have methods to update, delete or add records. The connection is kept open for exclusive use by the DataReader until the DataReader is closed.

To understand the use of SqlCommand objects and DataReaders, you will create the form, Employee_Reader1, shown in Figure 12-65, that performs three different tasks:

1. Find the phone number of an employee by specifying the EmpId. This involves using the SqlCommand object to execute an SQL to find a single value.
2. Find the last name, first name and salary by specifying the EmpId. This involves using the SqlCommand object to execute an SQL, and using a DataReader to access the row retrieved by the SQL.
3. Update the salary.

Figure 12-65: Use of SqlCommand and SqlDataReader

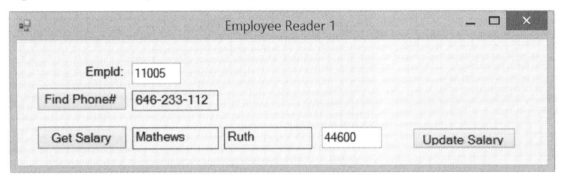

Creating the SqlCommand Object

The SqlCommand object, cmdEmployee, is created from the SqlCommand class, as follows:
    SqlCommand cmdEmployee = new SqlCommand()
where cmdEmployee could be any valid variable name. Because the Command object is used in multiple methods on this form, it is created at the class level.

You need to add the using directive, *using System.Data.SqlClient*, to specify the namespace that contains the class.

After you create the Command object, you need to specify the connection string and the SQL. To specify the connection, you create the SqlConnection object and assign it to the Connection property, as shown in line 26 of Figure 12-66. The connection string then is stored in the ConnectionString property of the Connection object.

Figure 12-66: Specifying the connection information

```
20      // Create a SqlCommmand object
21      SqlCommand cmdEmployee = new SqlCommand();
22
23      private void Employee_Reader1_Load(object sender, EventArgs e)
24      {
25          // Create a Connection object and assign it to the Connection property:
26          cmdEmployee.Connection = new SqlConnection();
27          // Specify the connection string for the Connection object
28          cmdEmployee.Connection.ConnectionString = @"Data Source = (LocalDB)\MSSQLLocalDB;
29                                      AttachDbFilename = C:\VC\HR2.mdf";
30      }
```

Next, you specify the SQL to find the phone number for a specified EmpId. The SQL is stored in the CommandText property of the SqlCommand object, as follows:

cmdEmployee.CommandText = "SELECT Phone FROM Employee WHERE EmpId = " + txtEmpId.Text;

EmpId is a numeric field; so, the value of EmpId is not enclosed in apostrophes, as in the case of string data. If EmpId is of string type, its value should be enclosed within apostrophes, as follows:

"SELECT Phone FROM Employee WHERE EmpId = '" + txtEmpId.Text + "'";

ExecuteScalar method of SqlCommand

The SqlCommand object has different methods for different types of SQL statements. An SQL with a single value in the result set can be executed using the ExecuteScalar method of the Command object, which returns the value. If the result set has more than one value, the method returns the first column of the first row.

The process of finding the phone includes opening the connection, calling the ExecuteScalar method to execute the SQL, and closing the connection. Figure 12-67 shows the code to find the phone number.

Figure 12-67: Using ExecuteScalar method to find phone number

```
32          private void btnPhone_Click(object sender, EventArgs e)
33          {
34              // Specify the SQL for the Command object
35              cmdEmployee.CommandText = "SELECT Phone FROM Employee WHERE EmpId = "
36                                      + txtEmpId.Text;
37              cmdEmployee.Connection.Open();  // open the connection
38              // Call the ExecuteScalar method to execute the SQL, and
39              // store the returned value in the variable, phone
40              string phone = cmdEmployee.ExecuteScalar().ToString();
41
42              lblPhone.Text = phone;
43              cmdEmployee.Connection.Close();
44          }
```

Line 40 shows the statement to call the ExecuteScalar method that returns the result (phone number) from executing the SQL specified in line 35. The returned value is stored in the variable, phone, and displayed in the Label, lblPhone.

Tutorial 7: SqlCommand and SqlDataReader to Work with Tables

This tutorial develops the Employee_Reader1 form, shown in Figure 12-65, and another form named Employee_Reader2 that retrieves and displays multiple Employee records.

Step 7-1: Open the project, Ch12_Databases_DataReader. Open the form, Employee_Reader1.

Add the code from Figure 12-66 to create the SqlCommand object and specify the connection information. Make sure that you specify the correct path for the database file.

Add the code from Figure 12-67 to specify the SQL and display the phone number.
Run the form and test the code using EmpId 11005.

ExecuteReader Method of SqlCommand

The ExecuteReader method executes an SQL, creates and populates a DataReader with the result set, and returns the DataReader that lets you access the rows, sequentially. So, this method is useful when the SQL result set includes multiple values from one or more rows. In the current form, you will use this method to execute an SQL that retrieves a single row containing the LastName, FirstName and Phone for a specified EmpId.

The DataReader returned by ExecuteReader may be assigned to a variable of DataReader type, as in the following statement:
   SqlDataReader drEmployeeReader = cmdEmployee.ExecuteReader();

where drEmployeeReader could be any valid variable name.

Read Method of the DataReader

The rows in the result set of an SQL are accessed using the Read method of the DataReader, as follows:
   drEmployeeReader.Read();

The Read method advances the reader to the next row, thus letting you access the values within the row. Initially, the position of the Reader is before the first record. To read multiple rows, you use a loop.

Figure 12-68 shows the code that uses the Read method and performs the following tasks:
- Specifies the SQL
- Calls the ExecuteReader method
- Calls the Read method of DataReader
- Displays column values in Labels

Figure 12-68: Code to display column values

```
46      private void btnGetSalary_Click(object sender, EventArgs e)
47      {
48          // Assign the SQL to the CommandText property
49          cmdEmployee.CommandText = "SELECT LastName, FirstName, Salary FROM Employee WHERE EmpId = "
50                                                  + txtEmpId.Text;
51          cmdEmployee.Connection.Open();
52          // Call the ExecuteReader method to execute the SQL.
53          SqlDataReader drEmployee = cmdEmployee.ExecuteReader();
54
55          // Move the DataReader position to the first (and only) row.
56          drEmployee.Read();
57
58          // Get column values from the current row of the DataReader, and display in Labels:
59          lblLastName.Text = drEmployee["LastName"].ToString();
60          lblFirstName.Text = drEmployee["FirstName"].ToString();
61          txtSalary.Text = drEmployee["Salary"].ToString();
```

Line 53 shows the statement that calls the ExecuteReader method and assigns the DataReader returned by the method to the variable, drEmployeeReader, which is of SqlDataReader type.

The statement in **Line 56** advances the DataReader position to the first and only row of the DataReader.

If the result set of the SQL has multiple rows, you will call the Read method inside a loop to access the rows, as discussed later in the Employee_Reader2 form.

Getting Column Values from a Row Using the DataReader

You can get values from a column of the current row by specifying the name of the column, as shown in lines 59–61 in Figure 12-68. The data retrieved using this method is object type, which is converted to string type for display in Labels.

Step 7-2: Add the code from Figure 12-68 to display the name and phone number.
Run the form, and test the code.

Alternate Methods to Get the Column Values

You also may use the ordinal (the position) of the column, like an index, in place of the column name to get the value. Line 65 in Figure 12-69 uses the ordinal, 2, of the Salary column. The ordinal would be 0 for the first column, 1 for the second column, and so on. Salary is the third column in the result set of the SQL, so its ordinal is 2. In general, it is more efficient to retrieve data using the ordinals. However, this is prone to errors due to using the wrong ordinals.

Figure 12-69: Alternate methods to get column values

```
62
63          // Alternate methods to get the salary
64          // Use the ordinal, 2, in place of the column name, Salary
65          txtSalary.Text = drEmployee[2].ToString(); // Salary is the 3rd column of the row
66          // or,
67          Int32 salaryOrdinal = drEmployee.GetOrdinal("Salary");
68          txtSalary.Text = drEmployee[salaryOrdinal].ToString();
69
70          // Get Salary in its original type, double
71          double salary = drEmployee.GetDouble(2);
72          txtSalary.Text = salary.ToString();
73
74          drEmployee.Close();
75          cmdEmployee.Connection.Close();
76      }
```

To use meaningful names for ordinals, you may get the ordinal using the column names, store it in a variable, and use the variable name in place of the literal values. The statement in line 67 uses the GetOrdinal method to get the ordinal, and stores it in a variable. Line 68 uses the variable name to get the salary. This method would help to minimize errors due to using literal values like 1 or 2 for ordinals.

Regardless of the type of fields in the database table, the column values retrieved using the above methods are object types. So, they need to be converted to appropriate types. For example, if Salary is to be used in calculations, it must be converted to a numeric type.

An alternative is to get values in their original types using methods like GetDouble, GetDateTime and GetString, shown in Table 12-1. This approach minimizes the need to convert data after it is retrieved. For example, if Salary needs to be used in calculations, it can be retrieved as Double, which is its type in the database table, using the following statement shown in line 71:

double salary = drEmployee.GetDouble(2);

This approach, however, requires knowledge of the exact type of the fields, when data are retrieved.

Step 7-3: Comment out line 61, and add the code from line 65 in Figure 12-69 to display salary using the ordinal. Run the form to test the code.

Comment out line 65, and add the code from lines 67 and 68, which use the GetOrdinal method. Run the form to test the code.

Comment out lines 67 and 68, and add the code from lines 71 and 72, which use the GetDouble method. Uncomment lines 74 and 75. Run the form to test the code.

The DataReader has several other methods and properties that help to work with it. Some commonly used methods/properties are shown in Table 12-1.

Table 12-1: Methods and Properties of SqlDataReader

Method/property name	Description	Example
Close	Closes the DataReader object	drEmployee.Close
Dispose	Releases resources used	drEmployee.Dispose
GetBoolean, GetDateTime, GetDecimal, GetDouble, GetFloat, GetInt32, GetString	Gets the value in its original type	drEmployee.GetString(0) drEmployee.GetDouble(2)
GetOrdinal(String)	Gets the ordinal	drEmployee.GetOrdinal["EmpId"]
GetSchemaTable	Returns a data table that contains field attributes	drEmployee.GetSchemaTable
GetValue	Gets the value as object	drEmployee.GetValue(0)
GetValues	Gets the values in the row as an array of objects	object[] colValues = new object[10]; drEmployee.GetValues(colValues);
Read	Advances to next row	drEmployee.Read()
Properties		
FieldCount	Gets the number of columns	int count = drEmployee.FieldCount
HasRows	Indicates whether the DataReader contains any rows	if (drEmployee.HasRows)

ExecuteNonQuery Method of SqlCommand

Previous sections used the ExecuteScalar method and ExecuteReader method of the SqlCommand to retrieve data from tables. The SqlCommand has a third method, **ExecuteNonQuery**, to execute queries that do not retrieve any data, but can be used to update, insert and delete records, and to do other database operations like create tables. For update, insert and delete, the method returns the number of records affected. For most other operations, the return value is -1.

You will use the ExecuteNonQuery to update the salary when the user enters the new salary into the TextBox, txtSalary, and clicks the Update Salary button on the Employee_Reader1 form presented earlier in Figure 12-65. The code is shown in Figure 12-70.

Figure 12-70: Use of ExecuteNonQuery to update record

```
77
78          private void btnUpdate_Click(object sender, EventArgs e)
79          {
80              cmdEmployee.CommandText = "UPDATE Employee SET Salary = " + txtSalary.Text
81                                      + " WHERE EmpId = " + txtEmpId.Text;
82              cmdEmployee.Connection.Open();
83              int count = cmdEmployee.ExecuteNonQuery();
84              MessageBox.Show(count + " record updated");
85              cmdEmployee.Connection.Close();
86          }
87      }
```

Line 80 stores the SQL statement in the CommandText property of the SqlCommand.

Line 83 calls the ExecuteNonQuery method to execute the SQL. The method returns the number of records updated, which is stored in the variable, count, although it is not used in this form.

You may use the same method to delete records and add new records.

Step 7-4: Add the code from Figure 12-70 to update the salary.

Accessing Multiple Rows Using a DataReader

Next, we will look at how to access multiple rows of an SQL result set. The code is essentially the same as that used to display a single row, except that we use a loop to call the Read method multiple times.

Consider an application that lists all employee records for a DeptNo selected from a ComboBox, as illustrated in the form Employee_Reader2, shown in Figure 12-71. A ListView is used to display the data.

Figure 12-71: Accessing multiple records using DataReader

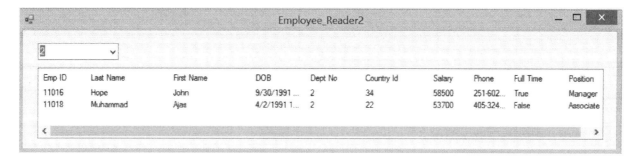

This application involves the following steps:
- Get DeptNo's from the Department table and add them to the ComboBox.
- Get the Employee records for the selected DeptNo.
- Store each record in a String array, and add it to the ListView.

Looping through Multiple Rows

This application involves accessing multiple rows of DeptNo's and Employee data. This is done by repeatedly calling the Read method of the DataReader in a while loop, as illustrated in Figure 12-72.

Figure 12-72: Using the while loop to access multiple rows

```
while (drEmployee.Read() == true)
{
        (statements to access the row)
}
```

The Read method in the while loop advances the reader to the next row, each time through the loop. The loop terminates when there are no more rows in the DataReader, as identified by the value, *false*, returned by the DataReader.

Getting DeptNo's from Department Table

As before, you will use the SqlCommand object and SqlDataReader objects to retrieve DeptNo's. Figure 12-73 shows the code to retrieve the data into a DataReader, and add them to the ComboBox.

You will use the same SqlCommand object, cmdEmployee, to retrieve Department and Employee data by changing the SQL assigned to the CommandText property.

Figure 12-73: Retrieving DeptNo's and adding to a ComboBox

```
19      // Create a SqlCommmand object
20      SqlCommand cmdEmployee = new SqlCommand();
21
22      private void Employee_Reader2_Load(object sender, EventArgs e)
23      {
24          // Create a Connection object and specify the connection string
25          cmdEmployee.Connection = new SqlConnection();
26          cmdEmployee.Connection.ConnectionString = @"Data Source = (LocalDB)\MSSQLLocalDB;
27                                          AttachDbFilename = C:\VC\HR2.mdf";
28          // Add DeptNo's from the Department table, to the Combobox
29          cmdEmployee.CommandText = "SELECT DeptNo FROM Department"; // specify SQL
30          cmdEmployee.Connection.Open();
31          // Call the ExecuteReader method to execute the SQL.
32          SqlDataReader drEmployee = cmdEmployee.ExecuteReader();
33
34          while (drEmployee.Read() == true)  // loop through the rows
35          {
36              cboDeptNo.Items.Add(drEmployee["DeptNo"]);  // add DeptNo to the ComboBox
37          }
38          cmdEmployee.Connection.Close();
39      }
```

Line 32 executes the SQL, and creates the DataReader, drEmployee, that contains multiple rows, each containing a single DeptNo.
The while loop in **line 34** processes the Read method repeatedly to access the rows.

Line 36 adds each DeptNo to the ComboBox.

Step 7-5: Open the form Employee_Reader2. Add the missing code from Figure 12-73. Run the form. Make sure that the ComboBox displays the DeptNo's.

Displaying Employee Records in a ListView

Displaying the data in a ListView involves the following steps:

- Retrieve Employee records, and access each row using a while loop.
- Use the **GetValues** method to get the entire current row, and store it in an object array.
- Convert the object array to a String array.
- Create a ListViewItem using the String array.
- Add the ListViewItem to the ListView.

Figure 12-74 shows the code to implement these steps.

Figure 12-74: Displaying Employee data in a ListView

```
41      private void cboDeptNo_SelectedIndexChanged(object sender, EventArgs e)
42      {
43          cmdEmployee.CommandText = "SELECT * FROM Employee WHERE DeptNo = "
44                                      + cboDeptNo.SelectedItem;
45          cmdEmployee.Connection.Open();
46          SqlDataReader drEmployee = cmdEmployee.ExecuteReader();
47
48          // display data in a ListView by storing the current row in an array
49          lvwEmployee.Items.Clear();
50          while (drEmployee.Read() == true)   // loop through the rows
51          {
52              int colCount = drEmployee.FieldCount;  // get number of columns
53              object[] colValuesObj = new object[colCount];  // create an object array
54
55              // Get values from the current row, and store in the array, colValuesObj
56              drEmployee.GetValues(colValuesObj);
57              // Convert objects to strings and store in the String array, colValues
58              String[] colValues = Array.ConvertAll(colValuesObj, element => element.ToString());
59
60              // Create a ListViewItem using the array, colValues
61              ListViewItem lviEmp = new ListViewItem(colValues);
62              lvwEmployee.Items.Add(lviEmp);  // add the ListViewItem to ListView:
63          }
64          cmdEmployee.Connection.Close();
65      }
```

Let's look at the steps in more detail.

Retrieve Employee records, and access each row using a while loop

Lines 43–46 retrieve Employee records for the selected DeptNo, and create the DataReader, as before.

Line 50 uses the Read method in a while loop to access the rows.

Use the GetValues method to get the entire current row, and store it in an object array

Lines 52 and 53 use the **FieldCount** property, which gets the number of columns, to create the object array.

```
int colCount = drEmployee.FieldCount;
object[] colValuesObj = new object[colCount];
```

Line 56 gets the values, and stores them in the object array, colValuesObj. Note that GetValues returns the set of values as an object array.

```
drEmployee.GetValues(colValuesObj);
```

Convert the object array to a String array

Line 58 uses the Array.ConvertAll method to convert the values from object to String:

```
String[] colValues = Array.ConvertAll(colValuesObj, element => element.ToString());
```

The **Array.ConvertAll** method converts the elements of an array of one type to another type, and returns an array of the new type.

The first argument, colValuesObj, is the input array.
The second argument "element => element.ToString()" can be thought of as a function with element as the input, and element.ToString() as the output. The word, element, in the second argument could be any word or notation.

Create a ListViewItem using the String array

Line 61 creates the ListViewItem, as discussed in Chapter 10:
```
ListViewItem lviEmp = new ListViewItem(colValues);
```

Add the ListViewItem to the ListView

Line 62 adds the ListViewItem to the ListView, lvwEmp:
```
lvwEmployee.Items.Add(lviEmp);
```

Accessing Individual Column Values

If it is necessary to work with individual column values rather than displaying data in a ListView, the values may be accessed directly from the DataReader using the column name or the ordinal, as shown in the following code:

```
while (drEmployee.Read() == true)
{
   double salary = (double) drEmployee["Salary"];
   // or,
   Int32 salaryOrdinal = drEmployee.GetOrdinal("Salary");
   double salary = drEmployee.GetDouble(salaryOrdinal);
   // (statements that use salary)
}
```

You also may store an entire row into an array, and access the individual columns from the array:

```
while (drEmployee.Read() == true)   // loop through the rows
{
   int colCount = drEmployee.FieldCount;   // get number of columns
   object[] colValuesObj = new object[colCount];   // create an object array

   // Get values from the current row, and store in the array, colValuesObj
   drEmployee.GetValues(colValuesObj);

   Int32 salaryOrdinal = drEmployee.GetOrdinal("Salary");
   double salary = (double) colValuesObj[salaryOrdinal];
   // (statements that use salary)
}
```

Step 7-6: Add the code from Figure 12-74. Run the form to test the code.

Review Questions

12.23 What is the function of a DataReader?

12.24 True or false: When you use a data table and data adapter to access records from a database table, the connection to the database table is maintained until the data table is closed.

12.25 True or false: A DataReader provides an efficient way to move forward and backward through the records retrieved by an SQL statement.

12.26 An SqlCommand object is to be used to execute an SQL statement that retrieves the salary for an employee, by specifying the employee's id. What method of the Command object should you use?

12.27 An SqlCommand object is to be used to execute an SQL statement that retrieves an employee's id, name and phone number using the employee's id. What method of the Command object should you use?

12.28 An SqlCommand object is to be used to execute an SQL statement that updates an employee's salary. What method of the Command object should you use?

12.29 What does the Read method of the SqlDataReader do?

Exercises

Exercise 12-1

Create a new project named Exercise12-1 and develop a form named DeptSearch that lets the user display records from Department table, one at a time using typed data table and data bound controls. Use a ComboBox for DeptNo field and TextBoxes for DeptName and DeptPhone. Use the ComboBox to let the user select a DeptNo from a list of all DeptNo's and display the corresponding record.

Exercise 12-2

Add a ComboBox named cboCountries to EmployeeSearch form within the project Ch12_Databases. Change the Text property of the ComboBox to "Filter By Country." The ComboBox should display a list of all unique CountryId's. When the user selects a CountryId from the ComboBox, filter the employee records using the selected CountryId. That is, the records displayed on the form should be limited to those for the selected CountryId.

Exercise 12-3

Add a ComboBox named cboDeptNo to EmployeeSearch form within the project Ch12_Databases. Change the Text property of the ComboBox to "Filter By Dept." The ComboBox should display a list of all DeptNo's from the Department table. When the user selects a DeptNo from the ComboBox, filter the employee records using the selected DeptNo; that is, the records displayed on the form should be limited to those for the selected DeptNo.

Next, modify the form so that the ComboBox displays a list of all department names instead of department numbers. When the user selects a name, use the corresponding DeptNo (ValueMember) to filter records.

Exercise 12-4

Create a new project named Exercise12_4 and develop a form named EmpSearch that lets the user display records from Employee table, one at a time, in TextBoxes. Use typed data table and data bound controls. Add a ComboBox named cboDeptNo to the form. Change the Text property of the ComboBox to "Fill By Dept." The ComboBox should display a list of all DeptNo's from the Department table. When the user selects a DeptNo from the ComboBox, load employee records for the selected DeptNo from the Employee database table to the Employee data table so that the user will be able to view only the selected records.

Exercise 12-5

Create a new project named Exercises12_Untyped, and develop a form named DeptDisplay that lets the user display records from Department table, one at a time using untyped data table and unbound controls. Allow the user to move between records. Use TextBoxes for all fields.

Exercise 12-6

Add a ComboBox named cboCountries to Employee_Select form within the project Ch12_Databases_UntypedDatasets. Change the Text property of the ComboBox to "Filter By Country." Create a data table, load the unique CountryId's into the table and display the CountryId's in the ComboBox. When the user selects a CountryId from the ComboBox, filter the employee records using the selected CountryId; that is, the records displayed on the form should be limited to those for the selected CountryId.

Exercise 12-7

Create a new project named Exercises12_Untyped, and develop a form named EmployeeSelect that lets the user display records from Employee table, one at a time, in TextBoxes. Use typed data table and data bound controls. Add a ComboBox named cboDeptNo to the form. Change the Text property of the ComboBox to "Fill By Dept." The ComboBox should display a list of all DeptNo's from the Department table. When the user selects a DeptNo from the ComboBox, load employee records for the selected DeptNo from the Employee database table to the data table, dtEmployee, so that the user will be able to view only the selected records.

Exercise 12-8

Create a new project named Exercises12_Untyped, and develop a form named DeptDAddEdDel that lets the user display records from Department table, one at a time using untyped data table and unbound controls. Allow the user to add, edit and delete records. Use TextBoxes for all fields.

Exercise 12-9

Open the project, Employee_DataReader. Open the form, Employee_Reader1. Add a button to this form so that the user can delete the currently displayed record by clicking the button.

Exercise 12-10

Create a new project named Exercises12_DataReader, and develop a form named DeptSelect that uses SqlCommand and SqlDataReader objects to let the user do the following tasks:

1. Type in a DeptNo into a TextBox and display the department's name in a Label.

2. Select a DeptNo from a ComboBox. Display the department's name and phone number in a ListBox.

Programming Assignment 3

The main goal of this assignment is to give you experience in working with database tables, text files, DataGridView and ListView.

This assignment is a modified version of assignment 2. A major change is the use of database tables (instead of sequential files) to store information. You are to develop an application for "Ace Auto Rentals," which is a small business that rents automobiles.

The **Rentals** database used in this assignment can be found in Tutorial_Starts/DataFiles folder. This database includes two tables: the Rental table that stores information on car rentals, and AutoTypes that contains the daily rental rates for different types of automobiles. Copy the database file to a local folder.

The following is a sample of the records in the **Rental** table:

RentalId	CustNumber	CustName	RentalDate	NumOfDays	AutoTypeId	DiscCategory	TotalCost
1001	10138	Josh Philip	11/5/2015	2	3	Business	81
1002	10150	Mini George	10/10/2015	1	2	Favorite	48
1003	10144	Jeffrey Mathews	12/22/2015	3	1	State	179.5
1004	10130	Cathy Arnold	12/21/2015	2	2	Standard	120
1005	10135	Gina Rodriguez	11/22/2015	5	4	Standard	200
1006	10126	Faith Abraham	11/23/2015	4	3	Favorite	112
1007	10115	Chris Cheng	12/22/2015	4	2	Standard	80
1008	10124	Sashi Nair	12/20/2015	2	6	Business	45

The following table shows the field types of Rentals. The data types char() and varchar() are text data.

Name	Data Type	Allow Nulls
RentalId	int	
CustNumber	char(5)	✓
CustName	varchar(40)	✓
RentalDate	date	✓
NumOfDays	int	✓
AutoTypeId	int	✓
DiscCategory	varchar(15)	✓
TotalCost	float	✓

The **AutoTypes** table appears as follows:

AutoTypeId	Description	DailyRate
1	SUV	70
2	mini-van	60
3	luxury	45
4	full-size	40
5	mid-size	35
6	compact	25

Part 1

Create a form named DisplayRentals that lets the user display **Rental** records, one at a time, as shown below.

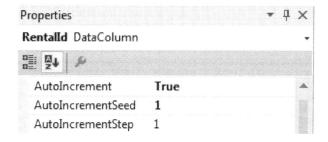

Note: If the version of SQL Server you use is incompatible with the database, when you create the data source, you will be asked whether you want to update the database. Answer "Yes."
Again, make sure that you answer **"No"** when the DataSource ConfigurationWizard asks you whether you want to copy the database to the project folder. (If you answer "Yes," the changes you make to the table will be lost the next time you run the project.)

The primary key, RentalId, is an auto-increment field. To view the fields and their properties, right click Rental table in Server Explorer window, and select *Open Table Definition*. The T-SQL window displays the SQL that creates the table. The key word **IDENTITY(1,1)** specifies that RentalId is an auto-increment field that increments by 1, starting with 1.

To view the data tables, open RentalDataSet.xsd. Select RentalId field and view its properties. Make sure that the auto-increment properties are set as follows:

Properties

RentalId DataColumn

AutoIncrement	True
AutoIncrementSeed	1
AutoIncrementStep	1

Part 2

Create a second form named EnterRentals that lets the user enter new records into the Rental table, as shown below.

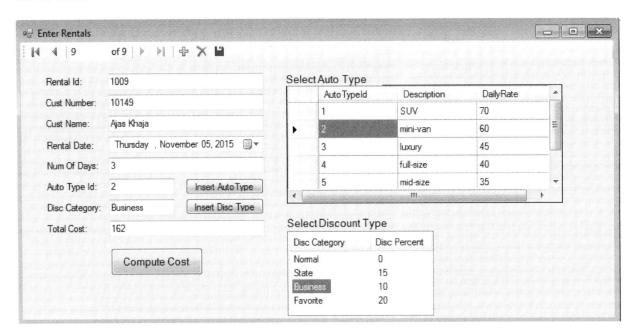

Use a DataGridView to display records from the table AutoTypes (in Rentals database), and to let the user select an AutoTypeId. When the user clicks the "Insert Auto Type" button, copy the selected AutoTypeId to the corresponding TextBox in the new record.

Similarly, use a ListView to display records from the DiscCategories.txt file, shown below, and to let the user select a discount category. Keep DiscCategories file in the Bin/Debug folder of the project. When the user clicks the "Insert Disc Type" button, the selected discount category is copied to the corresponding TextBox in the new record.

```
Normal, 0
State, 15
Business, 10
Favorite, 20
```

When the user clicks "**Compute Cost**" button, the total cost is computed and inserted into the TextBox for Total Cost in the new record.
   Total cost: number of days * Daily Rate *(1- discPercent/100))

To compute total cost, use the daily rate from the selected row in the DataGridView (see EmpDirectory form in Ch12_Database). Similarly, use the discount percent from the selected row in the ListView (see ListViewApp form in Ch10_GUIadditionalControls).

Part 3

Create a third form, **Rentals**, that serves as a switchboard with buttons and **menus** that allow the user to go to EnterRentals and DisplayRentals forms. Make Rentals the startup form.

Appendix A

Binary Files

A.1 Writing to Binary Files

The key differences between text files and binary files are

1. Text files store all data as text, whereas binary files store data in its original form without converting to text using the standard internal representation of numbers, dates, and so on. Therefore, the contents of text files can be viewed and edited in a text editor like Notepad, but binary files are not readable in a text editor.
2. Data typically is written to text files by combining an entire record into a single string and writing it in a separate line using the WriteLine() method of StringWriter. Data is written to binary files individually using the **Write()** method of BinaryWriter. Data is not grouped into rows.
3. The StreamWriter object to write data to text files can be created without explicitly creating a FileStream object. However, creating a BinaryWriter to write data to a binary file requires creating the FileStream object that is used by the BinaryWriter, as follows:

    FileStream variableName = newFileStream("FileName", FileMode);
    StreamWriter variableName = newBinaryWriter(FileStream object);

    Example
    FileStream examsStream = newFileStream("Exams.bin", FileMode.Append);
    StreamWriter examsWriter = newBinaryWriter(examsStream);

FileMode.Append: Append specifies that new data is added to existing data. If the file doesn't exist, it will be created.

FileMode.OpenOrCreate would overwrite existing data when new data is written. If the file doesn't exist, it will be created.

Tutorial: Working with Binary Files

Step A-1: Open the project, Ch8_FilesAndArrays. Open the WriteToBinaryFile form that has the same user interface as the WriteToTextFile form that you used in Chapter 8. Open the code window. Figure A-1 shows the complete code to write data from the TextBoxes to a specified file.

Step A-2: Add the code in lines 19 and 20, and in lines 30 and 32 from Figure A-1. Run the project. Select a folder and enter Exams.bin for file name.

Step A-3: Enter the name, score1 and score2, for five students as shown below, save them, and exit the form:

Mike Burns	75	92
Cathy Arnold	86	95
Wesley Mathews	82	90
Jerry Luke	70	89
Gina Rodriguez	77	94

Open Exams.bin with Notepad. (Right click the file, select *Open with,* and select *Notepad.*) Note that the two scores are not readable because they are not stored as text.

Figure A-1: Writing to binary file

```csharp
using System.Windows.Forms;
using System.IO;      // Add this line
namespace Ch8_FilesAndArrays
{
    public partial class WriteToBinaryFile : Form
    {
        public WriteToBinaryFile()
        {
            InitializeComponent();
        }
        FileStream examsStream;     // Declare variable to represent FileStream object
        BinaryWriter examsWriter;   // Declare variable to represent BinaryWriter object
        private void WriteToBinaryFile_Load(object sender, EventArgs e)
        {
            ExamsSaveDialog.ShowDialog();   // Display Save As Dialog box
            // Get the file name selected by the user and save in the variable fileName
            string fileName = ExamsSaveDialog.FileName;

            if (fileName.Length > 0)    // if the user selected a file
            {
                //Create FileStream object and assign to examsStream:
                examsStream = new FileStream(fileName, FileMode.Append);
                //Create BinaryWriter objects and assign to examsWriter:
                examsWriter = new BinaryWriter(examsStream);
            }
            else
            {
                MessageBox.Show("No file selected");
                btnSave.Enabled = false;
            }
        }
```

A.2 Reading Binary Files

Data is read from a Binary file using the BinaryReader object. Similar to the BinaryWriter object, the BinaryReader object uses the FileStream object, created as follows:

> FileStreamvariableName = newFileStream(@"FileName", FileMode);
> BinaryReadervariableName = newBinaryReader(FileStream);

Example:
> FileStream examsStream = newFileStream(@"E:\Exams.bin", FileMode.Open);
> BinaryReader examsReader = newBinaryReader(examsStream);

FileMode.Open: Opens the file if it exists; if not, an exception is thrown.

The code to read Exams.bin file (which was created in the form WriteToBinaryFile), and compute the average score for each student, is shown in Figure A-2. Exams.bin contains the name, score1 and score2, for multiple students.

Figure A-2: Reading from Binary Files

```
21      private void btnComputePercent_Click(object sender, EventArgs e)
22      {
23          string name, fmtStr;
24          float score1, score2;
25          float percentScore;
26
27          FileStream examsStream = new FileStream("Exams.bin", FileMode.Open);
28          BinaryReader examsReader = new BinaryReader(examsStream);
29          fmtStr = "{0, 15}{1, 6}{2, 6}{3, 6:N1}";
30          lstDisplay.Items.Clear();
31          while (examsReader.PeekChar() != -1)
32          {
33              // Read the next string (name)from Exams.bin
34              name = examsReader.ReadString();
35              // Read the next number (float type) from Exams.bin
36              score1 = examsReader.ReadSingle();
37              score2 = examsReader.ReadSingle();
38              percentScore = 100*(score1 + score2) / 200;
39              //Display the data and average in the List  box
40              lstDisplay.Items.Add(string.Format(fmtStr, name, score1, score2, percentScore));
41          }
42          examsReader.Close();
43      }
```

Lines 27–28: Create the FileStream and BinaryReader objects.

Line 31: The BinaryReader doesn't have an EndOfStream property. Instead, it has the PeekChar method that returns -1 when the end of file is reached.

Lines 34–37: Each data item is read separately. There is a separate method to read each type of data:

ReadString to read a string
ReadSingle to read a float type
ReadDouble to read a double type

An advantage of using binary files is that there is no need to convert data items like score1 and score2 from text to numeric types, as is necessary when reading from text files.

Step A-4: Open ReadBinaryFile form. Add the missing code from Figure A-3. Test the code to make sure you get the output as shown below:

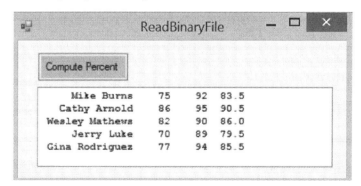

Appendix B

Creating a Database

SQL Server databases can be created in Visual Studio because the Express edition of SQL Server DBMS is automatically installed on your computer when you install Visual Studio.

Tutorial: Creating HR Database

This tutorial creates a database called HR (for human resources) that includes a table named Employee (EmpId, LastName, FirstName, DateOfBirth, Phone, DeptNo, CountryId, Salary, FullTime).

Step B-1: Create a new project named Ch11_Databases and rename the form to **EmpDirectory**.

Step B-2: Create a new database as follows:

Select: **Project, Add New Item, Service-based Database.**

Change the name of the database to: **HR.mdf,** andClick: **Add.**

If the **Data Source Configuration Wizard window** opens, asking you information about the data source, click Cancel because currently there are no tables in the database.
Notice that the Solution Explorer now includes the HR database file. In addition, you will see the **Server Explorer** (or Database Explorer if you are using Visual Studio Express) on the left side of the form will show the name of the database, shown in Figure B-1. If the Server Explorer is not visible, select *View, Server Explorer.*

Figure B-1: Server Explorer

Open the project directory, Ch12_Databases. You will see the database file, HR.mdf.

Step B-3: To add a table to the database, right click Tables and select **Add New Table**. You will see the Table Designer window, shown in Figure B-2.

Figure B-2: Table Designer window

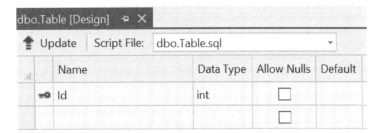

Step B-4: Specify the field names, data types and other attributes, and change the table name in the T-SQL statement at the bottom (CREATE TABLE [dbo].[Table]) to **Employee**, as shown in Figure B-3.

Figure B-3: Field names and attributes in Table Design window

Name	Data Type	Allow Nulls	Default
🔑 EmpId	int	☐	
LastName	nvarchar(50)	☐	
FirstName	nvarchar(50)	☑	
DateOfBirth	date	☑	
Phone	nvarchar(50)	☑	
DeptNo	int	☑	
CountryId	int	☑	
Salary	float	☑	((0))
FullTime	bit	☑	

CREATE TABLE [dbo].[Employee] (

Step B-5: Click the *Update* button at top left. On the *Preview Database Update* window, click *Update Database* button. You will see the message "Update completed successfully."

Step B-6: Click the Refresh button on the Server Explorer window and expand *Tables*.
   Right-click Employee and select *Show Table Data* from the menu.
   You now may enter records into the table, as shown in Figure B-4.

Figure B-4: Adding records to table

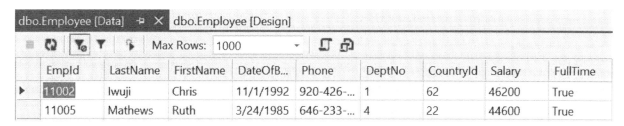

EmpId	LastName	FirstName	DateOfB...	Phone	DeptNo	CountryId	Salary	FullTime
11002	Iwuji	Chris	11/1/1992	920-426-...	1	62	46200	True
11005	Mathews	Ruth	3/24/1985	646-233-...	4	22	44600	True

You may copy HR.mdf file to a folder outside the current project directory for use in another project. But, before you can copy the file, you need to close the current project.

You may create additional tables within the HR database using the same process.

Appendix C

Answers to Review Questions

Chapter 1

1.1   Input: key word(s) that you type in.
         Process: find web pages related to the key word(s)
         Output: information on related web sites

1.2   Input: student Id, course number, section number
         Process: check availability of seats, prerequisites and conflicts with other classes; add student to class list.
         Output: student schedule, class list

1.3   Define the purpose and identify the input, process and output of the program.
Design and develop the Graphical User Interface (GUI).
Identify the components and logic of the program.
Design and develop files/databases, if any.
Write and test the program.

1.4   Syntax error

1.5   Logic error

1.6   False. There could be logic errors that produce wrong results

1.7   Machine Language

1.8   A complier is a special software that translates programs from a high level language to Machine Language or to an intermediate language to create an executable program.

1.9   An interpreter converts each statement in a program to Machine Language and runs it without producing an executable program.

1.10   Microsoft Intermediate Language Assembly is the executable program created by the compiler in Visual Studio.

1.11   The Common Language Runtime translates each statement from Microsoft Intermediate Language to Machine Language and runs it.

1.12   Text property

1.13   The form specified in the statement Application.Run(new ...()).

1.14   TextBox

1.15 Text

1.16 A namespace is a container for a group of classes in the .Net Framework Class Library

1.17 A using directive makes a group of classes from the .Net Framework Class Library available within a form.

1.18 A method is a program unit that performs a task.

1.19 An event handler is a method that executes in response to an event.

1.20 float

1.21 decimal

1.22 float and double types take less memory than decimal, and they are faster to work with.

1.23 The first letter of a variable is lowercase, and the first letter of subsequent words, if any, are capitalized.

1.24 A variable name must start with a letter. A variable can have only letters, numbers and underscore.

1.25 double orderQty = 4.5;

1.26 string lastName = "Jones";

1.27 string firstName = "Jack";

1.28 double totalAmount = 145.50;

1.29 double totalCost = 2145.00;

1.30 double unitPrice = 15.50;

1.31 int quantity;
      quantity = 15;

1.32 int quantity = int.Parse(txtQuantity.Text);

1.33 int units = int.Prase(txtUnits.Text);

1.34 currentSales: 0
      previousSales: 0

1.35   currentSales: 110
       previousSales: 110

1.36   17

1.37   4

1.38   4,525.46

1.39   variable

1.40   A constant must be initialized when it is declared. A constant cannot be assigned a value after it is declared.

1.41   A local variable is a variable that is declared inside a method.

1.42   A field is a variable that is declared at the class level and it is available in all methods within a form. A field exists until the form is closed.

Chapter 2

2.1   decimal

2.2   float and double types take less memory than decimal, and they are faster to work with.

2.3   double

2.4   decimal

2.5   no error

2.6   float cost = **(float)** 2.50;   or   float cost = 2.50F;

2.7   no error

2.8   **short** quantity = 200;

2.9   decimal price = 300;
      float currentPrice = **(float)** price;

2.10 no error

2.11 int qty1 = 1000000000;
      int qty2 = 2000000000;
      long total = (long) qty1 + qty2;

2.12 **double** total = 255.50 + 500;   // or, float total = (float)255.50 + 500

Appendix C: Answers to Review Questions

2.13   int totalScore = 25;
       int count = 4;
       float average = (float)total/count;

2.14   decimal price = 100;
       float quantity = **(float)**5.5;
       decimal netCost = price***(decimal)**quantity - 10

2.15   (fName.Substring(0, 2) + " " + lName).ToUpper()

2.16   id.Substring(studentId.Length-2, 2);

2.17   Both Today and Now includes today's date. Now also includes the current time, but Today includes only the default time 12:00:00 AM.

2.18   TimeSpan tsRental = DateTime.Now - rentalDate;
       lstDisplay.Items.Add(tsRental.Days);
       // or,
       TimeSpan tsRental = DateTime.Today - rentalDate;
       lstDisplay.Items.Add(tsRental.TotalDays);

2.19   TimeSpan tsRentalTime = new TimeSpan(rentalTime.Hour, rentalTime.Minute, rentalTime.Second);
       rentalDateTime = rentalDate.Add(tsRentalTime);

2.20   DateTime dueDate = invoiceDate.AddDays(30);

Chapter 3

3.1   5

3.2   0

3.3   Admit

3.4   Deny

3.5   Deny   -- ((ACT >= 24) || (GPA >= 3)) yields true. ! true yields false.

3.6   Admit -- ((ACT >= 24) && (GPA >= 3)) yields false. ! false yields true.

3.7   C

3.8   string grade;
       int totalScore = 85;
       if (totalScore >= 70)
               grade = "Pass";
       else
               grade = "Fail";

3.9   10

3.10

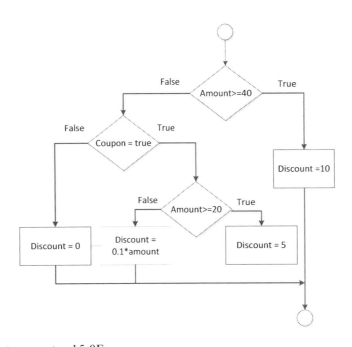

3.11   float amount = 15.0F;
       float discount;
       bool coupon = true;
       if (amount >= 40)
          discount = 10;
       else if (coupon = true)
          if (amount >= 20)
             discount = 5;
          else
             discount = 0.1F*amount;
       else
          discount = 0;

3.12   Integer, String and Boolean

3.13
```
int rating = 2;
double raise;
switch (rating)
{
  case 3:
    raise = 0.25;
    break;
  case 2:
    raise = 0.1;
    break;
  default:
    raise = -1.0;
    break;
}
```

3.14 false

3.15 0

3.16 true

3.17 When the user moves focus away from the control

3.18 The CauseValidation property of both controls must be set to true.

Chapter 4

4.1
```
int yearlyInvestment = 750;
int totalAmt = 0;
int year = 0;
while (year < 4)
{
  year = year + 1;
  totalAmt = totalAmt + yearlyInvestment;
}
```

4.2 750

4.3 2017   76000
     2018   77000

4.4 0

4.5 1

4.6 2016 66000
     2017 67000
     2018 68000

4.7   2018 64000
      2017 63000

4.8   The break statement lets you exit a loop. The continue statement lets you skip statements within the body of the loop but continue the iterations.

Chapter 5

5.1   A method is a program unit that performs a particular task

5.2   An event handler is a special type of method that is executed when a certain event of a control takes place.

5.3   True
5.4   True

5.5   ComputeCost();

5.6   An argument is a data item that is passed to a method by the calling program.

5.7   A parameter is a variable that receives an argument that is passed to a method.

5.8   An argument is associated with the parameter that is in the same position as the argument, unless the name of the parameter is specified along with the argument.

5.9   The scope of a parameter is the method in which it is declared.

5.10 False

5.11 Hi
      There

5.12 When an argument is passed by value, only its value is passed to the corresponding parameter.

5.13 0

5.14 When an argument is passed by reference, the memory location (address or reference) of the argument variable is passed to the parameter so that the called method can change the value of the argument by changing the value of the corresponding parameter.

5.15 150 100 250

5.16 0 0 25

5.17 Top-down design involves breaking down a larger problem to smaller subtasks, each performed by a separate method.

5.18 Makes software development simpler and easier.
      Makes the code more re-usable.
      Easier to get a high-level view of the code.

5.19 An enumeration is a set of named integer constants

5.20 An enumeration provides a convenient way to select from a limited set of numbers. In addition, an enumeration uses meaningful names in place of numbers, which helps minimize errors in making a selection.

Chapter 6

6.1   Maximum property should be set to 10.

6.2   lblUnits.Text=hsbUnits.Value.ToString

6.3   hsbUnits_ValueChanged

6.4   True

6.5   Checked

6.6   rdbDefault.Checked = true

6.7   Two

6.8   cboNames.SelectedItem (and cboNames.SelectedText, but, it is not available in event handlers).

6.9   cboNames.SelectedIndex = -1

6.10 A CheckBox

6.11 A ComboBox

6.12 A set of CheckBoxes

6.13 Basketball

6.14 Cricket

Chapter 7

7.1   A value type variable holds the actual value assigned to that variable, whereas a reference variable holds only the memory location of the object assigned to the variable.

7.2   12

7.3   string[] student = new string[20];

7.4   string[] student;
      student = new string[20];

7.5   string[] weekDay = {"Sun", "Mon", "Tue", "Wed", "Thu", "Fri", "Sat"};

7.6   weekDay[1] = "Monday";

7.7   lstDisplay.Items.Add(weekDay[1]);

7.8   for (int index = 0; index <= weekDay.Length-1; index = index + 1)
       lstDisplay.Items.Add(weekDay[index]);

7.9   The index 7 in testScore[7]) causes a runtime error because 7 is outside the valid range (0-6).

7.10   foreach (string day in weekDay)
       lstDisplay.Items.Add(day);

7.11   string[] workday = new string[5];
       Array.Copy(weekDay, 1, workday,0,5);

7.12   Monday

7.13   Mon

Chapter 8

8.1   System.IO

8.2   The method reads and returns the line at the current read position and then advances the read position to the next line. An exception is thrown if no data are found.

8.3   75648

8.4   extCost = fields[2]*fields[3]

8.5   orderNumbers[index] = fields[1];
       extCosts[index] = extCost;

8.6   foreach (string show in shows)
       cboShows.Items.Add(show);

8.7   int index = Array.IndexOf(months, txtMonthName.Text);
       lstDisplay.Items.Add (unitsSold[index]);

8.8   If ExamWriter is declared in the Load event handler, it won't be available in the btnSave_Click event handler, and in the FormClosing event handler, resulting in error.

8.9   If ExamWriter is declared and created in btnSave_Click event handler, it won't be available in the FormClosing event handler of the form.

8.10   If ExamWriter is declared, created and closed in btnSave_Click event handler, ExamWriter will be created and closed each time the Save button is clicked. This is less efficient than creating it once in the Load event handler, using it multiple times and then closing when the form is closed.

8.11   205

Chapter 9

9.1   You can add items to a collection without specifying the index position. Further, the size of a collection is increased automatically when you insert new items.

9.2   Generic collections are strongly typed; that is, they allow you to specify a single type that can be stored in the collection

9.3   prices.Add(fields);

9.4   string price;
foreach (string[] row in prices)
{
    if (row[0] == "1001")
        price = row[1];
}
// An alternative is to use the index of the row, as follows:
string[] row = prices[4];
string price = row[1];

9.5   A Dictionary requires a unique key, which makes it easier to find an item using its key. Further, a Dictionary is implemented as a hash table, whereas a List is implemented as an array.

Chapter 10

10.1   A ListView lets you add multiple data items to a row and lets you directly access each item from a row. A ListBox lets you add only a single string to a row.

10.2   A ListView is a GUI control. Each row of a ListView is a ListViewItem object

10.3   The Lion King

10.4   80

10.5   When you use multiple tab pages on a single form, you can access controls on any one tab page from another page by using the name of the control. By contrast, to access controls on one form from another form, you need to create an instance of the page,

Chapter 11

11.1   DisplayOrders displayOrders = new DisplayOrders();
displayOrders.Show();
float totalCost = float.Parse(displayOrders.lblTotalCost.Text);

11.2 Hiding a form keeps the form and its properties and settings in memory. You don't lose the data entered/displayed in controls at runtime.
Closing a form removes the form from memory. So, you lose all current settings and data.

11.3 The Show method opens the form and C# continues processing the statements that follow the statement. However, with the ShowDialog method, C# opens the form and stops processing the statements that follow the statement until the dialog window is closed.

11.4 Create a Public method within a class and call it from different forms.

11.5 A static method is not associated with a specific instance (object) of the class that contains the method. It is associated only with the class. So, it is invoked using the class name.

To call a nonstatic method, you have to create an object of the class and use the object name in place of the class name.

11.6 A context menu is associated with a control, and it is displayed by right clicking the control.

11.7 Option 3: Selecting an item from a Context menu

Chapter 12

12.1 data tables are temporary tables maintained in memory. Each data table may contain data from one or more database tables that are typically stored on the disk.

12.2 The table adapter connects the data table to the database, and retrieves data from and saves data to the database.

12.3 The binding source links the user interface controls on the form to a data table.

12.4 The Fill method of the table adapter copies data from the database table into the data table.

12.5 The Update method of the table adapter saves the data from the data table into the database table

12.6 A data bound control is a user interface control that is linked to a data source

12.7 The data in a data bound control is linked to the binding source, and the binding source is linked to the data table.

12.8 Set all DataBindings to **None**.
Set DataSource to CustomerBindingSource
Set DisplayMember to CustomerId

12.9 The user will be able to see only employee records with CountryId, 22.
It will have no effect on the records in the data table. The data table still will have all records from Employee table.

12.10 The Employee data table will have only records for employees from DeptNo 11.

12.11 Connection string, SQL

Appendix C: Answers to Review Questions

12.12   Creating data access objects at runtime using code makes the code more flexible, efficient and reusable.

12.13   Typed data tables contain fields that are of the same type as the fields in the database table, and they are created typically at design time. Untyped data tables do not contain information on its fields. They are created at runtime.

12.14   Data bound controls are bound to specific fields in the data table through the binding source. So, data on the form is tightly linked to the fields in the data table. The current row is automatically displayed in the controls. Unbound controls are not directly bound to any field.

12.15   Update method of the table adapter.

12.16   Set the DataSource property of the binding source to the data table.

12.17   Help move between rows, get the position/count of rows, and select rows using its Filter property.

12.18   Store the current row in a DataRowView object, and copy the data from the DataRowView to the controls on the form.

12.19   Both codes let the user view IS student records. Code segment #1 will load only IS student records into the data table. Code segment #2 will load all student records into the data table, but will allow the user to view only IS student records.

12.20   The EndEdit method applies the changes in the binding source rows to the data table it is linked to.

12.21   False

12.22   The Update method of the data adapter saves the changes in the data table to the database table.

12.23   A **DataReader** provides an efficient way to sequentially access the data retrieved by an SQL statement from a database

12.24   False

12.25   False

12.26   ExecuteScalar

12.27   ExecuteReader

12.28   ExecuteNonQuery

12.29   Advances the DataReader to the next row

Index

Access identifier, 145
Argument, 150
   default values, 155
   passing by reference, 157
   passing by value, 154
Arrays, 208, 238
   2-Dimensional arrays, 268
   Array initializer, 212
   Array names, 220
   Array.Copy method, 222
   Array.IndexOf method, 225
   Jagged arrays, 275
   Passing arrays to methods, 264
Autohide, 12
Binary Files, 429
Binding source, 353
Binding Source, 380
Boolean, 26
Boolean expression, 87
break statement, 140
Button, 16
Camel Case, 14, 29
Casting, 59
CauseValidation property, 115
Character data, 27
CheckBoxes, 180
Class, 6, 22
   Parial class, 22
Class Library, 6
Clearing a ListBox, 48
Collections, 279
ComboBox, 188, 250, 371
   ComboBox styles, 191
Common Language Runtime (CLR), 6
Compiler, 6
Concatenating Strings, 71
connectionString, 386
Console Application, 123
   Console.ReadLine, 124
   Console.WriteLine, 124
constant, 45, 151
continue statement, 141
Control, 3, 16
data bound control, 384
Data Flow Diagrams, 3

Data Source, 354
Data tables, 353
Data Validation, 106
Database Systems, 353
   Creating a SQL Server database, 432
   Data tables, 353
   Dataset, 353
   Parameter query, 381
   Searching a key field, 369
Data-bound controls, 360
DataGridView, 357, 364
DataReader, 413
DataRow, 388
DataRowView, 388
DateTime
   DateTime operations, 74
   Methods of DateTime, 79
   Today vs. Now, 78
DateTimePicker, 75
Debugging, 5
decimal type, 58
Decision Structure, 86
Default values of data types, 31
Dictionary, 279, 289
   Passing Dictionary as a parameter, 338
Directory, 257
Docking, 12
double type, 59
do-while Loop, 134
EndEdit method, 406
Enumerations, 170
ErrorProvider Control, 113
ExcecuteScalar method, 415
Exception handler, 51
ExecuteNonQuery method, 418
ExecuteReader method, 416
Fields, 49
Files, 230
   Append vs. Overwrite, 256
   Binary files, 230
   Directory class, 257
   SaveFileDialog, 260
   Sequential access files, 230
Filtering rows, 397
Floating point, 28, 57

Flowchart, 4
for Loop, 135, 216, 223
for vs. foreach, 219
foreach Loop, 142, 219, 287
Form
    Accessing from another form, 331
    Creating multiple instances, 333
    FormClosing event, 255
    Hiding vs. Closing, 333
    Load event, 254
    Login form, 342
    Sharing methods between forms, 335
    Show vs. ShowDialog, 334
Graphical User Interface, 3
High-level languages, 6
if else if Statement, 96
if else Statement, 87
Implicit cast, 64
Infinite Loop, 131
Initializing a variable, 31
Input, 3
int type, 27, 56
Integer, 26
Integer literals, 61
Integrated Development Environment (IDE), 6
Intermediate language, 6
Interpreter, 6
KeyValuePair collection, 294
Label, 16
    Autosize property, 16
Lifetime of a variable, 30
List<T> Collection, 281
ListBox, 198
    Items collection, 47
ListView, 302, 307, 316, 321
ListViewItem, 303, 317
Literals, 59
Local variables, 49, 62
Locals Window, 62
Logic error, 4
Logical operators, 93
Low-level, 5
Machine language, 6
Machine Language, 5
Menus, 345
    Context Menu, 348
MessageBox, 52
Method, 22, 144, 264
    access identifier, 145
    cohesive methods, 155
    passing values to methods, 148
    public, 145
    static methods, 338
    void method, 146
Methods that Return a Value, 168
Microsoft Intermediate Language (IL) Assembly, 6
Multiplication with large integers, 66
Namespace, 21
Nested if Statements, 98
Non-integer (real) types, 28
Output, 2
parameter
    out parameter, 159
Parameter, 149, 338
parameter query, 381
Parse, 35
PositionChanged event, 390
Precedence of Operators, 37
Prefixes for controls, 16
Pre-test loops, 132
Process, 3
Project, 8
Properties window, 10
Pseudo code, 4
RadioButtons, 182
Real number, 26
Reference types, 68
Relational operator, 87
Replace (string, string) method, 71
Runtime error, 4
SaveFileDialog control, 260
Scope of Variables, 49
Scroll Bars, 176
SelectCommand, 398
SelectedIndexChanged event, 195
Sequential Access Files, 230
Solution, 8
Solution Explorer, 10
SortedDictionary, 279
SortedList, 279
SQL Server, 432
SqlCommand, 413
SqlCommandBuilder, 403
Startup form, 15
StreamReader class, 231
StreamWriter class, 253
string Type, 28
    Concatenating strings, 71

    String operations, 69
    String.Format method, 47
    Substring method, 70
    ToLower method, 70
    Trim method, 70
String vs. Character, 72
Structure chart, 148
switch Statement, 102
Syntax error, 4
Tab Pages, 312, 326
Table adapter, 353, 380, 381
TextBox, 16
ToolStrip, 350
ToolTip, 13
Top-down design, 144, 161

Trim() method, 70
Try-catch, 51
TryParse method, 106
typed dataset, 384
UML diagram, 3
unbound control, 384
untyped datase, 384
Using **directive, 21**
Validating Input, 184
    validating event, 111
Value types, 68
Value Types vs. Reference Types, 209
while Loop, 121
Widening conversion, 64